Mathematical Concepts

Jürgen Jost

Mathematical Concepts

 Springer

Jürgen Jost
Max Planck Institute for Mathematics
in the Sciences
Max Planck Society
Leipzig
Germany

ISBN 978-3-319-20435-2   ISBN 978-3-319-20436-9   (eBook)
DOI 10.1007/978-3-319-20436-9

Library of Congress Control Number: 2015943831

AMS Codes: 00A05, 00A06, 01A65, 03-02, 03B45, 03G30, 18B25, 18-XX, 18-02, 18Axx, 18Fxx, 18Gxx, 20A05, 20B05, 20C05, 06Bxx, 08Axx, 14-XX, 14A15, 51K10, 53B05, 53B20, 53C21, 54A05, 55Nxx, 55P10, 55U10, 58A05, 92B99

Springer Cham Heidelberg New York Dordrecht London

Springer International Publishing AG Switzerland is part of Springer Science+Business Media (www.springer.com)

Mathematics translates concepts into formalisms and applies those formalisms to derive insights that are usually not amenable to a less formal analysis.

Preface

This book provides an introduction to and an overview of the most important concepts and structures of modern mathematics. Therefore, I try to motivate and explain those concepts and structures and illustrate them with examples, but I often do not provide the complete proofs of the basic results. Therefore, this book has a broader scope and is less complete and detailed than standard mathematical textbooks. Its main intention is to describe and develop the conceptual, structural, and abstract thinking of mathematics. Specific mathematical structures then illustrate that conceptual approach, and their mutual relationships and their abstract common features should also provide deeper insight into each of them.

The book thus could be used

1. simply as an overview of the panorama of mathematical structures and the relations between them, to be supplemented by more detailed texts whenever you want to acquire a working knowledge of some structure,
2. by itself as a first introduction to abstract mathematics,
3. together with existing textbooks, to put their results into a more general perspective,
4. after having studied such textbooks, in order to gain a new and hopefully deeper perspective.

Putting it differently, this means that the readers should use this book in the manner that best suits their individual needs and aims. It is not my intention to suggest or prescribe a particular way of reading this book. Of course, you could, in principle, try to read the book from the beginning to the end. You could also, however, browse through it and see what strikes you as especially interesting or useful. Or you could simply search for particular topics that you want to learn or understand better. Whichever way you approach it, I hope that this book will prove useful for you.

In any case, however, I should emphasize that mathematics can only be properly learned by going through and understanding the proofs of the main results in detail and by carrying out exercises and computing examples. This book does not systematically provide such proofs and exercises, so it needs to be supplemented by suitable other texts. (Some such textbooks are listed in the bibliography and are cited in the individual chapters of this book. Of course, there exist many more good textbooks than listed, and my selection may appear somewhat random to some experts in the respective fields).

After having dutifully expressed this warning, let me also issue a warning in an opposite direction. Beware of those self-appointed guardians of certain narrow subfields of mathematics who claim that you cannot understand what their subfield and work is about unless you study it with the insiders over many years. Often, such people are just afraid of competition from outsiders. On the contrary, I maintain that every important conceptual mathematical idea can be understood with some effort. Perhaps it requires somewhat more effort than the present book might offer to you, and in any case you will also need some ability in flexible and abstract thinking. But do not let this deter you. This book is intended to offer you some help and guidance on your path toward a deeper understanding of mathematics. This book wants to convey a positive, encouraging message, and not a negative, discouraging one.

Let me now describe the contents and topics. The first chapter gives an informal overview. Chapter 2 introduces some basic structures, like graphs, monoids, and groups, rings and fields, lattices and Boolean and Heyting algebras, as well as the basic notions of category theory. In Chap. 3, we first treat relations in an abstract manner and then discuss graphs as the mathematical structures encoding binary relations. As a more concrete example of mathematical reasoning, we discuss the representation theory of finite groups and apply this to illustrate the space of all graphs on the same set of vertices. In Chap. 4, we introduce topological spaces, as well as the more general class of pretopological spaces. A topological structure on a set distinguishes a subclass of the Boolean algebra of its subsets. The members of this subclass are called the open sets of the topology, and they constitute a Heyting algebra. (Since in general, the complement of an open set need not be open itself, they no longer form a Boolean algebra.) On a topological space, we can define sheaves and cohomology groups and thereby obtain algebraic invariants. Also, we introduce measures and with their help supplement the algebraic invariants by geometric ones. In the next Chap. 5, we analyze the concept of space from the points of view of topology, differential geometry, and algebraic geometry. In differential geometry, we identify the basic notion of curvature, whereas the algebro-geometric approach is based on the concept of a scheme. In the next Chap. 6, we turn to algebraic topology in more detail and discuss general homology theory. We illustrate this for simplicial complexes, and this also allows us to develop a dynamical picture of topology. This can be seen as a discrete analog of Conley theory, the extension of Morse theory to dynamical systems. I then insert Chap. 7, where the generation of structures through specified operations, perhaps obeying certain constraints, is discussed. In Chap. 8, we return to categories and provide an introduction to the basic results of category theory, like the Yoneda lemma and its applications. In Chap. 9, devoted to topoi, we combine algebraic structures such as Boolean and Heyting algebras, the geometric perspective of categorial concepts like presheafs, with the abstract approach of mathematical logic. The last chapter is something of an anticlimax. It reviews the various structures that can be imposed upon or induced by the simplest examples, the sets with no, one, or two elements. While much of this is, of course, rather trivial, it should give the reader the opportunity to review the basic concepts. Of course, to study these examples, one does not have to wait until the end, but can also utilize them while reading about a certain type of structure in another chapter. In the other chapters, a box in the page margin indicates that one of the recurring standard examples is discussed in the text. Occasionally, I use the abbreviation "iff" for "if and only if," which has become customary in mathematical texts.

Also, at certain places, I point out—possible or already existing—connections with theoretical biology. A systematic conceptual framework for theoretical biology still awaits to be developed, but I believe that some of the concepts presented in this book could yield important ingredients.

While some aspects of the present book may be new, like the discrete Conley theory or some items in the discussion of the concept of space, most of this book simply condenses and illustrates what can be found in existing textbooks. The motivation, as already mentioned in the beginning, is to provide a comprehensive overview and an orientation among the many important structures of modern mathematics. Of course, there are many omissions in this book. In particular, the most fundamental concepts of analysis like compactness are not treated, nor are such important structures as Banach spaces. Also, no mention is made of number theory beyond the elementary concept of a prime number.

Since this book covers a wide range of mathematical topics, some conflicts with established notation in certain fields are inevitable, because the different notational conventions are not always compatible. One point that the reader should be alerted to is that I use the symbol 1 for the truth value *true*, although some other texts use 0 instead.

Clearly, in many of the special fields discussed here, I am not an expert in the technical sense. I wanted to share the understanding that I do possess, however, with my students and collaborators, and to guide them into the powerful realm of abstract modern mathematical concepts, and therefore, I lectured on these topics in graduate courses at Leipzig. I hope that the present book can likewise serve to share that understanding with its readers.

Also, the style of the book leads to inconsistencies, in the sense that certain mathematical concepts are presupposed, but not explained, at various places. Foremost among them is the core concept of analysis, that of differentiation. We need this because some of the most important concepts and examples depend on differential calculus. This is clearly inconsistent, because I shall explain more basic principles, like that of continuity, carefully. The reason is that, to understand the essential thrust of this book and its examples, we need not go into the conceptual foundations of calculus, but can hopefully get away with what the reader knows from her basic calculus course. In any case, all the required material from calculus can be found in my textbook [59]. Also, at certain places, I use constructions and results from linear algebra without further explanation.

Moreover, the style of this book is not uniform. Some passages are rather elementary, with many details, whereas others are much denser and technically more difficult. You don't have to read this book linearly. It might be most efficient to first select those parts that you find easiest to understand and only subsequently proceed to the more technical sections.

I thank Nils Bertschinger, Timo Ehrig, Alihan Kabalak, Martin Kell, Eckehard Olbrich, Johannes Rauh, and other participants of my course for useful questions, insightful comments, and helpful suggestions. Over the years, of course, the conceptual point of view presented here has also been strongly influenced by the work of or discussions with friends and colleagues. They include the mathematicians Nihat Ay, Paul Bourgine, Andreas Dress, Tobias Fritz, Xianqing Li-Jost, Stephan Luckhaus, Eberhard Zeidler, and the late Heiner Zieschang, the theoretical biologists Peter Stadler and the late Olaf Breidbach, and the organization scientist Massimo Warglien. Oliver Pfante checked the manuscript and found some typos and minor inconsistencies. I am also grateful to several reviewers for constructive criticism. Pengcheng Zhao provided some of the figures. For creating most of the diagrams, I have used the latex supplement DCpic of Pedro Quaresma.

I thank the IHES for its hospitality during the final stages of my work on this book. I also acknowledge the generous support from the ERC Advanced Grant FP7-267087, the Volkswagenstiftung and the Klaus Tschira Stiftung.

Contents

Notational Conventions

Often, when a set is equipped with some additional structure which is clear from the context, that structure is not included in the notation. For instance, we may speak of a topological space X and thus write X in place of $(X, \mathcal{O}(X))$ where $\mathcal{O}(X)$ is the collection of open subsets of X defining the topological structure.

Overview and Perspective

1

1.1 Overview

This book is concerned with the conceptual structure of mathematics. Leaving largely aside the concepts of infinity and limit,[1] we shall cover those aspects that we can perhaps summarize under the headings

1. Distinctions
2. Relations
3. Abstractions
4. Generation

In this introductory chapter, I shall attempt to describe the above aspects in general and informal terms. While this may be somewhat vague and abstract, I hope that it can serve as some motivation for the more formal developments in subsequent chapters.

1.1.1 Properties and Distinctions

For concreteness, we consider here a set S and its elements. Subsequently, however, we shall consider more general structures and try to talk about them in more intrinsic terms.

When we consider the elements of a set, we want to distinguish them. Or more precisely, it is rather the other way around. In order to be able to identify some individual element, item, or object, we need to be able to distinguish it from others. Such a distinction can come from a property that some elements possess in contrast to others. While any particular property may only separate one collection of objects from another one, an element might then become identifiable through the combination of several properties that together are not shared by any other object.

---

[1] In the original sense of mathematical analysis; unfortunately, also certain constructions in category theory have been named "limits", and we shall deal with those in this text.

© Springer International Publishing Switzerland 2015
J. Jost, *Mathematical Concepts*, DOI 10.1007/978-3-319-20436-9_1

In the most basic case, we have a binary distinction, that is, an object can have a property or not. When the objects are elements of a set, we then have the subset of those elements that possess the property, and the complementary subset of those that don't. We can then also turn things around and start with a specific subset. The property would then be that an element belongs to that subset. Thus, if we can identify arbitrary subsets of a given set, we have as many distinctions as there are subsets, and conversely, if we have sufficiently many distinctions, we can identify arbitrary subsets. A finite set with n elements has 2^n subsets. When every subset corresponds to a distinction, we may not only distinguish a single element from the rest, but we can also distinguish any collection of elements from the rest, that is, those not in that particular collection. Thus, there are exponentially more distinctions than elements. As Cantor discovered, this principle also applies to infinite sets.

It turns out that often a more refined perspective is needed. A property may not hold absolutely, but perhaps only under certain circumstances, or in certain possible worlds, an old idea of Leibniz as revived by the logician Kripke. The question then is not what is actually the case, but which possibilities are compatible with each other, or compossible, using Leibniz' terminology. Also, properties need not be independent of each other, but there might exist relationships or correlations between their values at one instance or in different possible worlds. We shall describe the formal structures appropriate for handling these issues.

1.1.2 Relations

When we discuss properies as a tool to make distinctions, this leads to the question of where such properties come from. A property should relate to the structure of an object. This in turn leads to the question of what the structure of an object is. From a formal point of view, as adopted here, *the structure of an object consists of its relations with other objects*. This, as we shall explore in this book, is a very powerful principle. In particular, this applies to large classes of formal objects, and not only to the elements of sets. Nevertheless, there is an important difficulty. An object may also entertain relations with itself. These can be viewed as symmetries of that object.

In the diagram below, when we reflect the left figure across the dashed line, we get a different figure, as depicted on the right.

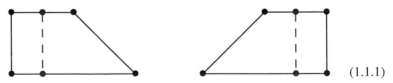

$$(1.1.1)$$

When, however, we reflect the figure in the next diagram, we get a figure that is undistinguishable from the first one.

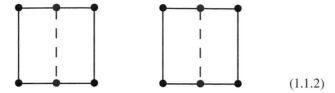

(1.1.2)

One says that the figure is symmetric about the dashed axis. The important point here is that we can describe that symmetry as the invariance under a certain *operation*, in this case a reflection. In fact, this particular figures possesses further symmetries. It is also invariant under reflection about a vertical axis, or under a counterclockwise rotation about its center by 90, 180 or 270°.

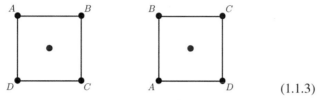

(1.1.3)

In this diagram, we have labelled the vertices in order to indicate the effect of the counterclockwise rotation by 90°. Again, these symmetries are described in terms of operations. In contrast, the figure on the left in (1.1.1) is not invariant under any such operation. So, that figure is related by the operation of reflection about the vertical line to a different figure, that on the right. Thus, applying that operation here leads to another figure, whereas the square in (1.1.2) is related to itself, and the operation does not produce a new figure. Such an operation that relates an object to itself is called an *automorphism* of that object. Such automorphisms can be composed. For instance, we may first reflect the object about the vertical axis and then about the horizontal one or apply a rotation of a multiple of 90° to it. Still, the figure stays invariant. Or, one can reverse the effect of an automorphism. We could first rotate the figure counterclockwise by 90°, and then rotate it clockwise by 90° (or equivalently, rotate it counterclockwise by 270°). One expresses these facts, that automorphisms can be composed and reversed, by saying that the automorphisms form a *group*. Thus, the square in (1.1.2) possesses a symmetry group consisting of reflections and rotations and their compositions. Other objects may, of course, have other symmetry groups. Among the plane figures, the circle is that with the largest symmetry group. In contrast, the figure in (1.1.1) does not possess any non-trivial symmetries. Only the identity operation, which consists in doing nothing, leaves that figure invariant. (This identity operation, although it may appear completely trivial, is very important for the mathematical structure of a group. Just think of the role of 0 in arithmetic. Adding 0 does not change any number, and so the addition of 0 is the identity operation of arithmetic. Although, or perhaps rather because this is so obvious, it took a long time to discover 0 as an arithmetic object.)

Once we understand the structure of the symmetries, that they form a group, we possess a general tool. For the example just analyzed, this group is generated by a reflection and a rotation. But the principle is general and applies to other geometric shapes and helps to organize the pattern of the symmetries that such a shape might have. For instance, without the concept of a group, it might be a rather formidable task to understand the symmetries of, say, an icosahedron, but with a little group theory, this becomes quite easy.

Thus, when we want to characterize an object through its relations with other objects, we also have to take the automorphism group of that object into account. When an object possesses a nontrivial automorphism group, this in particular indicates that it has some internal structure. We can then try to resolve that structure through a more fine-grained perspective. Instead of considering the category (a term to be formally defined in Chap. 2, but perhaps already intuitively meaningful at this stage) of figures in the plane, within which the square is a single object, we rather focus on the category consisting only of the four vertices of the square. Each of these vertices has no nontrivial internal structures, and so cannot be further resolved. The automorphisms of the square (except for the trivial one, the identity), while leaving the square as a whole invariant, then permute those points. Nevertheless, there is some structure between the points that remains invariant. Whenever two vertices are connected by an edge, then their images under any automorphism of the square will still be connected by an edge. For instance, no automorphism of the square can just interchange the positions of A and B, but leave the other two vertices, C and D, invariant. When, in contrast, we forget those relations and consider only a set consisting of four elements A, B, C, D, then we can freely permute them without changing that set itself. Thus, the automorphism of the set of four elements is the group of all permutations of those elements. It is larger than the automorphism group of the square which consisted only of certain permutations, those corresponding to reflections and rotations. The square as a collection of four vertices with four edges between them, that is, where four out of the six possible pairs have a particular relation, as expressed by an edge, has more structure than the simple set of four elements. And since there is more structure that needs to be preserved, the automorphism group is correspondingly smaller. This is a general principle, that additional structure reduces the number of symmetries.

Let us return once more to the figure in (1.1.1)

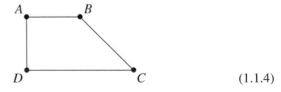

$$(1.1.4)$$

The connectivity pattern as expressed by the edges is the same as for the square. When we consider (1.1.4) as a figure in the plane, it is different from (1.1.3) because the distances between the vertices are different. Denoting the distance between two points P, Q by $d(P, Q)$, we would have $d(A, B) = d(A, D) < d(B, D) = d(B, C) < d(C, D) < d(A, C)$. In particular, any vertex can be distinguished from another one by its specific collection of distances. For this reason, there cannot be any symmetry that preserves all distances, in contrast to the situation for the square, where the arrangement $d(A, B) = d(B, C) = d(C, D) = d(A, D) < d(A, C) = d(B, D)$ still permits some, but not all, of the vertex permutations to be symmetries.

In any case, the preceding already suggests to interpret relations in terms of formal operations. Thus, by considering the structure of objects we are led to consider the structure of sets or classes of operations. The principle can then be iterated, taking such classes of operations as objects and describing their structure in terms of new operations, that is, considering operations on operations.

Objects or structures can also often be turned into operations. For instance, in arithmetic, a number might be considered as an object, but it can also be seen as the operations of adding or multiplying by that number. In that way, an object translated into an operation can gain a generative power. For instance, through the repeated addition of 1, one can obtain any natural number.

The preceding has been argued from the perspective of a single object. But we can also assume the perspective that a relation takes place between two or more objects. We can then identify those collections of objects that stand in specific relations with each other. This suggests a geometric representation. For simplicity, we discuss here only symmetric relations, that is whenever A stands in a relation with B, then B also stands in a relation with A. When the relation is binary, that is, involves only two objects at a time, we can represent it by a graph,

$$A \bullet\!\!\!-\!\!\!-\!\!\!-\!\!\!-\!\!\!-\!\!\!-\!\!\!\bullet B \qquad\qquad (1.1.5)$$

that is, that if A and B are related, then, representing A and B as vertices, the relation is depicted by an edge connecting them. And if C is not related to A or to B, it is not connected to either of them.

Now, when it is also possible that three points stand in a relation, we need to distinguish the situation where we only have pairwise relations between, say, A, B and C,

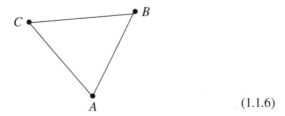

$$(1.1.6)$$

from that where also the triple A, B, C has a relation.

We depict the latter by filling the triangle enclosed by the three edges. Of course, this scheme also allows us to represent more complicated configurations of relations geometrically, by a so-called *simplicial complex*. The concepts of algebraic topology will then enable us to systematically distinguish between qualitatively different situations. For that purpose, we shall need and construct more refined invariants than the Euler characteristic introduced in the next Sect. 1.1.3, the (co)homology groups.

$$(1.1.7)$$

There can be different types of relations. For simplicity, we restrict ourselves here to the case of relations between two objects. They could be

- *discrete*, as in the case of the graph (1.1.5) where a relation is expressed by the presence of an edge
- *qualitative*, like nearness, or
- *quantitative*, like the distance between two points.

1.1.3   Abstraction

Abstraction ignores or forgets irrelevant details and identifies the essential aspects or properties. One such possibility is to describe a collection of objects by their number, that is, count how many objects there are, regardless of their individual or shared properties. Thus, a finite set is characterized by the number of its elements, that is, by a positive integer, taken from $\mathbb{N}$. But in many cases, we can not only *count* objects, but also *measure* them. Measurements usually yield real numbers instead of integers. Since not every real number is rational, that is, described as a fraction of two integers, as already the old Greeks discovered, or in any other way by finitely many integers, measuring cannot be reduced to counting.[2]

There also exist coarser distinctions than enumeration. For instance, we could simply look at the parity of a number, whether it is odd or even. When we use 1 for an odd and 0 for an even number, and when we then restrict the operations that we know from the integers, addition and multiplication, to those parities, we obtain the rule $1 + 1 = 0$ (the sum of two odd

---

[2]The Pythagorean principle that nature reduces to numbers was inspired by the discovery that musical harmonies are based on rational frequency relations. More generally, physically interacting elements usually get into some state of resonance that is expressed by rational relations. This may be the general reason why measurements can typically be reduced to counting after all, notwithstanding the fact that roots of algebraic equations with rational coefficients are usually not rational themselves, and numbers like π and e are not even algebraic, that is, cannot be obtained as roots of such algebraic equations.

numbers is even). With that, we get an algebraic structure on the set with the two elements 0 and 1. Despite its simplicity, this algebraic structure is fundamental.

Both counting and measuring utilize only positive numbers. When we also admit negative numbers, we gain the possibility of cancellation. In other words, we can count with signs. For instance, when one has an object like a simplicial complex that contains constituents of different dimension, one can count the even-dimensional constituents with a positive sign and the odd-dimensional constituents with a negative sign.

$$
\begin{matrix}
\bullet\ +1 \\
\big|\ -1 \\
\bullet\ +1
\end{matrix}
\tag{1.1.8}
$$

When we sum the contributions in this figure, we obtain $+1 - 1 + 1 = 1$. This is the so-called Euler characteristic (this idea had already been discovered by Descartes, in fact). It is the same as that of a single point,

$$
\bullet\ +1
\tag{1.1.9}
$$

When the single point in (1.1.9) stands for the bottom vertex in (1.1.8), we could express this by saying that the upper vertex and the edge in (1.1.8), being in adjacent dimensions, have cancelled each other. In fact, we can turn this cancellation into a process,

$$
\begin{matrix}
\bullet & & \bullet & & \bullet & & \bullet \\
\big| & & \big| & & \big| & & \\
\bullet & & \bullet & & \bullet & & \\
t = 0 & & t = \frac{1}{3} & & t = \frac{2}{3} & & t = 1
\end{matrix}
\tag{1.1.10}
$$

which takes place in continuous time from $t = 0$, when the full configuration is there, until time $t = 1$, when it has shrunk to a point. In other words, an edge with two vertices can be continuously deformed into a single vertex. That both configurations have the same Euler characteristic, 1 in this case, is a necessary (but not sufficient) condition for such a deformation to be possible. On the other hand, the single vertex, as in (1.1.9) stays there and cannot be deformed to nothing, because "nothing" would have Euler characteristic 0 instead of 1.

In the same vein, the triangle in (1.1.6), consisting of three vertices and three edges, has Euler characteristic $3 \cdot (+1) + 3 \cdot (-1) = 0$, whereas the filled triangle below it has an additional two-dimensional piece that will count as $+1$, so that the Euler characteristic becomes $3 \cdot (+1) + 3 \cdot (-1) + 1 = 1$. Therefore, this simple count with cancellations tells us that the unfilled and the filled triangle are "topologically different". In particular, one cannot be deformed into the other, because (as we shall prove) topological invariants have to stay constant under continuous deformations. Of course, for this simple example, this may appear as completely evident, but the power of the formalism will be so general that it will also allow for such conclusions in situations which are not evident at all.

1.1.4   Generation

In many cases, it is extremely cumbersome to describe a structure by listing all details, and in other cases, this is even impossible. One cannot list all positive integers simultaneously, as such a list would have to be infinite. But this is not necessary anyway. One can simply prescribe the principle of generation. Starting with 1, by repeated addition of 1, one can obtain any positive integer. In the terminology of Aristotle, instead of an actual infinity, we have a potential one, as the process of adding 1 needs never end. Thus, we only need a generator, the integer 1 in this case, and a rule of generation, addition in this example, for a condensed description of the structure, the set of positive integers here. This principle applies to other mathematical structures as well, be they finite or infinite. For instance, the square in (1.1.3) can be generated from a single vertex, say A, by applying rotations about the center to generate the other vertices. Thus, the generator here would be the vertex A, and the operations would be those rotations. (If we had also wanted to include the edges, we should have started with the edge from A to B, say, and would have obtained the others by rotating that one.)

At a more abstract level, proofs of mathematical statements are generated from the premises by applying the rules of logical inference. Once one understands how to do it, one no longer has to carry out all the details. As emphasized by Hermann Weyl [115], what is important is rather the creative power of mathematics, that is, the construction or discovery of rich structures in which suitable premises lead to contentful results, and the identification of those lines of reasoning among the multitude of possible ones that can deduce such results. One finds a proof of a mathematical result not by mechanically trying out one possible formal scheme after the next, but rather through structural insights.

The possibility of generation depends on structural regularities. This can be quantified. Kolmogorov's concept of algorithmic complexity assigns to each structure the length of the shortest program that can generate it. The more regularities a structure possesses, the more it can be condensed, and the shorter the corresponding program. Thus, in the algorithmic sense, regular structures are simple, and random ones, without apparent regularities, are algorithmically complex. Often, structures that appear rather complicated can be generated by simple rules. For instance, some simple dynamical rules can generate rather intricate patterns, like chaotic attractors.

1.2   A Brief Historical Sketch

This book is conceptual, not historical. Historically, mathematical discoveries have not always been made in their logical order. Perhaps, it was often rather the reverse of the logical order. While in hindsight some of the general principles may appear rather simple, typically only after some concrete, but complicated structures had been investigated and understood in great detail, the underlying abstract principle emerged. In other words, there is the tension or contrast between the universal validity, or so mathematicians think, of mathematical structures and results, and their historically contingent genesis or discovery. Therefore, a conceptual and a historical perspective are necessarily different. Nevertheless, I shall attempt here a—very brief and perhaps rather superficial—sketch of the history of mathematical principles and structures. Of course, this should not and cannot replace a more systematic investigation of the history of mathematics.[3]

Abstract mathematics began in classical Greek antiquity. Euclid (fl. ca. 300 B.C.) developed a formal deductive scheme for the geometry of the plane and of three-dimensional space. In fact, after the bible, Euclid's "Elements" is probably the most translated text in the intellectual history of mankind, and that with the most printed editions and copies. The Euclidean system of deductive geometry still forms an essential component of mathematical high school curricula. Aristotle (384–322 B.C.) isolated the basic formal rules for logical inference, and this was similarly influential in the history of Western thought.

Gottfried Wilhelm Leibniz (1646–1716) then put forward the powerful idea of the formalization of thinking via a symbolic language or scheme. His version of infinitesimal calculus won the day over Newton's because of his superior symbolism. In contrast, his ideas about geometry as an abstract science of spatial relationships (which he called "Analysis situs", the analysis of position) were not sufficiently developed and formalized in order to have a similarly sweeping impact (see [97] for a detailed analysis). But he also discovered the binary system, that is, the rules for computation with the two symbols 0 and 1 only.

Leonhard Euler (1707–1783), inspired by Leibniz' idea of an Analysis situs, could translate a combinatorial problem into a graph theoretical one, that is, construct a geometric representation, and solve the latter. In a rather different direction, but also inspired by Leibniz' vision, Hermann Grassmann (1809–1877) developed what is now called linear algebra, the rules for the algebraic operations with vectors and matrices. The science of space then took a decisive new turn with Bernhard Riemann's (1826–1866) concept of a manifold equipped with a notion of distance derived from an infinitesimal structure [96]. A manifold is an abstract object, in contrast to a geometric shape in three-dimensional Euclidean space. In any case, the conceptual approach of Riemann, as opposed to a computational or algorithmic one, had a large impact. Georg Cantor (1845–1918) went on to create abstract set theory [20], and the axiomatic version of Ernst Zermelo (1871–1953) and Abraham Fraenkel (1891–1965) is still considered to be the most basic foundation of mathematics, even though this has been chal-

---

[3]Of course, there exist many books on the history of mathematics, for instance the very comprehensive [21] and [119]. Perhaps naturally, most of them develop a historical instead of a conceptual perspective, and therefore are not so useful for our present enterprise. The most systematic and comprehensive history of mathematical thought that I am aware of is [68]. Studies of the historical development of mathematical concepts that combine a historical perspective with that of mathematical research are contained in [67, 27]. Besides general treatises on the history of mathematics, there also exist studies devoted to the development of mathematical structures, like [37] on set theory or [25] on algebraic structures.

lenged by proposals from category theory. Felix Hausdorff (1868–1942) then developed set theoretical topology from the basic notion of an open set, and thereby came up with the abstract notion of space that was generally adopted [48].

Although Riemann's original approach had been conceptual and not algorithmic, the tensor calculus of Riemannian geometry, as created in particular by Gregorio Ricci-Curbastro (1853–1925) ultimately became one of the most powerful algorithmic tools of mathematics, and, in particular, the mathematical basis of Einstein's theory of general relativity, as well as of modern quantum field theory. The concept of a connection, that is, how to pass from the infinitesimal geometry at one point of a Riemannian manifold to that at another point, was first introduced by Tullio Levi-Civita (1873–1941) and then in particular developed by Hermann Weyl (1885–1955) (for a general perspective, see [114]). The notion of a fiber bundle, where a geometric structure like that of a vector space or a Lie group is attached to every point of a manifold or, from a another perspective, where we have a family of such fibers indexed by the points of a manifold, was clarified by Charles Ehresmann (1905–1979).

Riemann's work also led to a new direction in topology—which was a new name for Leibniz' Analysis situs—the development of homology and cohomology theory. This was achieved by the combinatorial approach to topology of Henri Poincaré (1854–1912) and Luitzen Brouwer (1881–1966). Poincaré also emphasized the qualitative aspects of topology in his work on dynamical systems.

Modern algebra, and in fact, even mathematics as such as an autonomous science, distinct from the study of natural phenomena, is said to have begun with Carl Friedrich Gauss' (1777–1855) "Disquisitiones arithmeticae". The notion of a group emerged from the work of Gauss and, in particular, of Évariste Galois (1811–1832). Felix Klein (1849–1925) then defined geometrical structures through the groups of transformations that leave their constitutive relations invariant. Sophus Lie (1842–1899) systematically developed the theory of symmetries in group theoretic terms.

Riemann's conceptual ideas found their way into the work of Richard Dedekind (1831–1916), and what is now called "modern algebra" was developed by Emmy Noether (1882–1935) and her school, see in particular [111]. This also provided a new foundation of algebraic geometry, in the hands of André Weil (1906–1998) and many others, culminating in the work of Alexandre Grothendieck (1928–2014) (see for instance [4]). This work unified algebraic geometry and number theory and made spectacular developments in the latter field possible. This was achieved through a higher level of abstraction than had hitherto been considered feasible.

Such a level of abstraction had been prepared in the work of David Hilbert (1862–1943) on the axiomatic foundation of geometry [50] and other areas of mathematics, and then in the very systematic approach of the Bourbaki group to the structural thinking of mathematics. In particular, the important notion of a (pre)sheaf was introduced by Jean Leray (1906–1998) which then became a basic tool in the hands of Grothendieck, who developed concepts like schemes and topoi, see [4].

Working on the foundations of algebraic topology (see [31]), Samuel Eilenberg (1913–1998) and Saunders Mac Lane (1909–2005) invented the structural approach of category theory [32]. Sheaves, schemes, and topoi find their natural place in this framework. This even extends to logic and provides a common foundation for geometry and logic. This not only covers classical, but also intuitionistic logic where the law of the excluded middle no longer needs to hold and propositions may be only locally, but not necessarily globally true. The language of topoi can deal with such structures of contingent truths. Taking up another fundamental idea of Leibniz, that of possible worlds, the logician Saul Kripke (1940-) had constructed a possible world semantics for logic [72], and it was then discovered that the notion of a topos provides an abstract tool for investigating this aspect.

The preceding sketch of a conceptual history of mathematics was necessarily very cursory and perhaps rather one-sided, with many omissions. It sets the stage, however, for the topics to be developed and discussed in this book. In Sect. 5.1, when we discuss the concept of space, we shall return to the history of geometry.

Foundations

2

2.1 Objects, Relations, and Operations

2.1.1 Distinctions

A *set* is a collection of distinct or distinguishable objects, its elements. But how can these elements be distinguished? Possibly, by certain specific intrinsic properties that they possess in contrast to others. Better, by specific relations that they have with other elements.

This leads to the concept of *equivalence*, of "equal" versus "different", or of "nondistinguishable" versus "distinguishable". Objects that have the same properties or that stand in the same relation to all other elements are equivalent, as they cannot be distinguished from each other. Therefore, one might want to identify them, that is, treat them as the same and not as different objects. (One should be aware of the possibility, however, that the identification need not be unique because the objects may possess certain internal symmetries or automorphisms, a concept to be explained below.)

So, when we have a set of points, we may not be able to distinguish these points from each other, and therefore, we should identify them all. The resulting set would then consist of a single point only. However, strangely, we can distinguish different sets by their number of points or elements. That is, as soon as we can distinguish the elements of a set, we can also distinguish different sets. As we shall see, however, two sets with the same number of elements cannot be distinguished from each other, unless we can distinguish the elements themselves between the two sets.

Returning to the intrinsic properties, an element could have some internal structure, for instance be a set itself, or some space (see below), that is, it could be an object with structure. Or, conversely, we could say that an object consists of elements with relations between them. In any case, however, this seems to be some kind of higher level element. But the basic mathematical formalism ignores this kind of hierarchy. A collection of sets can again be treated as a set (so long as certain paradoxes of self-reference are avoided). More abstractly, we shall introduce the notion of a category below.

© Springer International Publishing Switzerland 2015
J. Jost, *Mathematical Concepts*, DOI 10.1007/978-3-319-20436-9_2

2.1.2 Mappings

In this section, we start with an elementary concept that most likely will be familiar to our readers. We utilize this also to introduce diagrams as a useful tool to visualize examples for some of the general concepts that we shall introduce in this chapter, and perhaps also to compensate our readers a little for the steepness of some of the subsequent conceptual developments.

Let us consider two sets S and T. We consider a special type of relation between their elements.

Definition 2.1.1 A *map*, also called a *mapping*, $g : S \to T$ assigns to each element $s \in S$ one and only one element $g(s) \in T$, also written as $s \mapsto g(s)$. Such a map $g : S \to T$ between sets is called *injective* if whenever $s_1 \neq s_2 \in S$, then also $g(s_1) \neq g(s_2)$. That is, different elements in S should have different images in T. The map $g : S \to T$ is called *surjective* if for every $t \in T$, there exists some, in general not unique, $s \in S$ with $g(s) = t$. Thus, no point in T is omitted from the image of S. Of course, if g were not surjective, we could simply replace T by $g(S)$ to make it surjective. A map $g : S \to T$ that is both injective and surjective is called *bijective*.

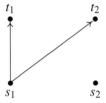

This does not define a map from the set $\{s_1, s_2\}$ to the set $\{t_1, t_2\}$ because s_1 has two images, instead of only one, whereas s_2 has none.

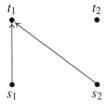

This is a map, but it is neither injective, because t_1 has two different preimages, s_1 as well as s_2, nor surjective, because t_2 has no preimage at all.

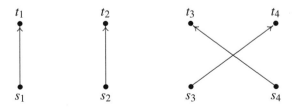

This represents a bijective map from $\{s_1, s_2, s_3, s_4\}$ to $\{t_1, t_2, t_3, t_4\}$.

Let us consider the set $\{1\}$ consisting of a single element only. Let S be another non-empty set. We can then specify an element s of S by a map

$$f : \{1\} \to S \text{ with } f(1) = s. \tag{2.1.1}$$

Thus, the set S corresponds to the maps $f : \{1\} \to S$. The set $\{1\}$ serves as a universal spotlight that can be directed by a map f to any particular element of the set S.

When S' is a subset of S, $S' \subset S$, then we have the inclusion map

$$i : S' \to S$$
$$s \mapsto s \text{ for every } s \in S'. \tag{2.1.2}$$

This map i is injective, but not surjective, unless $S' = S$.

More generally, a binary relation between the sets S and T is given by a collection of ordered pairs $R = \{(s, t)\}$ where $s \in S, t \in T$. While a relation is more general than a map, we can represent every such relation R by a map

$$r : S \times T \to \{0, 1\}$$
$$r((s, t)) = 1 \quad \Leftrightarrow \quad (s, t) \in R, \tag{2.1.3}$$

and hence, equivalently, $r((s, t)) = 0$ iff $(s, t) \notin R$. We might say here that 1 stands for "true", that is, the relation R holds, whereas 0 stands for "false", that is, R does not hold for the pair (s, t).

Mappings can be composed. That is, if $f : S \to T$ and $g : T \to V$ are maps, then the map $h := g \circ f$ is defined by

$$s \mapsto f(s) \mapsto g(f(s)), \tag{2.1.4}$$

i.e., s is mapped to the image of $f(s)$ under the map g, $g(f(s))$. We note that for this procedure to be possible, the target T of f, also called its *codomain*, has to agree with the *domain* of the map g.

Lemma 2.1.1 *The composition of maps is associative. This means that when $f : S \to T, g : T \to V, h : V \to W$ are maps, then*

$$h \circ (g \circ f) = (h \circ g) \circ f =: h \circ g \circ f. \tag{2.1.5}$$

Proof Under either of the variants given in (2.1.5), the image of $s \in S$ is the same, $h(g(f(s)))$. $\square$

Let us also explain the use of the brackets (.). The expression $h \circ (g \circ f)$ means that we first compute the composition $g \circ f$—let us call it η, and then the composition $h \circ \eta$. For $(h \circ g) \circ f$, it is the other way around. We first compute the composition $h \circ g =: \phi$ and then $\phi \circ f$. (2.1.5) tells us that the two results agree. In general, brackets (.) are used to specify the order in which the various operations in a formula have to be carried out. In some cases, there exists a general convention for such an order, and in those cases, brackets will not be needed. For instance, in an expression $a \cdot b + c$, we first compute the product $a \cdot b$ and then add c to it. Also, when two expressions are connected by an equality or inequality sign, like $a + b \leq c \cdot d$, then the expressions on the left- and on the right-hand side of that sign are each first computed and then the corresponding results are compared. But probably, every reader knows that. We shall repeatedly make use of such implicit conventions.

2.1.3 Power Sets and Distinctions

We consider a set S with finitely many elements, say $S = \{s_1, s_2, \ldots, s_n\}$. And we assume that there is some property that may either apply or not to each element of S. We write $P(s) = 1$ when s satisfies this property, and $P(s) = 0$ if it does not. According to the separation principle of Zermelo-Frankel set theory (see Sect. 2.2), each such P specifies the subset PS of those s that satisfy $P(s) = 1$, i.e.,

$$PS = \{s \in S : P(s) = 1\}. \qquad (2.1.6)$$

Conversely, for any subset S' of S, we can define such a property $P_{S'}$ by

$$P_{S'}(s) = 1 \text{ if and only if } s \in S'. \qquad (2.1.7)$$

Thus,

$$P_{S'}S = S'. \qquad (2.1.8)$$

We call the set of all subsets of S its *power set* $\mathcal{P}(S)$. Thus, each *subset* $S' \subset S$ becomes an *element* $S' \in \mathcal{P}(S)$. We can also say that each element of $\mathcal{P}(S)$ corresponds to a *distinction* that we can make on the elements of S, whether they possess a property P or not.

We can also interpret P as a map

$$P : S \to \{0, 1\} \qquad (2.1.9)$$

from S into the set $\mathbf{2} := \{0, 1\}$, and we shall therefore also write

$$\mathcal{P}(S) = \mathbf{2}^S. \qquad (2.1.10)$$

Now, when S is the empty set $\emptyset$, then its power set, the set of all its subsets, is

$$\mathcal{P}(\emptyset) = \{\emptyset\}, \qquad (2.1.11)$$

because the empty set is a subset of every set, hence also of itself. Thus, the power set of the empty set is not empty, but contains $\emptyset$ as its single element. That is, the power set of the empty set contains one element. This is the trivial distinction, but it is a distinction nonetheless.

Next, when $S = \{1\}$ contains a single element, which we now denote by 1, then its power set is

$$\boxed{\{1\}}$$

$$\mathcal{P}(\{1\}) = \{\emptyset, \{1\}\}, \qquad (2.1.12)$$

because we now have two possible properties or distinctions. When 1 does not satisfy the property, we get $PS = \emptyset$, but when 1 does satisfy it, we have $PS = \{1\}$.

$$\boxed{\{1, 2\}}$$

Moving on to a set $S = \{1, 2\}$ with two elements, we have

$$\mathcal{P}(\{1, 2\}) = \{\emptyset, \{1\}, \{2\}, \{1, 2\}\}, \qquad (2.1.13)$$

because now a property can be satisfied by none, both or either of the elements. That is, on two elements, we can make four different distinctions.

Pursuing this pattern, we see that the power set of a set with n elements has 2^n elements. Thus, the size of the power set grows exponentially as a function of the size of the original set.

We next turn to infinite sets. We start with the positive integers $\mathbb{N} = \{1, 2, 3, \dots\}$ and call a set S *infinite* if there exists an *injective map*

$$i : \mathbb{N} \to S. \tag{2.1.14}$$

$\mathbb{N}$ possesses infinite subsets that do not contain all the elements of $\mathbb{N}$ itself. For instance, the set of even positive integers, $2\mathbb{N} = \{2, 4, 6, \dots\}$, is infinite because we have the injective map $i : \mathbb{N} \to 2\mathbb{N}$ with $i(n) = 2n$ for $n \in N$. In fact, in this case, the map i is even bijective. There exist other sets S which may appear larger than $\mathbb{N}$ from a naive perspective for which we nevertheless have a bijection $i : \mathbb{N} \to S$. For example, let us consider the set $S = \{(n, m)\}$ of all pairs of positive integers n, m. We construct a bijection by the following procedure:

$$1 \mapsto (1, 1),$$
$$2 \mapsto (1, 2), \quad 3 \mapsto (2, 1),$$
$$4 \mapsto (1, 3), \quad 5 \mapsto (2, 2), \quad 6 \mapsto (3, 1),$$
$$7 \mapsto (1, 4), \dots,$$

that is, for every $k \in \mathbb{N}$ we enumerate the finitely many pairs with $n + m = k$ and after that move on to $k + 1$. Similarly, we can construct a bijection between $\mathbb{N}$ and the set of all N-tuples of elements of $\mathbb{N}$, for every $N \in \mathbb{N}$.

However, this is not possible between $\mathbb{N}$ and its power set $\mathcal{P}(\mathbb{N})$. Every element of $X \in \mathcal{P}(\mathbb{N})$ corresponds to a distinction that we can make in $\mathbb{N}$, that is, to a property that we can check for every $n \in \mathbb{N}$ and then assign those n that satisfy it to the set X. We can also express this via a binary sequence, like

$$100110100\dots \tag{2.1.15}$$

which means that the integers $1, 4, 5, 7, \dots$ satisfy the property whereas $2, 3, 6, 8, 9, \dots$ don't. We now describe Cantor's famous diagonal argument that the set of all such binary sequences $\sigma = (\sigma_1, \sigma_2, \sigma_3, \dots)$ cannot be put in a bijective correspondence with $\mathbb{N}$. The argument proceeds by contradiction. Suppose that there were such a correspondence, $i : n \mapsto \sigma(n)$. We then consider the sequence σ' constructed as follows. When $\sigma_k(k) = 1$, we put $\sigma'(k) = 0$, and if $\sigma_k(k) = 0$, we put $\sigma'(k) = 1$. Thus, at the kth position, σ' is different from $\sigma(k)$. Therefore, for every k, there is some position for which σ' is different from $\sigma(k)$. Thus, σ' is different from all the sequences $\sigma(k)$. But that means that the correspondence i is not surjective, and this is the contradiction.

In general, the power set $\mathcal{P}(S)$ of a set S is always "larger" than S itself, in the sense that on a set of elements, we can make more distinctions than there are elements.

Cantor's argument showed that there is no surjective map

$$i : \mathbb{N} \to 2^{\mathbb{N}}. \tag{2.1.16}$$

We shall now generalize the argument and the result (and the reader may want to skip the rest of this section upon a first reading) and show that only under very special circumstances can there be a surjective map

$$f : S \to \Lambda^S$$
$$x \mapsto f_x : S \to \Lambda \tag{2.1.17}$$

where

$$\Lambda^S := \{\lambda : S \to \Lambda\} \tag{2.1.18}$$

is the set of all maps from S to Λ. Each map $f : S \to \Lambda^S$ (whether surjective or not) yields

$$\tilde{f} : S \times S \to \Lambda$$
$$\tilde{f}(x, y) = f_x(y), \tag{2.1.19}$$

that is, for x, we have the map $f_x : S \to \Lambda$ which we can apply to $y \in S$. We then have

Lemma 2.1.2 *If there is a surjective map*

$$g : S \to \Lambda^S \tag{2.1.20}$$

then every map

$$\lambda : \Lambda \to \Lambda \tag{2.1.21}$$

has a fixed point, that is, there exists some $\ell \in \Lambda$ with

$$\lambda(\ell) = \ell. \tag{2.1.22}$$

Proof We consider the diagonal embedding

$$\Delta : S \to S \times S$$
$$x \mapsto (x, x) \tag{2.1.23}$$

and, recalling (2.1.19), the map

$$\phi := \lambda \circ \tilde{g} \circ \Delta : S \to \Lambda. \tag{2.1.24}$$

Anticipating the notations of category theory to be introduced in Sect. 2.3, we can represent this by a diagram

$$
\begin{array}{ccc}
S \times S & \xrightarrow{\ \ \tilde{g}\ \ } & \Lambda \\
\Delta \big\uparrow & & \big\downarrow \lambda \\
S & \xrightarrow{\ \ \phi\ \ } & \Lambda
\end{array}
\quad . \tag{2.1.25}
$$

Now, if g is surjective, there has to be some $x_0 \in S$ with

$$g(x_0) = \phi, \text{ or equivalently, } \tilde{g}(x_0, y) = \phi(y) \text{ for all } y \in S, \tag{2.1.26}$$

and hence in particular, and this is the crucial diagonal argument,

$$\tilde{g}(x_0, x_0) = \phi(x_0). \tag{2.1.27}$$

But then

$$\phi(x_0) = \lambda \circ \tilde{g} \circ \Delta(x_0) = \lambda \circ \tilde{g}(x_0, x_0) = \lambda(\phi(x_0)),$$

that is,

$$\ell = \phi(x_0)$$

satisfies (2.1.22), i.e., is a fixed point. □

Now, of course, for $\Lambda = \{0, 1\}$, the map λ with $\lambda(0) = 1, \lambda(1) = 0$ does not have a fixed point. Therefore, there can be no surjective map $g : S \to \{0, 1\}$ for any S, and in particular not for $S = \mathbb{N}$. Thus, Cantor's result holds. Note that Cantor's idea is translated into the diagonal operator Δ and the formula (2.1.27) in the above proof.

$\boxed{\{0, 1\}}$

More generally, on any set Λ with more than one element, we can permute the elements to construct a map without fixed points. Therefore, the above argument can be translated into a proof by contradiction. For any set Λ with more than one (for instance two) elements, the existence of a surjective map (2.1.20) would lead to a contradiction.

2.1.4 Structures

2.1.4.1 Binary Relations

We now look at a single set S. A *structure* consists of relations between the elements of the set S. Often, these relations are conceived or imagined as spatial relations. This leads to the concept of a *space*, to be defined below. This brings us into the realm of *geometry*.

A *relation* on a set S is given by a map

$$F : S \times S \to R \tag{2.1.28}$$

for some range set R. (In Chap. 3, we shall also consider relations involving more than two elements of a set S.)

Definition 2.1.2 When we have two sets with relations, (S_1, F_1) and (S_2, F_2), with the same range R, then we call a map $\phi : S_1 \to S_2$ a *homomorphism* (i.e., structure preserving) if for all $s, s' \in S_1$

$$F_1(s, s') = r \text{ implies } F_2(\phi(s), \phi(s')) = r \tag{2.1.29}$$

for all $r \in R$. We shall then also write this as $\phi : (S_1, F_1) \to (S_2, F_2)$.

Definition 2.1.3 Let $F : S \times S \to R$ be a relation and $\phi : S' \to S$ a map. We then define the *pullback relation* $\phi^* F : S' \times S' \to R$ by

$$\phi^* F(s', s'') = F(\phi(s'), \phi(s'')) \text{ for } s', s'' \in S'. \tag{2.1.30}$$

In particular, when $S' \subset S$, we can pull back a relation from S to S' via the inclusion (2.1.2).

With this definition,

$$\phi : (S', \phi^* F) \to (S, F) \tag{2.1.31}$$

becomes a homomorphism. We record this observation as a principle.

Theorem 2.1.1 *Relations can be pulled back by mappings, and the corresponding mappings then become homomorphisms.*

The simplest relation is a binary one, as explained at the end of Sect. 2.1.2. That is, two elements either stand in a relation, or they don't. When S is our set, according to (2.1.3), this can be expressed as

$$F : S \times S \to \{0, 1\}. \tag{2.1.32}$$

This is also known as a *directed graph* (sometimes also called a *digraph*) with vertex set S and with an ordered pair $(s_1, s_2) \in S \times S$ being an edge when $F(s_1, s_2) = 1$. We also call this an edge from s_1 to s_2.

This depicts a digraph with edges (s_1, s_2), (s_2, s_3), (s_3, s_2).

When F is *symmetric*, that is $F(s_1, s_2) = F(s_2, s_1)$ for all s_1, s_2, then this yields an undirected graph, usually simply called a *graph* for short. Here are some pictures of graphs.

Some graphs

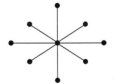

The second and the third are highly symmetric whereas the first does not exhibit any symmetries. Here, a *symmetry* would be a bijective homomorphism h from the vertex set to itself. That it is a homomorphism refers to the edge relation. It means that (s_1, s_2) is an edge precisely if $(h(s_1), h(s_2))$ is. Anticipating our discussion of automorphism groups, we see that such symmetries can be composed, that is, if h_1 and h_2 are symmetries, then so is $h_2 \circ h_1$. As an exercise, the reader might determine all the symmetries of the latter two graphs and their composition rules. The answer for the last graph is that the vertex in the center has to stay fixed under any symmetry whereas any permutation of the other 8 vertices yields a symmetry.

We now introduce some important types of relations.

Definition 2.1.4 When the binary relation F is

  reflexive: $F(s, s) = 1$ for all s

  transitive: if $F(s_1, s_2) = 1$ and $F(s_2, s_3) = 1$, then also $F(s_1, s_3) = 1$

  symmetric: $F(s_1, s_2) = F(s_2, s_1)$,

then we say that F defines an *equivalence relation* on S. In this case, one usually writes $s_1 \equiv s_2$ for $F(s_1, s_2) = 1$. We then define a quotient S/F of

S by the equivalence relation F whose elements are the equivalence classes

$$[s] := \{s' \in S : s' \equiv s\} \tag{2.1.33}$$

for $s \in S$.

In this section, we shall draw a number of graphs or digraphs to illustrate algebraic structures geometrically. We shall usually omit edges from the vertices to themselves, to simplify our drawings. That is, when depicting reflexive relations, the reflexivity condition is always assumed, but not drawn.

Here, two vertices s, s' are connected by an edge iff $F(s, s') = 1$. The first and the third graph here then represent equivalence relations, whereas the second one does not, as it is not transitive.

One can view an equivalence relation F on S as a partitioning of S into the equivalence classes. In the quotient S/F, equivalence is transformed into equality, $[s_1] = [s_2]$ iff $s_1 \equiv s_2$. We also obtain an induced relation F_q on S/F by putting $F_q([s], [s']) = 1$ if $[s] = [s']$ and $= 0$ otherwise. The map

$$q : S \to S/F$$
$$s \mapsto [s] \tag{2.1.34}$$

is then a homomorphism. Thus, an equivalence relation F induces a map (2.1.34) from S to its quotient S/F. Conversely, a map $\phi : S \to S'$ defines an equivalence relation by

$$F(s_1, s_2) = 1 \quad :\Leftrightarrow \quad \phi(s_1) = \phi(s_2), \tag{2.1.35}$$

that is, we identify elements of S that have the same image under ϕ. The target set S' thus becomes the quotient S/F.

There always exists the trivial equivalence relation F_0 on S with $F_0(s_1, s_2) = 1$ only if $s_1 = s_2$.

Definition 2.1.5 When the relation F is

      reflexive: $F(s, s) = 1$ for all s

      transitive: if $F(s_1, s_2) = 1$ and $F(s_2, s_3) = 1$, then also $F(s_1, s_3) = 1$

antisymmetric: if $F(s_1, s_2) = 1$ and $F(s_2, s_1) = 1$, then $s_1 = s_2$,

then we say that (S, F) is a *partially ordered set*, or *poset* for short. One usually writes $s_1 \leq s_2$ in place of $F(s_1, s_2) = 1$ in this case. A partial order provides some (partial) ranking of the elements of S.

Again, the second graph here is not transitive, nor is the fourth, and therefore they do not represent posets, whereas the other two do. Also, any graph with two arrows in opposite directions between the same two vertices would not be antisymmetric and hence would not represent a poset.

Definition 2.1.6 A *lattice* is a poset $(S, \leq)$ for which any two elements s_1, s_2 have a unique greatest lower bound, that is, there exists some $\underline{s}$, also written as $s_1 \wedge s_2$ and called the *meet* of s_1 and s_2, with

$$\underline{s} \leq s_1, \ \underline{s} \leq s_2, \text{ and } s \leq \underline{s} \text{ whenever } s \leq s_1, s \leq s_2, \qquad (2.1.36)$$

and a unique least upper bound $\bar{s}$, also written as $s_1 \vee s_2$ and called the *join* of s_1 and s_2, with

$$s_1 \leq \bar{s}, \ s_2 \leq \bar{s}, \text{ and } \bar{s} \leq s \text{ whenever } s_1 \leq s, s_2 \leq s. \qquad (2.1.37)$$

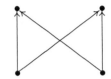

This poset is not a lattice as it neither has a unique greatest lower nor a unique least upper bound.

We leave it to the reader to check that the operations $\wedge$ and $\vee$ are associative and commutative. Here, associativity of, for instance, $\wedge$ means that always

$$(s \wedge s') \wedge s'' = s \wedge (s' \wedge s''), \qquad (2.1.38)$$

and commutativity that always

$$s \wedge s' = s' \wedge s. \qquad (2.1.39)$$

These notions will be taken up in Definitions 2.1.12 and 2.1.13 below.

Definition 2.1.7 We say that the lattice possesses 0 and 1 (not to be confused with the values in (2.1.32)), if it contains elements 0, 1 with the property that for all $s \in S$

$$0 \leq s \leq 1. \qquad (2.1.40)$$

Equivalently, as we leave for the reader to check, a lattice with 0 and 1 is a set with two binary associative and commutative operations $\wedge$ ("and"), and $\vee$ ("or") and two distinguished elements 0, 1, satisfying

$$s \wedge s = s, \ s \vee s = s \qquad (2.1.41)$$

$$1 \wedge s = s, \ 0 \vee s = s \qquad (2.1.42)$$

$$s \wedge (s' \vee s) = s \ = (s \wedge s') \vee s \qquad (2.1.43)$$

for any elements s, s'. The ordering can be recovered from these conditions by stipulating that $s \leq s'$ iff $s \wedge s' = s$, or equivalently, iff $s \vee s' = s'$. The exercise then amounts to verifying that these properties imply that $\leq$ defines an ordering in the sense of Definition 2.1.5, from the properties of $\wedge$ and $\vee$. Thus, here, we can recover a structure from operations; this aspect will be taken up in Sect. 2.1.6 below.

2.1.4.2 Metrics

When the range of F is larger, we obtain more general types of relations. When

$$F : S \times S \to \mathbb{R}, \tag{2.1.44}$$

we obtain the structure of a weighted and directed graph, with $F(s_1, s_2)$ being the weight of the edge from s_1 to s_2.

When we require that

$$F : S \times S \to \mathbb{R}^+ \text{ (the nonnegative real numbers),} \tag{2.1.45}$$

be symmetric, i.e. $F(s_1, s_2) = F(s_2, s_1)$ for all s_1, s_2, and satisfy the triangle inequality

$$F(s_1, s_3) \leq F(s_1, s_2) + F(s_2, s_3) \text{ for all } s_1, s_2, s_3, \tag{2.1.46}$$

we obtain a *pseudometric*.

When the points s, s' satisfy $F(s, s') = 0$, then, by (2.1.46), also $F(s, \sigma) = F(s', \sigma)$ for all other σ. Therefore, s and s' cannot be distinguished by their relations with other elements in terms of the pseudometric. Therefore, according to the general principle described above, they should be identified. (Above, in the definition of an equivalence relation, we had identified elements with $F(s_1, s_2) = 1$, but, of course, it amounts to the same when we identify elements with $F(s_1, s_2) = 0$ instead. It is an exercise for the reader to check that, for a pseudometric, this does indeed define an equivalence relation.) When we then identify all such equivalent points, we obtain a new set $\bar{S}$, a quotient of the original one, with an induced metric $\bar{F}$. Here, using the standard notation d in place of $\bar{F}$ for a metric, we have

$$d(x_1, x_2) > 0 \text{ whenever } x_1 \neq x_2, \text{ for all } x_1, x_2 \in \bar{S}. \tag{2.1.47}$$

When these conditions are satisfied, $d(., .)$ is called a *metric*, and we also say that (S, d) is a metric space (the notion of space will be defined below).

A metric provides us with a quantitative notion of nearness, in the sense that we can not only say that for instance y is closer than z to x if

$$d(x, y) < d(x, z), \tag{2.1.48}$$

but we can also quantify that difference.

Examples

1. On the real line $\mathbb{R}$, we have the Euclidean metric

Euclidean metric

$$d(x, y) = |x - y| \text{ for } x, y \in \mathbb{R}. \tag{2.1.49}$$

2. More generally, on $\mathbb{R}^d$, the set of d-tuples $(x^1, \ldots, x^d)$, $x^i \in \mathbb{R}$ for $i = 1, \ldots, d$, we have the Euclidean metric

$$d(x, y) = \sqrt{\sum_{i=1}^{d} (x^i - y^i)^2} \text{ for } x = (x^1, \ldots, x^d), y = (y^1, \ldots, y^d)$$

(2.1.50)

which, of course, reduces to (2.1.49) for $d = 1$.

Trivial metric

3. On any set S, we can define a metric by

$$d(s_1, s_2) = \begin{cases} 0 & \text{if } s_1 = s_2 \\ 1 & \text{if } s_1 \neq s_2. \end{cases}$$

(2.1.51)

Thus, any two different points have the same distance from each other. For a set with three points, this looks like

This metric is trivial in the sense that it does not allow any further distinction beyond whether two points are the same or different.

4. A metric d on the set S defines a connected graph if for any $s \neq s' \in S$ there exist $s_0 = s, s_1, \ldots, s_n = s' \in S$ with

$$d(s_{i-1}, s_i) = 1 \text{ for } i = 1, \ldots, n, \quad \text{and } d(s, s') = n \quad (2.1.52)$$

and we can then let the pair (s_1, s_2) define an edge iff $d(s_1, s_2) = 1$.
The first part of the condition then says that any two elements can be connected by a chain of edges. In that sense, the graph is connected. The second part of the condition then specifies that the distance between two vertices of the graph equals the minimal number of edges needed to get from one to the other.

5. On the set of binary strings of some fixed length n, that is, on objects of the form $(b_1 b_2 \ldots b_n)$ with $b_i \in \{0, 1\}$, we have the Hamming distance that counts in how many positions two strings $b = (b_1 b_2 \ldots b_n)$, $b' = (b_1' b_2' \ldots b_n')$ differ, that is,

$$d(b, b') = \sum_{i=1}^{n} |b_i - b_i'|.$$

(2.1.53)

6. Whenever $S' \subset S$, a metric on S induces a metric on S' by pullback under the inclusion $i : S' \to S$, see (2.1.2).

When d is a metric on S and $\phi : S' \to S$ a map, then $\phi^* d$ is a metric on S' only if ϕ is injective. Otherwise, it is only a pseudometric, and we need to pass to the quotient where points with the same image under ϕ are identified to obtain a metric, as just described.

The next definition will be formulated for metric spaces only, although it would also work for pseudometrics. The reason for this restriction is that its content for metric spaces is more interesting and useful than for general pseudometrics.

Definition 2.1.8 Let (S, d) be a metric space. We say that the point $z \in S$ is *between* the points x and y if

$$d(x, y) = d(x, z) + d(y, z), \tag{2.1.54}$$

that is, if the triangle inequality (2.1.46) becomes an equality.

A subset C of S is called *convex* if whenever $x, y \in C$, then also all points that are between x and y are contained in C as well.

The points z_1 and z_2 are between x and y with respect to the Euclidean metric in the plane $\mathbb{R}^2$, but w is not.

For the trivial metric (2.1.51), no point z is between two other points x, y. Consequently, any subset of a set S equipped with that metric is convex.

Definition 2.1.9 Let $p_1, \ldots, p_n \in S$ where (S, d) is a metric space. Then a point $b \in S$ with

$$b = \mathrm{argmin}_q \sum_{i=1,\ldots,n} d^2(p_i, q) \tag{2.1.55}$$

is called a barycenter of $p_1, \ldots, p_n$. Also, a point $m = m(p_1, p_2) \in S$ with

$$d(p_1, m) = d(p_2, m) = \frac{1}{2} d(p_1, p_2) \tag{2.1.56}$$

is called a midpoint of p_1 and p_2.

In particular, a midpoint $m(p_1, p_2)$ is between p_1 and p_2 in the sense of Definition 2.1.8. A midpoint m, if it exists, is also a barycenter of p_1 and p_2. This is easily seen. Let a be any point in S, and let $d(p_1, a) = \lambda d(p_1, p_2)$. (We may assume $\lambda \leq 1$, as otherwise p_2 would yield a lower value in (2.1.55) than a.) By the triangle inequality, $d(p_2, a) \geq (1 - \lambda) d(p_1, p_2)$. Thus,

$$d^2(p_1, a) + d^2(p_2, a) \geq \lambda^2 d^2(p_1, p_2) + (1 - \lambda)^2 d^2(p_1, p_2)$$
$$\geq \frac{1}{2} d^2(p_1, p_2) = d^2(p_1, m) + d^2(p_2, m),$$

and a midpoint thus indeed yields the smallest possible value in (2.1.55).

There is another characterization of midpoints that will become relevant in Sect. 5.3.3. Let

$$B(p, r) := \{q \in S; d(p, q) \leq r\} \text{ for } p \in S, r \geq 0 \tag{2.1.57}$$

be the *closed ball* with center p and radius r. Given $p_1, p_2 \in S$, we then ask for the smallest radius $r = r(p_1, p_2)$ with

$$B(p_1, r) \cap B(p_2, r) \neq \emptyset. \tag{2.1.58}$$

We then observe

Lemma 2.1.3 *The following are equivalent*

(i) $p_1, p_2 \in S$ *possess a midpoint.*
(ii) $r(p_1, p_2) = \frac{1}{2}d(p_1, p_2)$.

When S is finite, barycenters always exist, but midpoints need not. Neither barycenters nor midpoints need be unique. For the metric (2.1.51), there are no midpoints (unless $p_1 = p_2$), but any of the p_i is a barycenter of $p_1, \ldots, p_n$. On a connected graph as characterized by (2.1.52), s and s' possess a (not necessarily unique) midpoint iff their distance is an *even* integer.

2.1.5 Heyting and Boolean Algebras

In this section, we shall consider particular classes of lattices, the Heyting and Boolean algebras. These will play an important role in our discussion of topoi in the last chapter, and they will also arise in our discussion of topologies. Nevertheless, a reader who is primarily interested in the general and abstract aspects of structures may wish to skip this section upon a first reading and only return to it at some later stage.

Definition 2.1.10 A lattice with 0 and 1 is a *Heyting algebra* if for any elements s, s', there exists a (unique) element, called the *implication* $s \Rightarrow s'$, satisfying

$$t \leq (s \Rightarrow s') \text{ iff } t \wedge s \leq s'. \tag{2.1.59}$$

The element

$$\neg s := (s \Rightarrow 0) \tag{2.1.60}$$

is called the *pseudo-complement* of s.

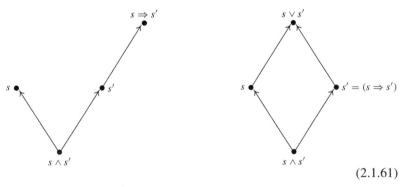

$$\tag{2.1.61}$$

In the diagrams (where the arrows from the lowest to the highest vertex that are required by transitivity are not shown), we see that $s \Rightarrow s'$ sits above s', but cannot sit higher above s than s', that is, it still has to satisfy $s \wedge (s \Rightarrow s') = s \wedge s'$, and in fact, it is the highest such element. For the pseudo-complement, we draw the following diagrams (where again some arrows required by transitivity are not shown)

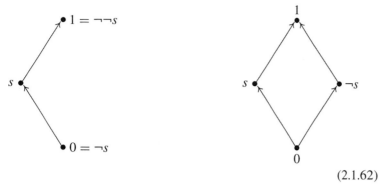

$$(2.1.62)$$

In a Heyting algebra, we have

$$(((s \Rightarrow s') \wedge s) \Rightarrow s') = 1, \qquad (2.1.63)$$

because $t \leq (((s \Rightarrow s') \wedge s) \Rightarrow s')$ iff $t \wedge (s \Rightarrow s') \wedge s \leq s'$ (by (2.1.59) and the associativity of $\wedge$) iff $t \wedge (s \Rightarrow s') \leq (s \Rightarrow s')$ (by (2.1.59) again), and this is satisfied for all t by the definition of $\wedge$, and finally, any element σ with $t \leq \sigma$ for all t has to be 1, as follows from (2.1.40). In the terminology of elementary logic, (2.1.63) says that the "modus ponens" is always valid. As an exercise for this formalism, you may wish to check that $t \Rightarrow (s \Rightarrow s') = t \wedge s \Rightarrow s'$.

In order to develop the concept of a Heyting algebra further, it is useful to reformulate the definition (2.1.10) of a Heyting algebra without invoking the order relation $\leq$.

Lemma 2.1.4 *A Heyting algebra is a set with two binary associative and commutative operations $\wedge, \vee$ and two distinguished elements $0, 1$, satisfying*

$$x \wedge x = x, \quad x \vee x = x \qquad (2.1.64)$$

$$1 \wedge x = x, \quad 0 \vee x = x \qquad (2.1.65)$$

$$x \wedge (y \vee x) = x = (x \wedge y) \vee x \qquad (2.1.66)$$

for any elements x, y, and a binary operation $\Rightarrow$ characterized as follows. For any elements y, z, there exists a (unique) element $y \Rightarrow z$ satisfying

$$x \wedge (y \Rightarrow z) = x \; iff \; x \wedge y \wedge z = x \wedge y \qquad (2.1.67)$$

for all x.

Proof The relations (2.1.64)–(2.1.66) are simply (2.1.41)–(2.1.43), and as explained there, from these operations, we can then recover the order relation $\leq$; it is characterized by $x \leq y$ iff $x \wedge y = x$, or equivalently, iff $x \vee y = y$. $\qquad \square$

Therefore, (2.1.67) can also be written as

$$x \leq (y \Rightarrow z) \text{ iff } x \wedge y \leq z \qquad (2.1.68)$$

for all x.

From the symmetry of $\wedge$, we get the symmetry

$$x \leq (y \Rightarrow z) \text{ iff } y \leq (x \Rightarrow z). \tag{2.1.69}$$

In the sequel, we leave out the brackets and write, for instance,

$$x \vee y \leq w \Rightarrow z \text{ in place of } (x \vee y) \leq (w \Rightarrow z). \tag{2.1.70}$$

That is, the operations $\vee, \wedge, \Rightarrow$ are carried out before the relation $\leq$ is applied. Similarly for $=$ in place of $\leq$.

Lemma 2.1.5 *A Heyting algebra satisfies the distributive law*

$$(x \vee y) \wedge z = (x \wedge z) \vee (y \wedge z) \text{ for all } x, y, z. \tag{2.1.71}$$

Proof For the proof strategy, we observe that

$$x = y \text{ iff for all } w : (x \leq w \text{ iff } y \leq w). \tag{2.1.72}$$

Indeed, this holds in any poset, by inserting $w = x$ and $w = y$ into (2.1.72) and using the antisymmetry of the ordering.

Now $(x \vee y) \wedge z \leq w$ iff $x \vee y \leq z \Rightarrow w$ (by 2.1.68) iff $(x \leq z \Rightarrow w$ and $y \leq z \Rightarrow w)$ iff $(x \wedge z \leq w$ and $y \wedge z \leq w)$ iff $(x \wedge z) \vee (y \wedge z) \leq w$. □

Similarly, we have distributive laws for $\Rightarrow$.

Lemma 2.1.6 *A Heyting algebra satisfies*

$$(x \vee y) \Rightarrow z = (x \Rightarrow z) \wedge (y \Rightarrow z) \tag{2.1.73}$$

and

$$x \Rightarrow (y \wedge z) = (x \Rightarrow y) \wedge (x \Rightarrow z) \tag{2.1.74}$$

for all x, y, z.

Note that we have both $\vee$ and $\wedge$ in (2.1.73), but only $\wedge$ in (2.1.74).

Proof By (2.1.69), $w \leq (x \vee y) \Rightarrow z$ iff $x \vee y \leq w \Rightarrow z$ iff $(x \leq w \Rightarrow z$ and $y \leq w \Rightarrow z)$ iff, using (2.1.69) again, $(w \leq x \Rightarrow z$ and $w \leq y \Rightarrow z)$ iff $w \leq (z \Rightarrow z \wedge y \Rightarrow z)$. This shows (2.1.73). We leave the proof of (2.1.74) to the reader. □

There are some further rules that we shall employ below.

$$x \Rightarrow x = 1 \tag{2.1.75}$$
$$x \wedge (x \Rightarrow y) = x \wedge y \tag{2.1.76}$$
$$y \wedge (x \Rightarrow y) = y \tag{2.1.77}$$
$$(x \Rightarrow (y \wedge x)) = x \Rightarrow y \quad \text{for all } x, y. \tag{2.1.78}$$

For instance, (2.1.78) follows directly from (2.1.71) and (2.1.75), and we leave the other equations as an exercise.

(2.1.60) and (2.1.73) imply one of De Morgan's laws,

$$\neg(x \vee y) = \neg x \wedge \neg y. \tag{2.1.79}$$

(The other De Morgan law, $(2.1.84) \neg (x \wedge y) = \neg x \vee \neg y$, does not hold in all Heyting algebras, but only in Boolean algebras. The reader can check this with the example of $\mathcal{O}(X)$, the algebra of open subsets of a topological space X, introduced in Sect. 4.1 below.)

We conclude our treatment of general Heyting algebras with

Lemma 2.1.7 *When a lattice L with 0 and 1 carries a binary operation $\Rightarrow$ satisfying (2.1.75)-(2.1.78), then this defines a Heyting algebra structure on L.*

Thus, the Heyting algebra structure can be recovered from the two binary operations $\wedge$ and $\Rightarrow$.

Proof We need to show that the two sides of (2.1.67) are equivalent when the conditions (2.1.75)–(2.1.78) hold. Thus, for going from left to right, assume that

$$x \wedge (y \Rightarrow z) = x, \qquad \text{hence}$$
$$x \wedge y = x \wedge y \wedge (y \Rightarrow z)$$
$$= x \wedge y \wedge z \text{ by (2.1.76)}$$

so that the right-hand side of (2.1.67) holds. Conversely,

$x = x \wedge (y \Rightarrow x)$ by (2.1.77)
$= x \wedge (y \Rightarrow (y \wedge x))$ by (2.1.78)
$= x \wedge (y \Rightarrow (x \wedge y \wedge z))$ if the r.h.s. of (2.1.67) holds
$= x \wedge ((y \Rightarrow z \wedge y) \wedge (y \Rightarrow x))$ by (2.1.77) (with the roles of x and y interchanged)
$= x \wedge (y \Rightarrow z)$ by (2.1.78) and (2.1.76)

which is the left-hand side of (2.1.67). $\qquad\square$

Definition 2.1.11 Finally, a *Boolean algebra* is a Heyting algebra in which the pseudo-complement $\neg s$ of every element s satisfies

$$s \vee \neg s = 1 \text{ and } s \wedge \neg s = 0. \tag{2.1.80}$$

$\neg s$ is then called the complement of s.

In particular, we then have in a Boolean algebra

$$\neg\neg s = s \text{ for every } s. \tag{2.1.81}$$

In a general Heyting algebra, however, this need not hold; we only have

$$s \leq \neg\neg s. \tag{2.1.82}$$

Thus, the right diagram in (2.1.62) is Boolean, but the left one is not.

Also, the implication and the pseudo-complement in a Boolean algebra are related by

$$s \Rightarrow s' = \neg s \vee s'. \tag{2.1.83}$$

Also, in a Boolean algebra, we have

$$\neg(s_1 \wedge s_2) = \neg s_1 \vee \neg s_2. \tag{2.1.84}$$

Therefore, in a Boolean algebra, the pseudo-complement together with $\vee$ determines $\wedge$ by applying (2.1.81)–(2.1.84)

$$s_1 \wedge s_2 = \neg(\neg s_1 \vee \neg s_2). \tag{2.1.85}$$

Similarly, in a Boolean algebra, $\neg$ and $\wedge$ determine $\vee$,

$$s_1 \vee s_2 = \neg(\neg s_1 \wedge \neg s_2). \tag{2.1.86}$$

A basic example of a Boolean algebra is $\{0, 1\}$ with the above operations, that is,

$$0 \wedge 0 = 0 \wedge 1 = 0, 1 \wedge 1 = 1, 0 \vee 0 = 0, 1 \vee 1 = 0 \vee 1 = 1, \neg 0 = 1. \tag{2.1.87}$$

As a more general example, let X be a set, and let $\mathcal{P}(X)$ be its power set, that is, the set of all its subsets. $\mathcal{P}(X)$ has an algebraic structure with the following operations as $\neg, \vee, \wedge, \Rightarrow$ (for all $A, B \in \mathcal{P}(X)$):

$$\text{Complement:} \quad A \mapsto X \setminus A \tag{2.1.88}$$
$$\text{Union:} \ (A, B) \mapsto A \cup B \tag{2.1.89}$$
$$\text{Intersection:} \ (A, B) \mapsto A \cap B := X \setminus (X \setminus A \cup X \setminus B) \tag{2.1.90}$$
$$\text{Implication:} \ (A, B) \mapsto A \Rightarrow B := (X \setminus A) \cup B. \tag{2.1.91}$$

We note that $C \cap A \subset B$ iff $C \subset (A \Rightarrow B)$, that is, the condition required in (2.1.59).

We also have the relations

$$A \cup (X \setminus A) = X \tag{2.1.92}$$
$$A \cap (X \setminus A) = \emptyset \tag{2.1.93}$$

for all $A \in \mathcal{P}(X)$. Thus, $\emptyset$ and X assume the roles of 0 and 1, resp., that is,

$$\emptyset \subset A \quad \text{(and also } A \cap \emptyset = \emptyset) \tag{2.1.94}$$

and

$$A \subset X \quad \text{(and also } A \cup X = X) \tag{2.1.95}$$

for all $A \in \mathcal{P}(X)$.

The Boolean algebra $\{0, 1\} \cong \{\emptyset, X\}$ then arises as the power set of a set X with a single element.

Boolean algebra $\{0, 1\} \cong \{\emptyset, X\}$

However, when we take X as a set with 2 elements, say $X = \{0, 1\}$,[1] and put

$$\mathcal{O}(X) := \{\emptyset, \{0\}, \{0, 1\}\}, \tag{2.1.96}$$

then we have a Heyting algebra that is not Boolean (because $\{0\}$ has no complement).

Heyting algebra $\{\emptyset, \{0\}, \{0, 1\}\}$

Returning to the case of Boolean algebras, and the above example of $\mathcal{P}(X)$ as an example of such a Boolean algebra, this is in fact an instance of a general relationship as given in the representation theorem of Stone. (As that theorem will not be utilized or referred to in the sequel, and since its proof invokes concepts that may seem a bit technical, some readers may want to skip it on a first reading and continue after (2.1.107).)

---

[1]Please do not get confused by the different meanings of the symbols 0 and 1 here, as algebraic symbols on one hand and as members of a set on the other hand.

Theorem 2.1.2 (Stone) *For any Boolean algebra B with operations* $\vee, \wedge, \neg$, *there exist a set X and an injective homomorphism of Boolean algebras*

$$h : B \to \mathcal{P}(X). \tag{2.1.97}$$

Here, a homomorphism $\eta : B_1 \to B_2$ of Boolean algebras has to satisfy $\eta(s_1 \wedge s_2) = \eta(s_1) \wedge \eta(s_2)$ for all $s_1, s_2 \in B_1$ where on the left-hand side, we have the operation $\wedge$ in B_1, and on the right-hand side the operation in B_2, and analogous relations for the other operations $\vee, \neg$, and also $\eta(0) = 0, \eta(1) = 1$, where again on the left-hand sides, we have the elements 0 and 1 in B_1, whereas on the right-hand sides, the corresponding elements in B_2. (Again, this is an instance of the concept of a homomorphism as a structure preserving map; here, the structure is that of a Boolean algebra, but we have also encountered and will repeatedly encounter homomorphisms for other structure. In Sect. 2.3, the concept will be developed from an abstract perspective.)

Proof A filter $\mathcal{F}$ on B is defined to be a subset of B with the following properties

$$0 \notin \mathcal{F}, \quad 1 \in \mathcal{F}, \tag{2.1.98}$$

$$\text{if } s \in \mathcal{F}, s \leq s', \quad \text{then also } s' \in \mathcal{F}, \tag{2.1.99}$$

$$\text{if } s_1, \ldots, s_n \in \mathcal{F}, \quad \text{then also } s_1 \wedge \cdots \wedge s_n \in \mathcal{F}. \tag{2.1.100}$$

An ultrafilter $\mathcal{F}$ is defined to be a maximal filter, that is, whenever for some filter $\mathcal{G}$

$$\mathcal{F} \subset \mathcal{G}, \quad \text{then } \mathcal{F} = \mathcal{G}. \tag{2.1.101}$$

Equivalently, $\mathcal{F}$ is an ultrafilter iff

$$\text{for all } s \in B, \quad \text{either } s \in \mathcal{F} \text{ or } \neg s \in \mathcal{F}. \tag{2.1.102}$$

(Note that as a consequence of (2.1.98) and (2.1.100), s and $\neg s$ cannot both be contained in a filter $\mathcal{F}$.)

The idea of the proof then is to let X be the set of all ultrafilters on B and define

$$h : B \to \mathcal{P}(X)$$
$$s \mapsto \{\mathcal{F} : s \in \mathcal{F}\}, \tag{2.1.103}$$

and verify that this is an injective homomorphism of Boolean algebras. So, let's see how this goes. Constructing filters is easy. For $s \in B$, let $\mathcal{F}(s) := \{s' \in B : s \leq s'\}$. Such an $\mathcal{F}(s)$ is called a principal filter. In general, however, this is not an ultrafilter. We have $\mathcal{F}(s) \subsetneq \mathcal{F}(\sigma)$ for any $\sigma \leq s, \sigma \neq s$. We can then try to iterate this construction and obtain larger and larger filters that asymptotically yield an ultrafilter. In fact, in general, one has to appeal to the axiom of choice to ensure the existence of ultrafilters.

We observe, however, that by (2.1.98) and (2.1.100), we cannot augment a filter $\mathcal{F}(s)$ so that it contains two elements $s_1, s_2 \leq s$ with $s_1 \wedge s_2 = 0$. Thus, if $s \not\leq t$, then $\neg t \wedge s \neq 0$, and hence we can obtain an ultrafilter containing s, but not t. This yields the injectivity of h.

Among the conditions to check for a homomorphism of Boolean algebras, the most difficult one is

$$h(s_1 \vee s_2) = h(s_1) \cup h(s_2). \tag{2.1.104}$$

We shall check this one and leave the others to the reader. Thus, if $s_1 \vee s_2 \in \mathcal{F}$ for some ultrafilter, then we claim that also s_1 or s_2 is in $\mathcal{F}$. This will then yield $h(s_1 \vee s_2) \subset h(s_1) \cup h(s_2)$. If on the contrary, neither of s_1, s_2 were in $\mathcal{F}$, then by (2.1.102), both $\neg s_1, \neg s_2 \in \mathcal{F}$, hence also $\neg s_1 \wedge \neg s_2 \in \mathcal{F}$ by (2.1.100). But since also, by (2.1.84), $\neg(\neg s_1 \wedge \neg s_2) = s_1 \vee s_2$ is in $\mathcal{F}$ by assumption, this would contradict (2.1.102). For the other direction, that is, $\supset$ in (2.1.104), if $s_1 \vee s_2 \notin \mathcal{F}$, then $\neg s_1 \wedge \neg s_2 = \neg(s_1 \vee s_2) \in \mathcal{F}$, and hence neither s_1 nor s_2 can be in $\mathcal{F}$, using (2.1.98) and (2.1.100) again. □

Actually, an ultrafilter $\mathcal{F}$ on a Boolean algebra B is the same as a homomorphism η from B to the Boolean algebra $\{0, 1\}$; we have

$$\eta(s) = 1 \text{ iff } s \in \mathcal{F}. \tag{2.1.105}$$

We leave it to the reader to check that the properties of ultrafilters ensure that this is indeed a homomorphism of Boolean algebras.

Also, the definition of a filter in the above proof is meaningful for any Heyting algebra H, not necessarily Boolean. Moreover, a filter $\mathcal{G}$ (not necessarily ultra) in H yields a homomorphism $\eta : H \to K$ into another Heyting algebra, with $\eta^{-1}(1) = \mathcal{G}$; this is the natural generalization of (2.1.105). For a principal filter $\mathcal{F}(s)$, this Heyting algebra is simply the poset $H/s := \{s' \in H : s' \leq s\}$ with the Heyting algebra operations induced from H. Equivalently, we can obtain it as the space of equivalence classes for

$$s_1 \equiv s_2 \text{ iff } s_1 \wedge s = s_2 \wedge s. \tag{2.1.106}$$

This construction extends to a general filter $\mathcal{F}$ as

$$s_1 \equiv s_2 \text{ iff } s_1 \wedge s' = s_2 \wedge s' \text{ for some } s' \in \mathcal{F}. \tag{2.1.107}$$

Let us now consider a structure that is not a Heyting or Boolean algebra. | Euclidean space $\mathbb{R}^d$ | We take the Euclidean space $\mathbb{R}^d$,[2] that is, the space consisting of tuples $x = (x^1, \ldots, x^d)$ with components $x^i \in \mathbb{R}$. The elements of $\mathbb{R}^d$ can be added; with $y = (y^1, \ldots, y^d)$, we have

$$x + y = (x^1 + y^1, \ldots, x^d + y^d) \tag{2.1.108}$$

and multiplied by real numbers α,

$$\alpha x = (\alpha x^1, \ldots, \alpha x^d). \tag{2.1.109}$$

Also, we have the scalar product

$$\langle x, y \rangle = \sum_{i=1}^{d} x^i y^i. \tag{2.1.110}$$

---

[2] This is, of course, an example of a vector space, but that latter concept will only be introduced below. Therefore, we recall some details here.

A linear subspace of $\mathbb{R}^d$ is a subset of the form

$$V = \{\alpha^1 v_1 + \dots \alpha^m v_m : \alpha^i \in \mathbb{R}, i = 1, \dots, d\} \qquad (2.1.111)$$

for some elements $v_1, \dots, v_m \in \mathbb{R}^d$, called generators of V; given V, these v_j are not unique, but this does not matter for our purposes. For two linear subspaces V, W, as is easily checked, their intersection $V \cap W$ is again a linear subspace, and so is their direct sum

$$V \oplus W := \{v + w : v \in V, w \in W\}. \qquad (2.1.112)$$

Finally, for a linear subspace, we have the complementary subspace

$$V^\perp := \{w \in \mathbb{R}^d : \langle v, w \rangle = 0 \text{ for all } v \in V\}. \qquad (2.1.113)$$

This is where we need the scalar product. Again, the complementary subspace of a linear subspace is indeed itself a linear subspace.

When we then take the intersection $\cap$ of linear subspaces as $\wedge$, the direct sum $\oplus$ as $\vee$ and the complement $\perp$ as $\neg$, $\mathbb{R}^d$ itself as 1, and the trivial subspace $\{0\}$ (where 0 stands for the element $(0, \dots, 0)$ of $\mathbb{R}^d$ whose components are all 0) as 0,[3] the linear subspaces of $\mathbb{R}^d$ do not constitute a Heyting algebra. For instance, the distributive law of Lemma 2.1.5 is not satisfied. Some subspace $W \neq \{0\}$ may intersect two other subspaces V_1, V_2 only at 0, but may nevertheless be contained in the direct sum $V_1 \oplus V_2$. For example, consider $\mathbb{R}^2$ spanned by the vectors $e_1 = (1, 0)$, $e_2 = (0, 1)$ and let V_1, V_2, W be the one-dimensional subspaces spanned by $e_1, e_2, e_1 + e_2 = (1, 1)$, resp. Then $V_1 \oplus V_2 = \mathbb{R}^2$ and hence this space contains W while neither V_1 nor V_2 does. In such a case, $(V_1 \oplus V_2) \cap W = W$, whereas $(V_1 \cap W) \oplus (V_2 \cap W) = \{0\}$. For those readers who went through the proof of Theorem 2.1.2, it is also instructive to see why that proof does not apply to this example. In fact, an ultrafilter $\mathcal{F}$ would have to correspond to a smallest subspace $\neq \{0\}$, that is, a subspace W generated by a single vector $w \neq 0$. In other words, the linear filter $\mathcal{F}$ would then consist of all linear subspaces V containing that W. But then (2.1.102) does not hold in general. In fact, there are many subspaces W for which neither W nor $W^\perp$ contain V. For example, take W as above as the span of $(1, 1)$ in $\mathbb{R}^2$, and V as the span of $(1, 0)$. Then neither V nor $V^\perp$, the space spanned by $(0, 1)$, contain W.

In any case, returning to (2.1.32), as soon as we have a basic distinction between two values, 0 and 1 (as in (2.1.32)), false or true, out or in, or whatever, then we can use that distinction to define relations between the elements of a set S.

2.1.6   Operations

Instead of relations, one can also consider *operations*. These are transformations of the elements of a set. When we have a structure, the operations

---

[3]Carefully distinguish the different meanings of the symbol 0 employed here!

are required to preserve that structure, in a sense defined below. The operations themselves usually constitute a structure. Such a structure can operate on itself, as in the case of a group, or it can operate on another structure, as for the representation of a group. In any case, operations bring us into the realm of *algebra*.

Operations also offer a new perspective on the issue of equivalence. Given an operation of some structure on a set, one can consider two elements of the latter as equivalent when there is an operation moving one of them into the other. In this way, one can form the quotient of the set by the operation, by identifying two elements that are related by an operation, or as one also says in this context, that can be mapped to each other. In order that this be an equivalence relation, so that we can indeed construct such a quotient, we need to require that this operation be reflexive, that is, every element can be mapped to itself (that is, doing nothing counts as an operation), symmetric, that is, when a can be mapped to b, then also b can be mapped to a, and transitive, that is, when a can be mapped to b, and b to c, then we can also map a to c. In terms of the operations, this means that we have an identity operation and that operations can be inverted and composed. When this composition is associative, then the operations constitute a group.

In this sense, we have the principle that any structure should be divided by its automorphism group, the concept of automorphism being defined below.

Let us introduce or recall the basic structures that are defined in terms of operations.

Definition 2.1.12 A *monoid M* is a set each element of which defines an operation

$$l_g : M \to M$$
$$h \mapsto gh. \tag{2.1.114}$$

In fact, we shall usually write this as a binary operation

$$(g, h) \to gh \tag{2.1.115}$$

mapping a pair of elements g, h of M to their product gh. This product has to be *associative*

$$(gh)k = g(hk) \text{ for all } g, h, k \in M, \tag{2.1.116}$$

and there must exist a distinguished element e (called the *neutral element*) with

$$eg = ge = g \text{ for all } g \in M. \tag{2.1.117}$$

For instance, a lattice with 0 and 1 possesses two such binary operations, $\wedge$ and $\vee$. According to (2.1.42), the neutral element for $\wedge$ is 1, but the neutral element for $\vee$ is 0.

On the set $\{0, 1\}$, we have two monoid structures, with the operations denoted by $\cdot$ and $+$, resp.,

$$0 \cdot 0 = 0, \ 0 \cdot 1 = 0, \ 1 \cdot 0 = 0, \ 1 \cdot 1 = 1 \text{ and} \tag{2.1.118}$$
$$0 + 0 = 0, \ 0 + 1 = 1, \ 1 + 0 = 1, \ 1 + 1 = 0. \tag{2.1.119}$$

Both structures will be important. Of course, this is also a special case of the observation we have just made about lattices, because we can let · correspond to ∧ and + to ∨ in the lattice with 0 and 1 only.

Monoid structures on $\{0, 1\}$

The operation l_g is called the left translation by g. Equivalently, we could write (2.1.115) in terms of the right translation r_h by h. Expressed in terms of such translations, (2.1.117) means that the left and right translations l_e, r_e by the neutral element are the identity operations on M.

Definition 2.1.13 A *group* G is a monoid in which each $g \in G$ has to possess an *inverse* $g^{-1} \in G$ satisfying

$$gg^{-1} = g^{-1}g = e. \tag{2.1.120}$$

The element e and the inverse g^{-1} of a given element g are uniquely determined, as is easily verified.

Definition 2.1.14 A subset S of a group G is called a set of *generators* of the group G if every element of G can be expressed as a product of elements from S and their inverses. (Such a set of generators is not unique.) The group is called *free* if it does not possess nontrivial relations. This means that there exists a set S of generators such that any element of G can be written in a unique way as the product of elements of S and their inverses, apart from inserting trivial products of the form gg^{-1}. (Again, S itself is not unique here.) G is called *torsionfree* if $g^n \neq e$ for all $g \in G, n \in \mathbb{Z}, n \neq 0$.

Free groups are torsionfree, because if $g^n = e$ nontrivially, then e can be expressed in more than one way as a product in G, and hence the same would hold for any other element, e.g., $h = g^n h$.

Definition 2.1.15 The monoid M or the group G is called *commutative* or, equivalently, *abelian* if

$$gh = hg \text{ for all } g, h \in M \text{ or } G. \tag{2.1.121}$$

For a commutative group, the operation is often written as $g + h$, with $-h$ in place of h^{-1}, and the element e is denoted by 0.

Of course, there is the trivial group containing a single element e, with $e \cdot e = e$. The smallest nontrivial group is given by (2.1.119), and we now write this as $\mathbb{Z}_2 := (\{0, 1\}, +)$ with $0+0 = 0 = 1+1$, $0+1 = 1+0 = 1$. When we consider the same set with a different operation, (2.1.118), which we now write as $M_2 := (\{0, 1\}, \cdot)$ with $0 \cdot 0 = 0 \cdot 1 = 1 \cdot 0 = 0$, $1 \cdot 1 = 1$, we obtain a monoid that is not a group (because 0 has no inverse).

Commutative group $\mathbb{Z}_2 := (\{0, 1\}, +)$

Monoid $M_2 := (\{0, 1\}, \cdot)$

More generally, for $q \geq 2$, we can consider the cyclic group $\mathbb{Z}_q := (\{0, 1, \ldots, q - 1\}, +)$ with addition defined modulo q, that is, $m + q \equiv m$ for all m. Thus, for instance, $1 + (q - 1) = 0$ or $3 + (q - 2) = 1$. We can also equip this set with multiplication modulo q, obtaining again a monoid M_q which is not a group.

Cyclic group $(\mathbb{Z}_q, +)$

Monoid $(M_q, \cdot)$

| Monoid $\mathbb{N}_0$ |
| Group of integers $\mathbb{Z}$ |

| Group $(\mathbb{Q}_+, \cdot)$ |

The nonnegative integers with addition also constitute a monoid $\mathbb{N}_0$; this monoid, however, can be extended to the group $\mathbb{Z}$ of integers.

Also, the positive rational numbers $\mathbb{Q}_+$, and likewise the nonzero rationals $\mathbb{Q} \setminus \{0\}$, equipped with multiplication constitute a group.

Definition 2.1.16 A subgroup H of a group G is simply some $H \subset G$ that also forms a group law under the group operation of G. That is, whenever $h, k \in H$, then also $hk \in H$ and $h^{-1} \in H$. Thus, H is closed under the group operation of G, that is, whenever we apply the group operations of multiplication or take the inverse of elements in H, then the result is still in H.

| Subgroup $m\mathbb{Z}$ of $\mathbb{Z}$ |

| Subgroups of $\mathbb{Z}_q$ |

We shall discuss the more abstract Definition 2.3.5 below. Every group G has the trivial group $\{e\}$ and G itself as subgroups. As a nontrivial example, $m\mathbb{Z} := \{\ldots, -2m, -m, 0, m, 2m, \ldots\}$ is a subgroup of $\mathbb{Z}$. Also, when p divides $q \in \mathbb{N}$, then $\{0, p, 2p, \ldots\}$ is a subgroup of $\mathbb{Z}_q$. When q is a prime number (that is, if $q = mn$ with positive integers m, n, then either $m = q, n = 1$ or the other way around), however, then $\mathbb{Z}_q$ does not possess any nontrivial subgroups. Thus, here an arithmetic property, that q be prime, is translated into a group theoretical one, that $\mathbb{Z}_q$ has no nontrivial subgroups.

| Ring $(\mathbb{Z}, +, \cdot)$ |

In fact, the integers $\mathbb{Z}$ also possess another operation, namely multiplication. This leads us to the next

Definition 2.1.17 A *ring* R possesses the structure of a commutative group, written as $+$ (often called addition), and another operation (called multiplication) that is associative, see (2.1.116), and which is *distributive* over $+$,

$$g(h+k) = gh+gk \text{ and } (h+k)g = hg+kg \text{ for all } g, h, k \in G. \quad (2.1.122)$$

The ring is said to be *commutative* when the multiplication is also commutative, see (2.1.121). It is said to possess an *identity* or *unit* if there is an element, denoted by 1, satisfying

$$g1 = 1g = g \text{ for all } g \in R. \quad (2.1.123)$$

Since $0 + 0 = 0$, the distributive law (2.1.122) implies

$$g0 = 0g = 0 \text{ for all } g \in R. \quad (2.1.124)$$

A ring with identity thus possesses a group structure (addition) as well as a monoid structure (multiplication) that is distributive over the group structure.

| Ring $(\mathbb{Z}_q, +, \cdot)$ |

For instance when we equip $\mathbb{Z}_q$ both with addition $+$ and multiplication $\cdot$ modulo q, we obtain a ring. The simplest example is, of course, $\mathbb{Z}_2$ with the two operations given in (2.1.118) and (2.1.119).

More generally, we can also form amalgams of the operations of addition and multiplication.

Definition 2.1.18 A *module* M over a ring R is an abelian group (with group operation denoted by $+$) whose elements can be multiplied by elements of R, denoted by $(r, g) \mapsto rg$ for $r \in R$, $g \in M$, with the following distributive and associative rules.

$$r(g + h) = rg + rh \tag{2.1.125}$$

$$(r + s)g = rg + rs \tag{2.1.126}$$

$$(rs)g = r(sg) \tag{2.1.127}$$

for all $r, s \in R$, $g, h \in M$.

If R possesses an identity 1, then the R-module M is called *unitary* if

$$1g = g \text{ for all } g \in M. \tag{2.1.128}$$

Of course, each ring is a module over itself, as well as a module over any subring[4] of itself. In Definition 2.1.21 below, we shall also consider subsets of a ring that are closed under multiplication of that ring, hence also constitute modules. In those cases, the operation of multiplication is already internal to the structure, but the concept of a module also allows us to consider the multiplication by elements of the ring as something additional, superimposed upon the internal group structure of M. In particular, the elements of R are not considered as elements of M, but rather as operations on M. In the sequel, we shall often encounter such modules over rings. In particular, at several places, we shall construct structures from a ring on which that ring then operates.

In any case, it is an important principle that one structure can operate on another one. In this book, we shall mostly interpret such operations as multiplications, but in other contexts they might arise as translations, time shifts (as an operation by the (positive) real numbers or integers), or whatever. As we shall see below, there are many examples where not a ring, but only a monoid or group operates.

We now come to an important special class of rings.

Definition 2.1.19 A commutative ring R with identity $1 \neq 0$ for which $R \setminus \{0\}$ also is a group under multiplication, i.e., for which every $g \neq 0$ possesses a multiplicative inverse g^{-1} with

$$gg^{-1} = 1, \tag{2.1.129}$$

is called a *field*.

A unitary module over a field is called a *vector space*.

We have already seen an example of a vector space, Euclidean space $\mathbb{R}^d$, see (2.1.108) and (2.1.109).

| Euclidean space $\mathbb{R}^d$ |

---

[4]In Definition 2.1.16, we have explained what a subgroup is, and you should then easily be able to define a subring, if you do not already know that concept.

Field
$\mathbb{Z}_2 = (\{0, 1\}, +, \cdot)$

Vector space $\mathbb{Z}_2^n$

Field $(\mathbb{Z}_p, +, \cdot)$ for prime p

An example of a field is $\mathbb{Z}_2 = (\{0, 1\}, +, \cdot)$ with the operations as defined above.[5] We then have the vector space $\mathbb{Z}_2^n$ over the field $\mathbb{Z}_2$. That vector space consists of all binary strings of length n, like (1011) $(n = 4)$ with componentwise addition modulo 2. For instance, $(1100)+(0110) = (1010)$ in $\mathbb{Z}_2^4$. The operation of the field $\mathbb{Z}_2$ on this vector space is given by $0 \cdot a = a, 1 \cdot a = a$ for all $a \in \mathbb{Z}_2^n$. And we have the simple rule that $a + b = 0 \in \mathbb{Z}_2^n$ iff $a = b$.

More generally, $\mathbb{Z}_q$ with the above ring structure is a field if and only if q is a prime number. When q is not prime, there exist elements without multiplicative inverses, namely the divisors of q. For example, in $\mathbb{Z}_4$, we have $2 \cdot 2 = 0 \bmod 4$.

The topic of rings and fields and the relations between them will be taken up in detail in Sect. 5.4.1.

Finally,

Definition 2.1.20 An *algebra* A is a module over a commutative ring R that possesses a bilinear multiplication, that is,

$$(r_1 a_1 + r_2 a_2)b = r_1 a_1 b + r_2 a_2 b \text{ for all } a_1, a_2, b \in A, r_1, r_2 \in R$$
$$a(r_1 b_1 + r_2 b_2) = r_1 a b_1 + r_2 a b_2 \text{ for all } a, b_1, b_2 \in A, r_1, r_2 \in R$$

$$(2.1.130)$$

(Here, for instance, $a_1 b$ denotes the multiplication of the two elements of the algebra, whereas $r_1 a$ is the multiplication of the element a of A by the element r_1 of the ring R.)

Of course, every ring R is not only a module, but also an algebra over itself. In that case, multiplication in the algebra and multiplication by an element of the ring is the same.

Less trivial, but typical examples are algebras of functions. As those will play an important role later on, let us systematically go through the construction. When U is a set, then the functions from U to a monoid, group, or ring constitute a monoid, group, or ring themselves. (This will be discussed from a more general perspective in Sect. 2.1.7.) For instance, when M is a monoid, and $f : U \to M, g : U \to M$, then for $x \in U$, we can simply put

$$(fg)(x) := f(x)g(x), \tag{2.1.131}$$

as the latter multiplication takes place in the monoid M. Moreover, we can multiply such a function $f : U \to M$ by an element m of M,

$$(mf)(x) := mf(x). \tag{2.1.132}$$

---

[5]Thus, we have equipped the group $\mathbb{Z}_2$ now with an additional operation, multiplication, but still denote it by the same symbol. We shall, in fact, often adopt the practice of not changing the name of an object when we introduce some additional structure or operation on it. That structure or operation will then henceforth be implicitly understood. This is a convenient, but somewhat sloppy practice. Probably, you will not need to worry about it, but as a mathematician, I should at least point this out.

From this, we see that the functions on U with values in a commutative ring form an algebra. Whether the set U also possesses an algebraic structure is irrelevant here.

Here is another construction of an algebra. Let $\gamma : R \to S$ be a homomorphism of commutative rings. Then S becomes an algebra over R. We have the addition and multiplication in S, and $(r, s) \mapsto \gamma(r)s$ yields the multiplication by elements of R, that is, the module structure for S. The multiplication in S then clearly satisfies the bilinearity laws (2.1.130).

Let us also systematically go through the example that you probably know best: The positive integers $\mathbb{N} = \{1, 2, 3, \dots\}$ together with addition do not form a monoid, as the neutral element is missing. This is easily remedied by including 0 and considering the nonnegative integers $\mathbb{N}_0 = \{0, 1, 2, \dots\}$ which form an additive monoid. This monoid is not a group because its elements, except 0, do not have inverses. This is fixed by enlarging it to the additive group of integers $\mathbb{Z}$. If we also include the operation of multiplication of integers, then $\mathbb{Z}$ becomes a ring. This ring is not a field, because apart from 1 and -1, its non-zero elements do not possess multiplicative inverses. Again, we can enlarge it, this time to obtain the field $\mathbb{Q}$ of rational numbers. ($\mathbb{Q}$ can be further enlarged to the fields of the real, the complex, or the p-adic numbers, but this is not our present concern.) All this may appear rather easy in the light of the concepts that we have developed. We should, however, remember that each such extension was an important step in the history of mathematics, and that it provided a crucial motivation for the corresponding abstract concept.

> $\mathbb{N}$

> Monoid $\mathbb{N}_0$

> Ring $\mathbb{Z}$

> Field $\mathbb{Q}$

Definition 2.1.21 A (left) *ideal* I in a monoid M is a subset of M with

$$mi \in I \text{ for all } i \in I, m \in M. \qquad (2.1.133)$$

An *ideal* in a commutative ring R with identity is a nonempty subset that forms a subgroup of R as an abelian group and satisfies the analogue of (2.1.133) w.r.t. multiplication.

An ideal in a commutative ring is then also a module over that ring in the sense of Definition 2.1.18.

With this concept of an ideal, one can characterize the groups G as precisely those monoids for which $\emptyset$ and G itself are the only ideals. Analogously, a commutative ring R with identity is a field if its only left ideals are $\{0\}$ and R itself. This will be of fundamental importance in Sect. 5.4 below.

When M is the additive monoid $\mathbb{N}_0$ of nonnegative integers, then its ideals are $\emptyset$ and all sets $\{n \geq N\}$ for some fixed $N \in \mathbb{N}_0$. The set of left ideals of the monoid M_2 is $\Lambda_{M_2} := \{\emptyset, \{0\}, \{0, 1\}\}$. We note that this is the same as the above Heyting algebra $\mathcal{O}(X)$ for the 2-element set $X = \{0, 1\}$. More generally, for M_q, the ideals are $\emptyset$, $\{0\}$, M_q and all subsets of the form $\{0, m, 2m, \dots, (n-1)m\}$ when $nm = q$ for $n, m > 1$, that is, nontrivial divisors of q. Thus, when q is a prime number, M_q has only the three trivial ideals, but when q is not prime, there are more. These then are also the ideals of the ring $(\mathbb{Z}_q, +, \cdot)$

> Ideals of monoids M_2 and M_q

Ideals of ring $\mathbb{Z}$

The ideals of the ring of integers $\mathbb{Z}$ are of the form $\{nm : n \in \mathbb{Z}\}$ for fixed $m \in \mathbb{Z}$. (Note that while for the monoid $\mathbb{N}_0$, we considered ideals with respect to the operation of addition, for the ring $\mathbb{Z}$, we consider ideals with respect to multiplication.)

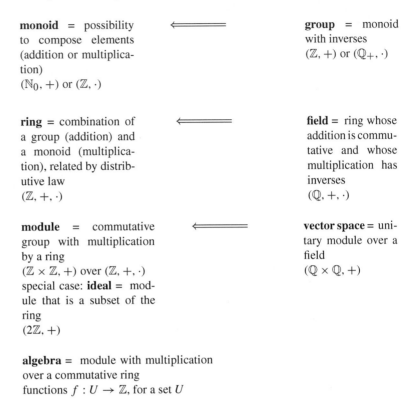

monoid = possibility to compose elements (addition or multiplication) $(\mathbb{N}_0, +)$ or $(\mathbb{Z}, \cdot)$ ⟵ **group** = monoid with inverses $(\mathbb{Z}, +)$ or $(\mathbb{Q}_+, \cdot)$

ring = combination of a group (addition) and a monoid (multiplication), related by distributive law $(\mathbb{Z}, +, \cdot)$ ⟵ **field** = ring whose addition is commutative and whose multiplication has inverses $(\mathbb{Q}, +, \cdot)$

module = commutative group with multiplication by a ring $(\mathbb{Z} \times \mathbb{Z}, +)$ over $(\mathbb{Z}, +, \cdot)$ special case: **ideal** = module that is a subset of the ring $(2\mathbb{Z}, +)$ ⟵ **vector space** = unitary module over a field $(\mathbb{Q} \times \mathbb{Q}, +)$

algebra = module with multiplication over a commutative ring functions $f : U \to \mathbb{Z}$, for a set U

The various algebraic structures and the relations between them, with examples. The arrow is an implication, in the sense that, e.g., every group is a monoid; structures build upon those above them.

So far, we have looked at the prototypes of abelian groups, $\mathbb{Z}_q$ and $\mathbb{Z}$. In many respects, however, the most important group is the symmet-

Symmetric group $\mathfrak{S}_n$

ric group $\mathfrak{S}_n$, the group of permutations of n elements, with the composition of permutations as the group operation. We think of these elements as ordered in the form $(1, \ldots n)$. A permutation of them is then written as $(i_1 i_2 \ldots i_n)$, meaning that the element $i_k \in \{1, 2 \ldots, n\}$ is put in the place of k, for $k = 1, 2, \ldots, n$. Of course, the i_j all have to be different, so as to exhaust the set $\{1, 2, \ldots, n\}$. The original ordering $(12 \ldots n)$ then stands for the identity permutation that leaves every element unchanged. The symmetric group $\mathfrak{S}_2$ consists of the two elements (12) and (21) and is therefore isomorphic to $\mathbb{Z}_2$ (we shall explain the term "isomorphic" only below in Definition 2.3.2, but you will probably readily understand its meaning in the present example). The group $\mathfrak{S}_3$

$\mathfrak{S}_3$

contains 6 elements, $(123), (213), (132), (321), (231), (312)$. (123) is the neutral element. $(213), (132)$ and (321) simply exchange two elements; for instance, (213) exchanges the first and the second element and leaves the third one in place. Each of these three permutations is its own inverse. In contrast, (231) and (312) permute the three elements cyclically, and they are inverses of each other, i.e., $(231) \circ (312) = (123)$. Moreover, $(231) = (132) \circ (213)$, that is, we first exchange the first two elements (note the reversal of order in our notation; here, the permutation (213) is carried out first, and then the permutation (132) is applied), and then the last two. After the first exchange, 1 is in the middle position, and the second exchange then moves it into the last position. Also, $(231) = (213) \circ (321)$, whereas $(213) \circ (132) = (312)$. In particular, $\mathfrak{S}_3$ is not abelian, since $(132) \circ (213) \neq (213) \circ (132)$.

In order to simplify the notation, one can leave out those elements that are not affected by the permutation. Thus, instead of (132), one simply writes (32) for the exchange of the second and the third element. With this notation, for example, $(21) \circ (32) = (312)$. Again, note the reverse order of the operation: We first exchange the last two elements, which brings 3 into the second position, and we then exchange the first two positions, which then brings 3 from the second into the first position.

We can also already make some general observations. The group $\mathfrak{S}_m$ is contained in $\mathfrak{S}_n$ for $m < n$—just take m of the n elements and permute them and leave the remaining $n - m$ elements alone. $\mathfrak{S}_m$ is a subgroup of $\mathfrak{S}_n$ in the terminology of Definition 2.1.16 or 2.3.5 below. This means that the inclusion that we have described defines $i : \mathfrak{S}_m \to \mathfrak{S}_n$ such that for all $g, h \in \mathfrak{S}_m$, we have

$$i(g \circ h) = i(g) \circ i(h) \qquad (2.1.134)$$

where the first $\circ$ is the multiplication in $\mathfrak{S}_m$ and the second one that in $\mathfrak{S}_n$. More generally, a map $i : G_1 \to G_2$ between groups or monoids satisfying (2.1.134) is called group or monoid homomorphism. Thus, a homomorphism is a map compatible with the group or monoid operations. Analogously, one may then define homomorphisms of rings or fields.

Returning to the example of the symmetric groups $\mathfrak{S}_n$, we see that, since $\mathfrak{S}_3$ is not abelian, this implies that neither are the groups $\mathfrak{S}_n$ for $n > 3$.

Also, when G is a group with finitely many elements, or a finite group for short, then left multiplication l_g by any $g \in G$ induces a permutation of the elements of G. For this, we simply observe that $l_g : G \to G$ is injective, because if $gh = gk$, then also $h = g^{-1}(gh) = g^{-1}(gk) = k$. That is, l_g maps different elements to different elements, and it therefore permutes the elements of g. Likewise, the assignment $g \to l_g$ is injective, in the sense that if $g_1 \neq g_2$, then also $l_{g_1} \neq l_{g_2}$, as for instance $l_{g_1} e = g_1 \neq g_2 = l_{g_2} e$.

A group G defines a graph. More precisely, take a group G with a set S of generators that is closed under inversion, that is, whenever $g \in S$, then also $g^{-1} \in S$ (one might simply start with any set S' of generators and enlarge S' to S by all the inverses of its elements). The so-called Cayley graph of the pair (G, S) then has as vertex set the elements of G, and there is an edge between $h, k \in G$ iff there is some $g \in S$ with $gh = k$. Since then also g^{-1} by our condition on S and $g^{-1}k = h$, this relation between h and k is symmetric, and the graph is therefore undirected. For instance, for the symmetric group $\mathfrak{S}_3$, we can take the generators (21), (32), (31) each of which is its own inverse. The resulting Cayley graph is then

$\boxed{\mathfrak{S}_n}$

$\boxed{\text{Cayley graph of } \mathfrak{S}_3}$

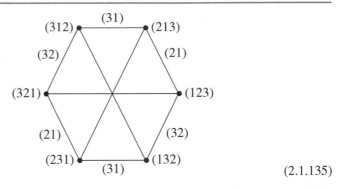

$$(2.1.135)$$

In this graph, in order to avoid clutter, we have labelled only some of the edges, and for the edges, we have used the abbreviated notation explained above. We also see that this graph is bipartite, that is, there are two classes of nodes, {(123), (231), (312)} and {(132), (321), (213)} with the property that there are only edges between nodes from different classes. In fact, the second class consists of permutations of two elements, in our shorthand notation {(32), (31), (21)}. Such permutations of two elements are called transpositions. The elements of the other class are products of an even number of such transpositions. In fact, each symmetric group $\mathfrak{S}_n$ contains the two subclasses of elements that can be written as products of an even or of an odd number of transpositions. (We can also formulate this slightly more abstractly and define the parity or sign $\mathrm{sgn}(g)$ of an element g of $\mathfrak{S}_n$ to be 1 (even) if it can be expressed as even number, and -1 (odd) if it can be expressed as an odd number of transpositions. Of course, we then have to verify that the parity is well defined, in the sense that there is no permutation that can be represented by both an even and an odd number of transpositions. The essential fact underlying this is that the product of two transpositions—which is even—can never be a single transposition itself—which would be odd. This, however, is readily checked. Also, $\mathrm{sgn}(gh) = \mathrm{sgn}(g)\mathrm{sgn}(h)$. In more abstract terminology, this means that sgn is a group homomorphism from $\mathfrak{S}_n$ to the group $\{1, -1\}$ with multiplication as the group law. In turn, the assignment $1 \to 0, -1 \to 1$ is a homomorphism of this group to the group $\mathbb{Z}_2$ with addition as the group law.)

The first subclass, the even products, actually forms a subgroup of $\mathfrak{S}_n$. This group is called the alternating group $\mathfrak{A}_n$.

In order to show the bipartiteness of this Cayley graph, we rearrange the

Cayley graph of $\mathfrak{S}_3$ positions of the nodes to get the following figure.

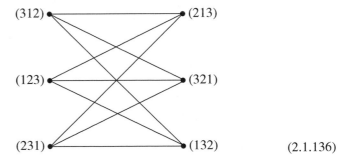

$$(2.1.136)$$

In fact, we have here a *complete* bipartite graph, meaning that every node in one class is connected with every node in the other class. Let us emphasize that (2.1.136) depicts the same graph as (2.1.135). Rearranging the positions of the nodes can create visualizations that look very different although they represent the same structure.

Here is another observation. The permutation (31) can also be expressed as the product $(31) = (21)(32)(21)$ of three transpositions of *adjacent* elements. Again, this is a general phenomenon. Any element of $\mathfrak{S}_n$ can be written as a product of transpositions of adjacent elements. This product representation is not unique (for instance, also $(31) = (32)(21)(32)$), but the parity (odd vs. even) is invariant. In particular, we can choose the set of transpositions of adjacent elements as a set of generators of $\mathfrak{S}_n$. This would change the above Cayley graph (2.1.135) to the following one

Cayley graph of $\mathfrak{S}_3$

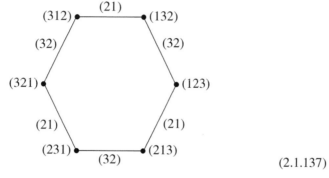

$$(2.1.137)$$

From this graph, for instance, we can directly read off the identity $(21)(32)(21) = (32)(21)(32)$.

Of course, we could also take $S = G \backslash \{e\}$, that is, let our set of generators consist of all nontrivial elements of G. The resulting Cayley graph would then be a complete graph, that is, a graph where each element is connected with every other element. In fact, the same is true for any finite group G, i.e., when we take all nontrivial elements of G as generators, the resulting Cayley graph is complete.

The symmetric groups will appear again in Sect. 3.4.4 below. In fact, there we shall see a converse to the preceding observation that we can obtain complete graphs as Cayley graphs of the symmetric (or any other) group: The automorphism group of a complete graph is a symmetric group.

Given such algebraic structures, you can of course combine them to form other such structures. For instance, we can form the product GH of two groups GH. The elements of GH are pairs $(g, h), g \in G, h \in H$, and the group operation is likewise assembled from those in G and H,

$$(g_1, h_1)(g_2, h_2) := (g_1 g_2, h_1 h_2). \qquad (2.1.138)$$

Subsequently, we shall utilize another important concept from group theory, that of a normal subgroup. To prepare for this concept, let us consider a group G with a subgroup N. We then consider the set G/N of the equivalence classes for the equivalence relation

$$g \sim h \text{ iff there exists some } n \in N \text{ with } h = ng. \qquad (2.1.139)$$

(2.1.139) defines an equivalence relation, indeed, because N is a subgroup, and not just a subset of G. For instance, for the transitivity of (2.1.139), let $g \sim h$ and $h \sim k$, that is, there exist $n, m \in N$ with $h = ng, k = mh$. But then $k = mng$ and since $mn \in N$ because N is a group, we obtain $g \sim k$. Nevertheless, it seems that this leads us outside the realm of groups, as so far, G/N is only a set, although we had started with groups G and N. In other words, we are led to the question of whether G/N can also be turned into a group. It turns out, however, that we need to impose an additional condition upon N for that purpose.

Definition 2.1.22 A subgroup N of a group G is called a *normal* subgroup if for every $g \in G$

$$g^{-1}Ng = N. \tag{2.1.140}$$

Lemma 2.1.8 *If N is a normal subgroup of G, then the quotient G/N can be equipped with a group structure.*

Proof We need to verify that the group multiplication in G passes to the quotient G/N in the sense that we can carry it out unambiguously on equivalence classes. That is, when $[g]$ denotes the equivalence class of $g \in G$, we need to verify that

$$[g][h] := [gh] \tag{2.1.141}$$

is well defined in the sense that it does not depend on the choice of elements in the equivalence classes. That means that when $g' \sim g, h' \sim h$, then we want

$$g'h' \sim gh, \tag{2.1.142}$$

that is,

$$[g'h'] = [gh]. \tag{2.1.143}$$

Now, when $g' \sim g, h' \sim h$, there exist $m, n \in N$ with $g' = mg, h' = nh$. Then

$$g'h' = mgnh = mgng^{-1}gh. \tag{2.1.144}$$

Now since $gNg^{-1} = N$ as N is a normal subgroup, the element gng^{-1} is again an element of N. Calling this element n', we obtain from (2.1.144)

$$g'h' = mn'gh, \tag{2.1.145}$$

and since $mn' \in N$ as N is a group, we conclude that (2.1.143) holds. $\square$

Putting it differently, the Lemma says that we have a group homomorphism

$$\iota : G \to G/N, \quad g \to [g]. \tag{2.1.146}$$

Let us consider some *examples* of normal subgroups.

- Any subgroup of an abelian group is normal, since $g^{-1}ng = n$ in that case.

- When $q = mn$ is not prime, that is, m, n, q are all integers > 1, then $\mathbb{Z}_m$ and $\mathbb{Z}_n$ are subgroups of $\mathbb{Z}_q$ ($\mathbb{Z}_m$ becomes the subgroup with elements $0, m, 2m, \ldots, (n-1)m$ of $\mathbb{Z}_q$, and analogously for $\mathbb{Z}_n$), and since $\mathbb{Z}_q$ is abelian, they are normal subgroups.

  Normal subgroups of $\mathbb{Z}_q$

- The alternating group $\mathfrak{A}_n$ is a normal subgroup of the symmetric group $\mathfrak{S}_n$, because the parity of $g^{-1}ng$ is the same as the parity of n. In fact, since we have observed above that sgn induces a group homomorphism $\mathfrak{S}_n \to \mathbb{Z}_2$, this also follows from the next example.

  Alternating group $\mathfrak{A}_n$

- If $\rho : H \to G$ is a group homomorphism, then $\ker(\rho) := \{k \in H : \rho(k) = e \in G\}$ is a normal subgroup. In fact, if $k \in \ker \rho$, then for any $h \in H$, we have $\rho(h^{-1}kh) = \rho(h)^{-1}\rho(k)\rho(h) = \rho(h)^{-1}e\rho(h) = e$, and hence also $\rho(h)^{-1}\rho(k)\rho(h) \in \ker \rho$.

Actually, the last example is not really an example, but a general fact. Every normal subgroup N of a group G is the kernel of a homomorphism, the homomorphism $\iota : G \to G/N$ of (2.1.146).

Definition 2.1.23 A group G is called *simple* if it is nontrivial and its only normal subgroups are the trivial group and G itself.

This is a very important concept, because simple groups are the basic building blocks of group theory. A group G that is not simple possesses a nontrivial normal subgroup N and therefore can be broken up into two smaller groups, the normal subgroup N and the quotient group G/N. When one or both of those are still not simple, the process can be repeated. When G is finite, the process then has to terminate after finitely many steps, and one has assembled the building blocks of G (by the Jordan-Hölder theorem, which we do not prove here, they are, in fact, unique). Thus, when one has a list of the simple finite groups, one can then construct all finite groups, and the subject of finite groups is mathematically completely understood. The complete classification of finite simple groups, however, turned out to be quite difficult and was only successfully completed around 2004.

The most important examples of finite simple groups are the cyclic groups $\mathbb{Z}_p$ for a prime p (as we have just seen in the examples of normal subgroups, $\mathbb{Z}_n$ is not simple when n is not prime) and the alternating groups $\mathfrak{A}_n$ for $n \geq 5$ (the latter fact is not completely trivial). There are such infinite series of finite simple groups, and 26 exceptional groups, among which the so-called monster group is the most complex. A recent textbook is [118]. The approach to the classification of finite groups as just described embodies an important principle of mathematical research. Whenever one wants to understand some class of mathematical structures, one first needs to identify the elementary building blocks that are no longer decomposable by some basic operation (here, taking the quotient by a normal subgroup), and then one has to classify those building blocks. All the structures in the class under consideration can then be composed out of those building blocks.

2.1.7 Parametrizing Constructions

We have defined or characterized algebraic structures like Boolean and Heyting algebras, monoids, groups, rings and fields in terms of operations. There were nullary operations—the distinguished elements 0, 1—, unary operations—the pseudo-complement $v \to \neg v$ or the inverse $g \to g^{-1}$— and binary operations—$(v, w) \to v \wedge w$ or $(g, h) \to gh$. Let S be such a structure, and let X be a set. Then, typically, the set S^X of maps

$$\phi : X \to S \qquad (2.1.147)$$

again carries the same type of structure. For instance, when S is a group, S^X is also a group, with group law

$$(\phi, \psi) \to \phi\psi$$
$$\text{with } \phi\psi(x) := \phi(x)\psi(x) \qquad (2.1.148)$$

where on the right hand side, we have the group operation of S applied to the elements $\phi(x)$ and $\psi(x)$. We may consider S^X as a family of groups parametrized by the elements of X, but the preceding construction tells us that this family is a group in its own right. So, if we want, we can iterate the construction and consider maps from some other set into S^X. Again, the result will be a group.

The same works for monoids or rings. It does not work, however, for fields. That is, the maps from a set X into a field F do not constitute a field; they only yield a ring (except for the trivial case when X has only one element). The reason is the exceptional role of the distinguished element 0 as being the only one without a multiplicative inverse. In F^X any ϕ that maps some, but not all elements of X to 0 then fails to have a multiplicative inverse, without being the neutral element for addition in F^X—the latter one being the map that maps *all* $x \in X$ to 0.

When we have a relation like in (2.1.32), $F : S \times S \to \{0, 1\}$, we have two immediate possibilities for defining such a relation on S^X,

$$F(\phi, \psi) := \sup_{x \in X} F(\phi(x), \psi(x)) \qquad (2.1.149)$$

or

$$F(\phi, \psi) := \inf_{x \in X} F(\phi(x), \psi(x)) \qquad (2.1.150)$$

In the first case, we have $F(\phi, \psi) = 1$ if there exists some $x \in X$ with $F(\phi(x), \psi(x)) = 1$. In the second case, this would have to hold for all $x \in X$.

We can do the same when the relation takes values in $\mathbb{R}$ or $\mathbb{R}^+$, provided the supremum or infimum stays finite.

Just as an aside, to put this into the perspective of analysis: If X carries a measure μ (see Sect. 4.4), we can also average this construction w.r.t. μ. For instance, when the relation is a metric $d(., .)$, and if $1 \leq p < \infty$, we obtain a metric on S^X via

$$d_p(\phi, \psi) := \left(\int_X d^p(\phi(x), \psi(x)) \mu(dx) \right)^{1/p} \qquad (2.1.151)$$

where $d^p(y, z)$ means $d(y, z)^p$; again, we need to assume that this expression is finite, of course. The case $p = 2$ is the most important one in practice.

The restriction $p \geq 1$ is needed for the triangle inequality to hold for the metric d_p on S^X. d_p is also called the L^p-metric on S^X. The analogue of (2.1.149) is

$$d_\infty(\phi, \psi) := \text{ess sup}_{x \in X} d(\phi(x), \psi(x)), \qquad (2.1.152)$$

where ess sup stands for the essential supremum (i.e., ess $\sup_{x \in X} f(x) :=$ $\inf\{a \in \mathbb{R} \cup \{\infty\} : f(x) \leq a$ for all $x \in X \setminus A\}$ for some $A \subset X$ with $\mu(A) = 0$, a so-called nullset, see Sect. 4.4). (For details, see e.g. [59]). Of course, when X is, for instance, finite, we may replace the essential supremum by the ordinary supremum.

2.1.8 Discrete Versus Continuous

Besides the distinction between geometry and algebra, another important distinction in mathematics is that between discrete and continuous structures. In abstract terms, this is the difference between *counting* and *measuring*. In its basic form, counting utilizes the positive integers $\mathbb{N}$, that is, the possibility of enumeration. Measuring depends on the real numbers $\mathbb{R}$, and the ancient Greeks have already discovered that measuring cannot be reduced to counting, because of the existence of irrational numbers, that is, numbers that cannot be expressed as a ratio of two integers. Real numbers can be expressed as sequences of integers, in fact, even as binary strings, but Cantor's diagonal argument displayed in Sect. 2.1.3 showed us that the reals cannot be enumerated by the integers.

Continuous structures constitute the realm of *analysis*. In analysis, fundamental concepts are limits and convergence, completion, and compactness.

The completely new aspect that analysis offers is that relations need not hold exactly, but only approximatively, and that this can be quantified in the sense that the approximation error can be controlled or estimated. For that purpose, analysis is based on continuous structures which enable us to define perturbations and variations. While such structures constitute the basis of analysis, the methods can then even be applied to discrete settings, as in numerical analysis, where continuous structures are utilized conceptually to interpolate between discrete values, and to prescribe the resolution of a discrete scheme by comparison with an idealized continuous one.

Returning to the above example of a pseudometric F, let us assume that there is also some metric d given on S. Then $F + \epsilon d$, for any positive real number ϵ, yields a perturbation of the pseudometric F that is positive definite, i.e., a metric on S. We may apply this even to the trivial case where $F \equiv 0$. In that case, the above quotient $\bar{S}$ obtained by identifying points s, s' with $F(s, s') = 0$ consists of one single point, whereas the quotient by $F + \epsilon d$ for any $\epsilon > 0$ is the entire set S. Therefore, the quotient construction in this case does not depend continuously on ϵ. In a certain sense, then the algebraic quotient construction and the analytic limit construction $\epsilon \to 0$ are not compatible with each other.

Some mathematical concepts connect two of the above domains, geometry, algebra, and analysis. The concept of a Lie group, for instance, connects all three of them, but this will not be treated in the present book.

2.2   Axiomatic Set Theory

This section will not be referred to in the sequel, and it can therefore be skipped. Its purpose is to provide a formal substrate for some notions that have been employed in the preceding in an intuitive sense. More precisely, we shall briefly describe the Zermelo-Frankel version of axiomatic set theory that started with [123]. There are many textbooks on axiomatic set theory, for instance [109].

The basic notions of a set and of membership in a set (expressed by the symbol $\in$) are not defined in axiomatic set theory, but they are assumed to exhibit certain properties that are stipulated in the axioms. Such axioms should be consistent (in the sense that it is not possible to derive from them both a statement and its negation), plausible, and rich enough to derive for instance the basic results of Cantor's theory of sets.

We now list 10 axioms of set theory that emerged from the work of Zermelo, Frankel, Skolem, von Neumann and others.

1. **Axiom of extension**: *Let A, B be sets. If for all x, $x \in A$ if and only if $x \in B$, then $A = B$.*
   Thus, sets with the same elements are equal. This axiom says that a set is determined by its elements.
2. **Axiom of the empty set**: *There exists a set $\emptyset$ with the property that for all x, $x \notin \emptyset$.*
   Here, $x \notin$ abbreviates "it is false that $x \in$". The empty set thus is simply the set without elements.
3. **Axiom of separation**: *Let A be a set. Then for every definite condition $P(x)$, there exists a set B such that for all x, $x \in B$ if and only if x satisfies $P(x)$.*
   Here, a definite condition is built from the so-called atomic formulas $x \in y$ and $x = y$ (where x and y are variables) through finitely many applications to formulas P, Q of connectives (*if P, then Q; P iff Q; P and Q; P or Q; not P*) and quantifiers (*for all x, P holds (abbreviated as $P(x)$); for some x, Q holds*). In a condition $P(x)$, the variable x must be free, that is, not under the scope of a quantifier. In contrast, for a condition as in the axiom, B must not be free, that is, it has be bound by a quantifier. The concepts of propositional and predicate logic that are employed will be taken up in Sect. 9.3. We have discussed this as a principle for specifying subsets of a given set already in Sect. 2.1.3.
4. **Axiom of pairing**: *If A, B are sets, there exists a set (A, B) that has A and B as its only elements.*
   (A, B) is called the unordered pair of A and B. This and the next two axioms ensure that the application of standard operations to sets produces sets again.
5. **Axiom of union**: *Let A be a set. Then there exists a set C with the property that $x \in C$ iff $x \in a$ for some $a \in A$.*
   This means that the union of all the sets that are elements of A is again a set. We also write $\bigcup A$ for this set. A more explicit notation would be $\bigcup_{a \in A} a$.

6. **Axiom of power set**: *Let A be a set. Then there exists a set $\mathcal{P}(A)$, called the power set of A, with the property that $B \in \mathcal{P}(A)$ whenever for all x, $x \in B$ implies $x \in A$.*

   The power set $\mathcal{P}(A)$ thus contains all the subsets of A as its elements. We have discussed the power set already in Sect. 2.1.3 where we have connected the axioms of separation and power set.

7. **Axiom of infinity**: *There exists a set N with the properties that $\emptyset \in N$ and whenever $x \in N$, then also $x \cup \{x\} \in N$.*

   Here, $\{x\}$ is the set having x as its only member, and $x \cup \{x\}$ means that we add to the set x x itself as a further member. The axiom of infinity can be seen as an abstract version of Peano's principle of induction that generates the natural numbers (positive integers). We consider $x \cup \{x\}$ as the successor of x. Such a set N might be written in an iterative fashion as

   $$N = \{\emptyset, \emptyset \cup \{\emptyset\}, \emptyset \cup \{\emptyset\} \cup \{\emptyset \cup \{\emptyset\}\}, \dots\}. \qquad (2.2.153)$$

   In fact, in place of $\emptyset$, we could have started with an arbitrary element x. Therefore, more succinctly, we introduce a symbol 1 and write $1' := 1 \cup \{1\}$. Then we obtain such a set N as

   $$N = \{1, 1', 1'', 1''', \dots\}. \qquad (2.2.154)$$

8. **Axiom of choice**: *Let A be a set whose elements are nonempty sets. Then there exists a mapping $f : A \to \bigcup A$ with $f(a) \in a$ for all $a \in A$.*

   Thus, for every $a \in A$, we can choose some element $f(a)$ of a.

9. **Axiom of replacement**: *Let A be a set and f a mapping defined on A. Then there exists a set B whose elements are precisely the $f(x)$ for $x \in A$.*

   Thus, the image of a set under a mapping is again a set. As an application, the map $1 \mapsto N$ from (2.2.154) produces the set

   $$\{N, N', N'', \dots\}. \qquad (2.2.155)$$

10. **Axiom of restriction**: *Every set A contains an element a with $A \cap a = \emptyset$.*

    Thus, A and its element a have no element in common. This last axiom is only introduced in order to rule out certain undesired models of the first nine axioms. More precisely, it serves to rule out infinite descending chains, that is, $\dots a_n \in a_{n-1} \in \cdots \in a_0$. In particular, according to this axiom, we must not have $a \in a$.

The preceding axioms are not all independent of each other. In fact, several of them can be left out without loss of scope, as they can be derived from the remaining ones. For instance, the axiom of the empty set can be omitted, and so can the axiom of separation which can be deduced from the axiom of replacement. Also, from the latter together with the axiom of power set one can deduce the axiom of pairing. Thus, the list of axioms reflects the historical development rather than their logical status. Also, some people do not accept the axiom of choice.

There is an alternative system of axioms, named after Bernays and Gödel; in some treatises, von Neumann is also included.

In the sequel, we shall assume that we have some fixed universe U of sets that satisfies the above axioms. Any set to be mentioned in the sequel will be assumed to be a member of U.

2.3   Categories and Morphisms

The concepts of *category* and *morphism* unify some (but not all) of the preceding.

Definition 2.3.1 A *category* $\mathbf{C}$ consists of *objects* $A, B, C, \ldots$ and *arrows* or *morphisms*

$$f : A \to B \tag{2.3.1}$$

between objects, called the *domain* $A = \mathrm{dom}(f)$ and *codomain* $B = \mathrm{cod}(f)$ of f. Arrows can be composed, that is, given $f : A \to B$ and $g : B \to C$, there is an arrow

$$g \circ f : A \to C. \tag{2.3.2}$$

(The requirement for the composition is solely that $\mathrm{cod}(f) = \mathrm{dom}(g)$.) This composition is *associative*, that is,

$$h \circ (g \circ f) = (h \circ g) \circ f \tag{2.3.3}$$

for $f : A \to B, g : B \to C, h : C \to D$.

For each object A, we have the *identity arrow* ("doing nothing")

$$1_A : A \to A \tag{2.3.4}$$

which satisfies

$$f \circ 1_A = f = 1_B \circ f \tag{2.3.5}$$

for all $f : A \to B$.

The associativity condition (2.3.3) can also be expressed by saying that whichever sequence of arrows we follow from A to D in the diagram below, the result is the same.

$$
\begin{array}{ccccccc}
 & & & h \circ g & & & \\
 & & \overset{\frown}{} & & & & \\
A & \overset{f}{\longrightarrow} & B & \overset{g}{\longrightarrow} & C & \overset{h}{\longrightarrow} & D \\
 & & \underset{\smile}{} & & & & \\
 & & & g \circ f & & &
\end{array}
$$

$$\tag{2.3.6}$$

We can either follow the red or the blue sequence or the middle one, $h \circ g \circ f$.

Somewhat sloppily, we shall occasionally write $C \in \mathbf{C}$ to say that C is an object of the category $\mathbf{C}$.

One might object here that Definition 2.3.1 is not really a mathematical definition because it is left undefined what an "object" or a "morphism" is or should be. Whenever we speak of a category, we therefore first need to specify what its objects and morphisms are. For the abstract language of categories, however, it is irrelevant what the objects and morphisms are. They only need to satisfy the rules laid down in Definition 2.3.1.

The idea is that the objects of a category share some kind of structure, and that the morphisms then have to preserve that structure. A category thus consists of objects with structure and directed relations between them. A very useful aspect is that these relations can be considered as operations.

Taking objects as vertices and arrows as edges, we can thus consider a category as a directed graph, with the property that each vertex stands in relation to itself, that is, has an edge from itself to itself. This graph might have multiple edges, as there could be more than one morphism between two objects.

In this sense, the arrows of a category are viewed as relations. One can also view them as operations, as mappings between the objects. An arrow from A to B thus maps A to B.

Viewing morphisms as operations may remind you of the notion of a group, but in contrast to what was required in the Definition 2.1.13 of a group for the left translation l_g by a group element, for the morphisms of the category we do not require that they can be inverted, nor that we can compose any two of them. Nevertheless, we have

Lemma 2.3.1 *A category with a single object is a monoid, and conversely.*

Proof Let M be a monoid, as in Definition 2.1.12. We consider the elements g of M as operations, $h \mapsto gh$, that is, as arrows

$$l_g : M \to M. \qquad (2.3.7)$$

Since they have to satisfy the associativity law, they define the morphisms of the category with the single object M. The neutral element e yields the identity morphism 1_M.

Conversely, the arrows of a category with a single object M can be considered as the left translations l_g of M, hence as elements of a monoid, as they satisfy the associativity law. The identity arrow 1_M yields the neutral element e of that monoid. □

Categories can be constructed and considered at different levels of abstraction, as we shall explore in the following. As a brief guide to things that will be explained in more detail shortly, let us point out the following principle: On one hand, the structures that we have considered in the preceding constitute categories. A set, a graph or digraph, a poset, lattice, Heyting or Boolean algebra, a monoid, group, ring, or field are all categories, with the objects being the elements of that structure, and the morphisms being given by the relations or operations within that structure.[6] On the other hand, however, at the next level, the ensemble of structures of a given type also constitute a category. Thus we shall have the category of sets, the category of

---

[6]Alternatively, as noted above, for an algebraic structure like a group, we could consider that structure as the single object, and its elements as the morphisms of that object.

posets, those of graphs and digraphs, of metric spaces, of monoids, groups, rings, or fields, etc. The morphisms then are structure preserving mappings between two such structures, e.g., between two groups. Thus, within the context of the corresponding category, we can consider all structures of a given type simultaneously and consider the structure preserving relations between them. We can then move to still higher levels of abstraction and consider categories of categories of categories and figure out what the morphisms in that case should be. Or we can consider categories of morphisms. And so on. This will be explored not only in the remainder of this section, but also throughout much of this book.

So, let us go into more detail now and develop the preceding abstract principle. Every set is a category, with the elements as the objects and the only arrows being the identity arrows of the elements. Thus, a set is a category with a most uninteresting structure, that is, there are no structural relations between different objects. In fact, the empty set $\emptyset$ also constitutes a category. This category has no objects and no arrows. This may strike you as the utmost triviality, but it turns out that for some formal constructions, it is quite useful to include this particular category.

However, reversely, we also have the category of sets, denoted by **Sets**, and also the category of finite sets. The objects of these categories are now sets, one of them being again the empty set $\emptyset$, and the morphisms are mappings

$$f : S_1 \to S_2 \tag{2.3.8}$$

between sets. In view of our above discussion of distinctions, this leads us to the concept of isomorphism:

Definition 2.3.2 Two objects A_1, A_2 of a category are *isomorphic* if there exist morphisms $f_{12} : A_1 \to A_2$, $f_{21} : A_2 \to A_1$ with

$$f_{21} \circ f_{12} = 1_{A_1}, \quad f_{12} \circ f_{21} = 1_{A_2}. \tag{2.3.9}$$

In this case, the morphisms f_{12}, f_{21} are called isomorphisms.

An *automorphism* of an object A is an isomorphism $f : A \to A$.

Of course, 1_A is an automorphism of A, but there may also exist others. Often, an automorphism is considered as a symmetry of A.

Since an automorphism can be inverted, the automorphisms of an object A of a category form a group, the automorphism group of A. In fact, this is how the group concept historically emerged. But we may then turn things around and consider a group as an abstract object that might be *represented* as the group of automorphisms of some object in a category. We'll return to that issue.

(2.3.9) means that isomorphisms are invertible morphisms. Isomorphic objects are then characterized by having the same morphisms, as follows from the associativity law. That is, for example when $f_{12} : A_1 \to A_2$ is an isomorphism, then a morphism $g : A_2 \to B$ corresponds to the morphism $g \circ f_{12} : A_1 \to B$, and similarly in other directions. In particular, 1_{A_2} then corresponds to f_{12}.

There may, however, exist more than one isomorphism between isomorphic objects A_1, A_2. In that case, the identification of the morphisms of these two objects is not canonical as it depends on the choice of such an isomorphism. In fact, we can precompose an isomorphism $f_{12} : A_1 \to A_2$ with any automorphism $f_1 : A_1 \to A_1$ or postcompose it with any automorphism $f_2 : A_2 \to A_2$ to obtain another isomorphism. Conversely, whenever $f_{12}, g_{12} : A_1 \to A_2$ are two isomorphisms, then $g_{12}^{-1} \circ f_{12}$ is an automorphism of A_1, and $g_{12} \circ f_{12}^{-1}$ is an automorphism of A_2. Thus, the identification of two isomorphic objects is only determined up to the automorphisms of either of them. In fact, the automorphism groups of two isomorphic objects are themselves isomorphic. This is an instantiation of the fact that isomorphic objects have the same relational structure with other objects, in this case with themselves. Again, however, the automorphism groups may in turn possess certain symmetries, making this identification noncanonical again.

As observed, since automorphisms can be inverted, the automorphisms of an object A of a category form a group. In that sense, the concept of a morphism is a generalization of that of an automorphism in two ways. Firstly, it need not be invertible, and secondly, it need not map an object A to itself, but can map it to another object B of the same category. Morphisms can be composed. In distinction to a monoid or group where any two elements can be freely composed, however, here we have the restriction that the domain of the second morphism has to contain the codomain of the first one in order that they can be composed. Since in a monoid or group, all elements have the monoid or group itself as their domain and codomain, there is no such restriction for the composition of monoid or group elements. Thus, we had observed in Lemma 2.3.1 that the categories with a single object are precisely the monoids.

In any case, the most fundamental of the monoid or group laws, associativity, has to be preserved for the composition of morphisms. In a certain sense, associativity is a higher law, as it is about the composition of compositions. It stipulates that such a composition of compositions does not depend on the order in which we compose the compositions. This has to be distinguished from the property of commutativity which requires that a composition of group elements be independent of the order of these elements. Commutativity does not hold for a general monoid or group. Commutative monoids or groups constitute a special subclass of all monoids or groups, with many additional properties that are not shared by other monoids or groups in general.

Thus, a category can be considered as a directed graph, as some kind of generalized monoid, or as a set with some additional structure of directed relations between its elements.

Again, within a category, we cannot distinguish between isomorphic objects. We may thus wish to identify them, but need to keep in mind that such an identification need not be canonical as it depends on the choice of an isomorphism, as explained above. The important point here is that the objects of a category are determined only up to isomorphism. The view of category is that an object B of a category $\mathbf{C}$ is characterized by its relations with other objects, that is, by the sets $\mathrm{Hom}_{\mathbf{C}}(., B)$ and $\mathrm{Hom}_{\mathbf{C}}(B, .)$ of morphisms $f : A \to B$ and $g : B \to C$, respectively, and as we have seen, for isomorphic objects B_1 and B_2, the corresponding sets can be identified, although not necessarily canonically. Thus, in a category, isomorphic objects cannot be distinguished by their relations with other objects.

In this sense, the category of finite sets contains just one single object for each $n \in \mathbb{N} \cup \{0\}$, the set with n elements, because any two sets with the same number of elements are isomorphic within the category of sets. Thus, the structure of the category of sets consists essentially in the cardinality. Again, this is notwithstanding the fact that the isomorphisms between sets of the same cardinality are not canonical as they can be composed with arbitrary permutations of the elements of the sets. In particular, the automorphism group of a set of n elements is the group $\mathfrak{S}_n$ of permutations of its elements introduced at the end of Sect. 2.1.6.

A poset becomes a category when we stipulate that there is an arrow $a \to b$ whenever $a \leq b$. In turn, we also have the category of posets, with arrows $m : A \to B$ between posets now given by monotone functions, that is, whenever $a_1 \leq a_2$ in A, then $m(a_1) \leq m(a_2)$ in B. Again, while we can consider a category as a graph, we can also consider the category of graphs. Morphisms are then mappings g between graphs $\Gamma_1 \to \Gamma_2$ that preserve the graph structure, that is, map edges to edges.

There can be categories with the same objects, but different morphisms. For instance, we can consider the category whose objects are sets, but whose morphisms are *injective* maps between sets. As another example, for a category with metric spaces as its objects, we could take the isometries as morphisms, that is, the mappings $f : (S_1, d_1) \to (S_2, d_2)$ with $d_2(f(x), f(y)) = d_1(x, y)$ for all $x, y \in S_1$. Alternatively, we can also take the more general class of distance nonincreasing maps, that is, those $g : (S_1, d_1) \to (S_2, d_2)$ with $d_2(g(x), g(y)) \leq d_1(x, y)$ for all $x, y \in S_1$. The isomorphisms of the category, however, are the same in either case. Algebraic structures also naturally fall into the framework of categories. Again, a single structure can be considered as a category, but we can also form the category of all structures of a given type. Thus, as already explained, for instance a monoid M or a group G yields the category with M or G as its only object and the multiplications by monoid or group elements as the morphisms. Thus, the monoid or group elements are not objects, but (endo)morphisms of this category.[7] In fact, for a group considered as a category, every morphism is then an isomorphism, because the group elements are invertible.

Considering monoid or group elements as morphisms, of course, reflects the general idea of a monoid or group as consisting of operations. We have already noted that the associativity law for monoids and groups is included in the definition of a category. In particular, the axioms for a category can also be considered as generalizations of the group axioms, as we do not require invertibility of the operations. Thus, the concept of a monoid is natural within category theory even though in general the concept of a group is more important than that of a monoid. In fact, a category with a single

---

[7]Alternatively, we could also consider the elements of a group or monoid as the objects of the corresponding category. The morphisms would again be the multiplications by elements. Thus, the classes of objects and morphisms would coincide.

object M is nothing but a monoid, where the composition of morphisms then defines the monoid multiplication. Thus, there are many morphisms from this single object to itself. Conversely, we have the categories of monoids, groups, finite groups, abelian groups, free (abelian) groups, Lie groups, etc. In such a category of groups, for instance, an object is again a group, but a morphism now has to preserve the group structure, that is, be a group homomorphism. We should be quite careful here. A monoid M or group G considered as a category is not a subcategory[8] of the category **Monoids** of monoids or **Groups** of groups, resp. The reason is that in those two cases, the notion of a morphism is different. For a single group as a category, the multiplication by any group element as an operation on the group itself is a morphism. Within the category of groups, however, a morphism $\chi : G_1 \to G_2$ between two objects has to preserve their group structure. In particular, χ has to map the neutral element of G_1 to the neutral element of G_2. Analogously, of course, for the case of monoids.

There is a generalization of the foregoing. Let M again be a fixed monoid, with its unit element denoted by e, and with the product of the elements m, n simply written as mn. By definition, the category $\mathbf{B}M = M - \mathbf{Sets}$ consists of all representations of M, that is, of all sets X with an operation of M on X, i.e.,

$$\mu : M \times X \to X$$

$$(m, x) \quad \mapsto mx$$

with $ex = x$ and $(mn)x = m(nx)$ for all $x \in X, m, n \in M$. (2.3.10)

A morphism $f : (X, \mu) \to (Y, \lambda)$ then is a map $f : X \to Y$ which is equivariant w.r.t. the representations, that is,

$$f(mx) = mf(x) \text{ for all } m \in M, x \in X \quad (2.3.11)$$

(where we also write $\lambda(m, y) = my$ for the representation λ). Expressed more abstractly,

$$f(\mu(m, x)) = \lambda(m, f(x)). \quad (2.3.12)$$

For instance, when L is a left ideal of M, then left multiplication by M on L yields such a representation.

Another interpretation of a category, which leads us into logic, a topic to be taken up in Sect. 9.3, is that of a deductive system. The objects of a deductive system are interpreted as formulas, the arrows as proofs or deductions, and the operations on arrows as rules of inference. For formulas X, Y, Z and deductions $f : X \to Y, g : Y \to Z$, we have the binary operation of composition, yielding $g \circ f : X \to Z$, as an inference rule. Thus, by stipulating an equivalence relation for proofs, a deductive system becomes a category. Or putting it the other way around, a category is a formal encoding of a deductive system. See [74] and Sect. 9.3.

We now develop some general concepts.

Definition 2.3.3 An arrow $f : A \to B$ between two objects of a category **C** is called

---

[8]An obvious definition: A category **D** is a subcategory of the category **C** if every object D and every morphism $D_1 \to D_2$ of **D** is also an object or a morphism, resp., of **C**.

- a *monomorphism*, or shortly, *monic*, in symbols,

$$f : A \rightarrowtail B, \text{ or } f : A \hookrightarrow B, \tag{2.3.13}$$

if for any morphisms $g_1, g_2 : C \to A$ in $\mathbf{C}$, $f g_1 = f g_2$ implies $g_1 = g_2$,
- an *epimorphism*, or shortly, *epic*, in symbols,

$$f : A \twoheadrightarrow B, \tag{2.3.14}$$

if for any morphisms $h_1, h_2 : B \to D$ in $\mathbf{C}$, $h_1 f = h_2 f$ implies $h_1 = h_2$.

These notions generalize those of injective and surjective mappings between sets, as introduced in Sect. 2.1.2.

An isomorphism is both monic and epic. In the category **Sets** the converse also holds, that is a monic and epic morphism is an isomorphism (in short jargon: In **Sets**, monic epics are iso). In a general category, this need not be true, however. For instance, in the category of free abelian groups, $f : \mathbb{Z} \to \mathbb{Z}, n \mapsto 2n$ is monic and epic, but not iso.

The above definition is an instance of a general principle in category theory, to define properties through relations with other objects or morphisms within a category. We shall systematically explore this principle below.

Definition 2.3.4 A morphism f is called an *endomorphism* if its codomain coincides with its domain A, in symbols

$$f : A \circlearrowleft . \tag{2.3.15}$$

Thus, an automorphism is an invertible endomorphism.

Definition 2.3.5 A *subobject* A of the object B of the category $\mathbf{C}$ is a monomorphism $f : A \rightarrowtail B$.

Often, the monomorphism f is clear from the context, and we then simply call A a subobject of B.

Thus, for instance, in the category **Sets**, we can speak of subsets, whereas in the category **Groups**, we have subgroups. For the set $\{1, 2, \ldots, n\}$ of n elements, any collection $\{i_1, i_2, \ldots, i_m\}$ for any distinct $i_k \in \{1, 2, \ldots, n\}$, $m < n$, yields a subset. And the group $\mathfrak{S}_m$, introduced at the end of Sect. 2.1.6, of permutations of those m elements is a subgroup of $\mathfrak{S}_n$. . The observation at the end of Sect. 2.1.6 that for a finite group G, the left translation l_g by any element $g \in G$ yields a permutation of the elements of G, with different elements inducing different permutations, means that we can consider G as a subgroup of the group of permutations of its elements. Slightly more abstractly: Every finite group is a subgroup of a symmetric group. This is known as Cayley's Theorem.

We can also consider the morphisms of one category $\mathbf{C}$ as the objects of another category $\mathbf{D}$. In other words, operations within one category can become the objects in another one. In particular, what we mean by an "object" in mathematics has little, if anything, to do with what an "object" is in ordinary language. In category theory, an object is anything on which we can perform systematic operations that relate it to other objects. And

$\boxed{\mathfrak{S}_n}$

when and since we can also operate on operations, they in turn can become objects.

But if we take operations as objects, what then are the operations of that category? The morphisms of **D** are arrows between morphisms of **C**, that is,

$$F : (f : A \to B) \to (g : C \to D), \qquad (2.3.16)$$

given by a pair

$$\phi : A \to C, \psi : B \to D \qquad (2.3.17)$$

of morphisms of **C** with

$$\psi \circ f = g \circ \phi. \qquad (2.3.18)$$

One also expresses this relation by saying that the diagram

$$
\begin{array}{ccc}
A & \xrightarrow{\ f\ } & B \\
\phi \downarrow & & \downarrow \psi \\
C & \xrightarrow{\ g\ } & D
\end{array}
\qquad (2.3.19)
$$

commutes. In short, the morphisms of a category of morphisms are commuting diagrams. As an example, when $f : A \to B$ is a morphism, the identity 1_f is obtained from the identities 1_A and 1_B through such a commutative diagram

$$
\begin{array}{ccc}
A & \xrightarrow{\ f\ } & B \\
1_A \downarrow & & \downarrow 1_B \\
A & \xrightarrow{\ f\ } & B.
\end{array}
\qquad (2.3.20)
$$

We shall now derive a simple condition for constructing a commutative diagram from two morphisms. This is the basic situation considered in [94]. Let X, Y be objects in some category **C**, and let $\pi : X \to Y$ be a morphism. When **C** = **Sets**, that is, X, Y are sets and π is a map between them, then π defines an equivalence relation,

$$x_1 \sim x_2 \text{ iff } \pi(x_1) = \pi(x_2). \qquad (2.3.21)$$

Let $f : X \to X$ be another morphism. We can then find a morphism $\tilde{f} : Y \to Y$ for which the diagram

$$
\begin{array}{ccc}
X & \xrightarrow{\ f\ } & X \\
\pi \downarrow & & \downarrow \pi \\
Y & \dashrightarrow{\tilde{f}} & Y
\end{array}
\qquad (2.3.22)
$$

commutes iff f commutes with the equivalence relation (2.3.21), that is,

$$\text{if } x_1 \sim x_2 \text{ then also } f(x_1) \sim f(x_2). \qquad (2.3.23)$$

When **C** is a category of sets with additional structure, then condition (2.3.23) becomes simpler to check, because we can utilize the additional structure. For instance, when **C** = **Groups** and $\pi : G \to H$ is a group homomorphism, then for a homomorphism $\rho : G \to G$, we get a commutative diagram

$$
\begin{CD}
G @>\rho>> G \\
@V\pi VV @VV\pi V \\
H @>\tilde{\rho}>> H
\end{CD}
\tag{2.3.24}
$$

iff

$$\rho(\ker \pi) \subset \ker \pi, \tag{2.3.25}$$

because in that case, (2.3.25) implies (2.3.23).

The condition (2.3.23) is not a categorical one. Therefore, we now reformulate it in a more abstract manner. Given the morphism $\pi : X \to Y$ in our category $\mathbf{C}$, consider the set of morphisms

$$K(\pi) = \{g : X \to Z \text{ morphism of } \mathbf{C} : \pi \text{ factors through } g\}, \tag{2.3.26}$$

that is, where there exists a morphism $\pi_g : Z \to X$ with $\pi = \pi_g \circ G$. Thus, $g \in K(\pi)$ iff there exists a commutative diagram

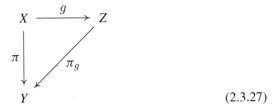

$$(2.3.27)$$

Lemma 2.3.2 *A necessary and sufficient condition for (2.3.22) to commute is*

$$K(\pi) \subset K(\pi \circ f). \tag{2.3.28}$$

Proof We consider the following diagram

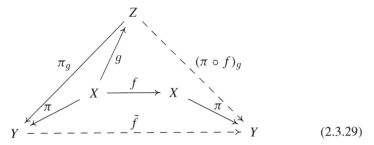

$$(2.3.29)$$

Let $g \in K(\pi)$. When (2.3.22) commutes, we have $\tilde{f}$ and can put $(\pi \circ f)_g = \tilde{f} \circ \pi_g$. Thus, $g \in K(\pi \circ f)$. For the other direction, we simply observe that $\pi \in K(\pi)$ with $Z = Y$ and $\pi_\pi = \mathrm{id}_Y$. Then, $\tilde{f} = (\pi \circ f)_\pi$ lets (2.3.22) commute. $\qquad \square$

In the case of groups, as in (2.3.24), (2.3.25), the crucial g is the projection onto the quotient group $G/\ker \pi$.

We can also fix an object C of $\mathbf{C}$ and consider the category $\mathbf{C}/C$ of all morphisms $f : D \to C$ from objects D of $\mathbf{C}$. This category is called a slice or comma category. A morphism $f \to g$ between two objects of this slice category, that is, from an arrow $f : D \to C$ to an arrow $g : E \to C$ is then a commutative diagram

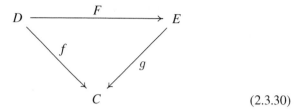

$$\text{(2.3.30)}$$

that is, an arrow $F : D \to E$ with $f = g \circ F$.

We can then also go ahead and form categories $\mathcal{C}$ of categories.[9] That is, the objects of $\mathcal{C}$ are categories $\mathbf{C}$, and the morphisms $F : \mathbf{C} \to \mathbf{D}$ of $\mathcal{C}$, called *functors*, then preserve the category structure. This means that they map objects and arrows of $\mathbf{C}$ to objects and arrows of $\mathbf{D}$, satisfying

$$F(f : A \to B) \text{ is given by } F(f) : F(A) \to F(B) \quad \text{(2.3.31)}$$

$$F(g \circ f) = F(g) \circ F(f) \quad \text{(2.3.32)}$$

$$F(1_A) = 1_{F(A)} \quad \text{(2.3.33)}$$

for all A, B, f, g. Thus, the image of an arrow under F is an arrow between the images of the corresponding objects (domain and codomain) under F, preserving compositions, and mapping identities to identities.

Functors play a very important role as typically one wants to assign to objects of a category with a perhaps complicated structure objects of a category with less structure that nevertheless capture the important qualitative features of the former. For instance, we can associate to a topological space its cohomology groups, as will be explained below. These groups are algebraic objects that encode qualitative topological properties of these spaces. The question that typically arises from such constructions is whether they capture all the relevant features. In the present example, this leads to the question to what extent the cohomology groups determine the topology of a space.

In general, a functor that maps one category to another one with less structure is called forgetful.

---

[9]In fact, these form more naturally so-called bicategories. We suppress this technical point here, however. See for instance [80]. More importantly, one has to be careful to avoid paradoxes of self-reference. Therefore, one respects the axioms of set theory as listed in Sect. 2.2 and considers only sets from a universe U. A category will be called small if both its objects and its arrows constitute a set from U (see Definition 8.1.2). One then looks only at categories of small categories.

Given two categories $\mathbf{C}, \mathbf{D}$, we can then also look at the category $\mathbf{Fun}(\mathbf{C}, \mathbf{D})$ of all functors $F : \mathbf{C} \to \mathbf{D}$. The morphisms of this category are called *natural transformations*. Thus, a natural transformation

$$\theta : F \to G \tag{2.3.34}$$

maps a functor F to another functor G, preserving the structure of the category $\mathbf{Fun}(\mathbf{C}, \mathbf{D})$. What is that structure, and how can it be preserved? Well, the defining property of a functor is that it maps morphisms of $\mathbf{C}$ to morphisms of $\mathbf{D}$. Thus, given a morphism $f : C \to C'$ in $\mathbf{C}$, we obtain morphisms $Ff : FC \to FC'$ and $Gf : GC \to GC'$ in $\mathbf{D}$. A natural transformation $\theta : F \to G$ then has to respect that relation. That means that for each $C \in \mathbf{C}$, it induces a morphism

$$\theta_C : FC \to GC \tag{2.3.35}$$

such that the diagram

$$
\begin{array}{ccc}
FC & \xrightarrow{\theta_C} & GC \\
{\scriptstyle Ff}\downarrow & & \downarrow{\scriptstyle Gf} \\
FC' & \xrightarrow{\theta_{C'}} & GC'
\end{array}
\tag{2.3.36}
$$

commutes.

As will be investigated in more detail in Sect. 8.3, in particular, we can consider functor categories of the form $\mathbf{Sets}^{\mathbf{C}}$, involving the category $\mathbf{Sets}$ of sets and some small category $\mathbf{C}$ ($\mathbf{C}$ is called small if its collections of objects and of arrows are both sets, see Definition 8.1.2). The objects of $\mathbf{Sets}^{\mathbf{C}}$ are functors

$$F, G : \mathbf{C} \to \mathbf{Sets}, \tag{2.3.37}$$

and its arrows are natural transformations

$$\phi, \psi : F \to G. \tag{2.3.38}$$

According to (2.3.35) and (2.3.36), this means that, for instance, $\phi : F \to G$ for each $C \in \mathbf{C}$ has to induce a morphism

$$\phi_C : FC \to GC \tag{2.3.39}$$

such that the diagram

$$
\begin{array}{ccc}
FC & \xrightarrow{\phi_C} & GC \\
{\scriptstyle Ff}\downarrow & & \downarrow{\scriptstyle Gf} \\
FC' & \xrightarrow{\phi_{C'}} & GC'
\end{array}
\tag{2.3.40}
$$

commutes.

We also need the opposite category $\mathbf{C}^{\mathbf{op}}$, obtained from $\mathbf{C}$ by taking the same objects, but reversing the direction of all arrows. This simply means that each arrow $C \to D$ in $\mathbf{C}^{\mathbf{op}}$ corresponds to an arrow $D \to C$ in $\mathbf{C}$. In some cases, this procedure is quite natural. For instance, when the category $\mathbf{C}$ is a poset, this simply amounts to replacing $x \leq y$ by $y \geq x$.

We can now consider the category $\mathbf{Sets}^{\mathbf{C}^{\mathbf{op}}}$ for some fixed small category $\mathbf{C}$. Here, for instance, the category $\mathbf{C}$ could be $\mathcal{P}(S)$, the objects of which are the subsets U of some set S and the morphisms the inclusions $V \subset U$. That is, $\mathcal{P}(S)$ has the structure of a poset, with the ordering relation $\leq$ given by inclusion $\subset$. This poset has a largest element, S, and a smallest one, $\emptyset$. (In fact, $\mathcal{P}(S)$ is a Boolean algebra, with the operations of intersection and union.) This category will be treated in Sect. 2.4.

From the preceding, we may form the impression that the most important example of a category is the category **Sets** of sets. Much of the terminology is oriented towards that example, for instance the representation of morphisms by arrows reminds us of maps between sets. The objects of many other categories are sets with some additional structure, such as a partial order or a group structure. We then have a natural functor from such a category to **Sets**, the so-called forgetful functor that simply forgets that additional structure. Also, it is a fundamental principle to be explored below in more detail that an object of a category can to a large extent be described or characterized by its hom-set, that is, by the set of morphisms into that object or the morphisms from that object.

Another basic example is the category of groups, **Groups**, which we have already discussed above. We have pointed out that the notion of a morphism is inspired by that of a homomorphism between groups. Homomorphisms between groups preserve the group structure, and a morphism between objects of some category has to preserve the characteristic structure of that category.

In order to show the applicability or limitations of the concepts introduced below, we shall often consider other categories. For that purpose, a particularly useful example will be the one consisting of a single poset. Here, the objects are the elements of that poset. As already emphasized, this should not be confused with the category of all posets, where the objects are posets themselves, instead of elements of posets.

The topic of categories will be taken up systematically below in Chap. 8. In order to appreciate the general approach of that chapter, it will be useful to first look at some more particular mathematical structures in depth. We shall do this in the following chapters.

Before moving on in this direction, let us insert a small warning. Even though category theory provides us with abstract principles and constructions that apply simultaneously to all categories, nevertheless, the concrete content of such constructions might be rather different according to the type of category under consideration. On one hand, we have the categories whose objects are simply elements of some set. These elements do not possess any internal structure. They may stand in binary relations F, and such relations then define the morphisms. In a set, there are no nontrivial such relations, that is, $F(s_1, s_2) = 1$ only for $s_1 = s_2$. In a poset, such a relation is denoted by $\leq$, that is, $s_s \leq s_2$ iff $F(s_1, s_2) = 1$. Such binary relations can also be represented geometrically as digraphs, that is, we draw an edge from s_1 to s_2 whenever $F(s_1, s_2) = 1$.

In contrast, at the next level, we have categories, like **Sets**, **Posets** or **Groups** whose objects do have some particular internal structure (although trivial in the case of **Sets**). Morphisms are required to preserve that structure, that is, be structure homomorphisms. For this type of categories, the constructions of category theory will turn out to much more useful than for the preceding ones, whose objects are elements without internal structure. While the preceding categories may serve to illustrate some of those constructions, as examples of the general thrust of the theory they may be somewhat misleading.

2.4   Presheaves

In this section, we shall take a first glimpse at the concept of a presheaf, which will be treated in more detail and depth in Sects. 4.5 and 8.4 below.

Here is a preliminary definition that will be refined in Sect. 4.5 where we shall work in the category of topological spaces instead of sets. A bundle over the set S is a surjective map $p : T \to S$ from some other set T. For $x \in S$, the preimage $p^{-1}(x)$ is called the fiber over x. S is called the base space, and T is the total space. A section of such a bundle is a map $s : S \to T$ with $p \circ s = 1_S$. Thus, a section associates to each element x of the base space an element of the fiber over x. Usually, for a given bundle, the space of sections is constrained; that is, not every such s with $p \circ s = 1_S$ represents a valid section of the bundle, but the sections need to satisfy certain restrictions or constraints.

Here is an interpretation. S could be a set of properties, observables or features of possible objects, like size, color, texture, material. The fiber over such a feature $x \in S$ then contains the possible values that this feature can assume. When x stands for color, the fiber over x might contain the values 'red', 'green', 'blue' etc., or if we desire, also more precise shades of color like 'azure blue', 'crimson', or 'yellowish pink'. A section then assigns to each object the values of its properties, that is, in the current example, its color, size, texture, material etc. The whole point now is that the values of the various properties in general are not independent of each other. A trivial example might elucidate this issue. When the object in question is a piece of gold, then the value 'gold' for the material constrains the color to be 'golden', and also its size and texture will obey certain restrictions. Deeper and more interesting examples come from theoretical biology, and lead, in fact, to the core of the field of morphology. Two hundred years ago, the biologist Cuvier (1769–1832), the founder of scientific paleontology and comparative anatomy, had already emphasized that a plant or animal does not just consist of an arbitrary collection of feature values, but that those are highly interdependent and determined by its mode of living. According to his principle of "Correlation of parts", the anatomical structures of the various organs of an animal are functionally related to each other and the structural and functional characteristics of the organs are all derived from the particular mode of living of the animal within its environment. Mammals are not only viviparous, but also typically possess fur and characteristic anatomical features, and carnivores not only have teeth and jaws adapted to catch and handle their pray, but also digestive tracts suited for their meat diets, and feet depending on the way they chase their prey, and so on. In fact, based on such correspondences, he could perform the stunning feat of reconstructing a particular dinosaur on the sole basis of a very incomplete fossil consisting only of a claw. Later on, when a more complete fossil skeleton was found, it agreed in remarkable detail with Cuvier's reconstruction which made him quite famous.[10] Translated into our language of bundles and sections, there exist strong correlations, constraints and restrictions between the values of the various features, and knowing such constraints, one can reconstruct much of a section from the values of particular features. The biological aspects will be explored in detail elsewhere. Here, we want to use this to introduce a key concept, that of a presheaf.

We recall the functor category $\mathbf{Sets}^{\mathbf{C^{op}}}$ for some small category $\mathbf{C}$ where $\mathbf{C^{op}}$ is obtained from $\mathbf{C}$ by reversing the directions of all arrows, and stipulate

Definition 2.4.1 An element P of $\mathbf{Sets}^{\mathbf{C^{op}}}$ is called a *presheaf* on $\mathbf{C}$.

For an arrow $f : V \to U$ in $\mathbf{C}$, and $x \in PU$, the value $Pf(x)$, where $Pf : PU \to PV$ is the image of f under P, is called the *restriction* of x along f.

Thus, a presheaf formalizes the possibility of restricting collections of objects, that is, the—possibly structured—sets assigned to subsets of S, from a set to its subsets.

We can put this into the more general context that will be developed in Chap. 8, but this will not be indispensable for understanding the current section. Anticipating some of Sect. 8.4 and also of Sect. 8.3, we consider $\mathrm{Hom}_{\mathbf{C}}(V, U)$, the set of morphisms in the category $\mathbf{C}$ from the object V to the object U. Each object $U \in \mathbf{C}$ then yields the presheaf yU defined on an object V by

$$yU(V) = \mathrm{Hom}_{\mathbf{C}}(V, U) \tag{2.4.1}$$

and on a morphism $f : W \to V$ by

$$yU(f): \quad \mathrm{Hom}_{\mathbf{C}}(V, U) \quad \to \mathrm{Hom}_{\mathbf{C}}(W, U)$$
$$h \qquad\qquad \mapsto h \circ f. \tag{2.4.2}$$

We shall also see in Sect. 8.3 that when $f : U_1 \to U_2$ is a morphism of $\mathbf{C}$, we obtain a natural transformation $yU_1 \to yU_2$ by composition with f, so that we get the Yoneda embedding (Theorem 8.3.1)

$$y : \mathbf{C} \to \mathbf{Sets}^{\mathbf{C^{op}}}. \tag{2.4.3}$$

The presheaf yU, with

$$yU(V) = \mathrm{Hom}_{\mathbf{C}}(V, U), \tag{2.4.4}$$

is also called the functor of points as it probes U by morphisms from other members V of the category. When we work with the category $\mathbf{Sets}$ and V is a single element set, then any morphism from such a single element set to a set U determines an element in U, a point of U. When V is a more general set, then this yields, in naive terminology, a family of points in U parametrized by V. The categorial approach thus naturally incorporates such generalized

---

[10] We refer to [44] for a conceptual analysis of Cuvier's position in the history of biology.

points. For instance, when we are in the category of algebraic varieties (to be defined below), we probe an algebraic variety U by considering morphisms from other algebraic varieties V, be they classical points or more general varieties.

When we go in the opposite direction and consider

$$zU(V) = \text{Hom}_{\mathbf{C}}(U, V), \qquad (2.4.5)$$

we obtain what is called the functor of functions. Here, classically, one would take as V a field such as the real numbers $\mathbb{R}$ or the complex numbers $\mathbb{C}$.

Of course, we can then also let U and V vary simultaneously, to make the construction symmetric.

We now return to the category $\mathcal{C} = \mathcal{P}(S)$ of subsets of some set S. For a presheaf $P : \mathcal{P}(S)^{\text{op}} \to \mathbf{Sets}$, we then have the restriction maps

$$p_{VU} : PV \to PU \text{ for } U \subset V \qquad (2.4.6)$$

that satisfy

$$p_{UU} = 1_{PU} \qquad (2.4.7)$$

and

$$p_{WU} = p_{VU} \circ p_{WV} \text{ whenever } U \subset V \subset W. \qquad (2.4.8)$$

Definition 2.4.2 The presheaf $P : \mathcal{P}(S)^{\text{op}} \to \mathbf{Sets}$ is called a *sheaf* if it satisfies the following condition. If $U = \bigcup_{i \in I} U_i$ for some family $(U_i)_{i \in I} \subset \mathcal{P}(S)$ and $\pi_i \in PU_i$ satisfies $p_{U_i, U_i \cap U_j} \pi_i = p_{U_j, U_i \cap U_j} \pi_j$ for all $i, j \in I$, then there exists a unique $\pi \in PU$ with $p_{UU_i} \pi = \pi_i$ for all i.

Thus, whenever the π_i are compatible in the sense that the restrictions of π_i and π_j to $U_i \cap U_j$ always agree, then they can be patched together to an element π of PU that restricts to π_i on PU_i.

The interpretation of a presheaf over $\mathcal{P}(S)$ that is most important for our current purposes takes place within the framework of fiber bundles developed at the beginning of this section. The set assigned to some $U \subset S$ by the presheaf P then would simply be the set of sections over U of a fiber bundle with base S. Such sections over a subset U are called local because they need not extend to all of S. Sections defined over all of S could be called global, and thus, local sections need not admit extensions to global sections. In contrast, by the presheaf condition, whenever we have a local section over some U, we can restrict it to any $V \subset U$. The sheaf condition, in turn, stipulates that locally compatible local sections can be patched together to a global section. Not every presheaf is a sheaf, and so, such an extension from local compatibility to what one may call global coherence need not always be possible.

For instance, when our fiber bundle is a Cartesian product $S \times \mathbb{R}$ or $S \times \mathbb{C}$, then all fibers are $\mathbb{R}$, and a local section over U is nothing but a real valued function on U. A presheaf then might stipulate further conditions or restrictions on such functions. In particular, when S carries some additional structure, like that of a metric, we could require the functions to respect that structure. In the case where S is equipped with a metric $D(.,.)$, we might request, for instance, that for every local section over U, that is, for every function $f : U \to \mathbb{R}$ that belongs to the presheaf, we have $|f(x) - f(y)| \leq d(x, y)$ for all $x, y \in U$. Or when S is a graph, we could only admit functions with values ± 1 and require that any such function on a subgraph U assign different values to neighboring vertices. That is, when $f(x) = 1$ for some vertex x, then $f(y) = -1$ for all neighbors y of x, and conversely. Such a function then cannot be extended to any subgraph that contains a triangle, that is, three vertices x, y, z each of which is a neighbor of the other two. More generally and precisely, a graph is called bipartite iff it admits such a function f. Bipartite graphs do not contain triangles, nor other cycles of odd length.[11] They consist of two classes of vertices, so that a function f with the above properties assigns the value $+1$ to one class and -1 to the other.

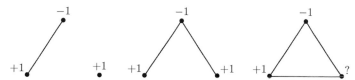

Here, we have sections for the first two graphs, but none for the third. This phenomenon is also called frustration. Of course, the first two graphs are subgraphs of the third, and so, we see here an example where a local section cannot be extended to a global section.

Returning to the biological principle of the "Correlation of parts", the sections of a presheaf would correspond to the different species, and the fact that only specific value combinations in the various fibers are realized by some section then reflects those correlations and constraints.

We now describe and further explore a different biological realization of presheafs, and discuss a proposal of Benecke and Lesne [11] to describe genomes in this formal framework. The DNA of a cell of a member of a biological species is a linear string composed of instances of four nucleotides, labelled by the letters A, T, C, G.[12] That is, we have a finite sequence (with about 3 billion elements in the case of humans) of positions, called genetic loci, each of which is filled by one of the letters A, T, C, G. Looking at this from the point of view of topology, we take as base space the space of genetic loci of a species, with a metric given by the distance between positions in the linear arrangement.[13] As the fiber, we take the possible nucleotide values, that is, A, T, C, and G. This fiber then carries a natural probability measure (see Sect. 4.4 below for the formal definition) given by the relative frequencies of the nucleotide values. In fact, first the fiber over each locus carries such a measure. We can then also compare fibers

---

[11]see Sect. 3.4.1 for the definition of a cycle.

[12]With certain exceptions that need not concern us here, all the cells of a given organism carry the same DNA sequence.

[13]We assume here that there is a one-to-one correspondence between the loci of different members of the species. That is, we assume that the only differences between individuals are given by point mutations, but not by insertions or deletions of nucleotide strings in the genome.

over different loci by the distance between those measures.[14] We can also simply look at the abstract fiber of the four nucleotide values and obtain a measure on it by averaging over all genetic loci.

An individual genome is then a section of this fiber bundle. The space of sections then yields a sheaf, even though, of course, not every section needs to be realized by the genome of some individual. Again, spaces of genomes then induce measures on the space of sections as well as on the individual fibers, or more generally, on the collection of fibers over any given set of loci. When the set of loci of two populations is in 1-1 correspondance, we can then look at the distance between the measures induced on the space of sections by the two populations. We can then define the genetic distance between the populations as such a distance between measures on the space of sections.

2.5   Dynamical Systems

Definition 2.5.1 A *dynamical system* is a homomorphism ϕ from the additive group $\mathbb{Z}$ of integers or real numbers $\mathbb{R}$ (or the semigroup of nonnegative integers or reals) to the group of (invertible) selfmaps $\mathcal{F}(S)$ of some set S. When the domain is the group (monoid) of (nonnegative) integers, we speak of a discrete-time dynamical system, and when we have the (nonnegative) reals, we talk about a continuous-time system.

In particular, $0 \in \mathbb{R}$ is mapped to id_S, and $\phi(t_1 + t_2) = \phi(t_1) \circ \phi(t_2)$. Often, there is more structure; for instance, S could be a topological space (see Definition 4.1.1) or a differentiable manifold (see Definition 5.3.3), and the selfmaps could be homeomorphisms (see Definition 4.1.11) or diffeomorphisms. Or, S could be a vector space, and the maps linear. We can write this as a diagram

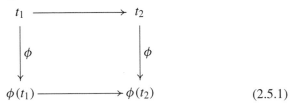

$$\phi(t_1) \longrightarrow \phi(t_2) \tag{2.5.1}$$

The variable t here is considered as time. The value $\phi(0)$ is called the *initial value* of the dynamical system (2.5.1). One usually writes $\phi(x, t)$ for $\phi(t)(x)$, the value of the map $\phi(t)$ applied to the initial value $x = \phi(0)$. This expresses the dependence on time t and the initial value x. The collection of points $\phi(x, t)$ as t varies is called the *orbit* of x.

---

[14]Here, we can use the distance induced by the Fisher metric on the space of measures. We can also utilize the Kullback-Leibler divergence, which is not quite a distance, in fact, because it is not symmetric. For the definition and for a geometric view of these distances, see [3, 6].

In the discrete-time case, we can simply take a map $F : S \to S$ (invertible when the system is to be defined on $\mathbb{Z}$) and consider its iterates, that is, put

$$\phi(x, n) = F^n(x), \qquad (2.5.2)$$

the n-fold iterate of F (when $n < 0$ and F is therefore assumed to be invertible, $F^n = (F^{-1})^{-n}$). Thus, $\mathbb{N}$ or $\mathbb{Z}$ operates on S by iteration of a self-map. As a diagram, this looks like

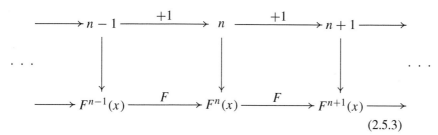

$$(2.5.3)$$

Such a discrete-time dynamical system, that is, a set S equipped with an endomorphism F, is also called an automaton.

Relations

3

3.1 Elements Standing in Relations

Relations take place between elements, and in order to conceptualize this, we can either take the elements or the relations as primary. Alternatively, we can try to combine these two dual approaches.

When we take the elements as fundamental, we start with a set V whose elements are denoted by v or $v_0, v_1, v_2, \ldots$. We assume that finite tuples $(v_0, v_1, \ldots, v_q)$ of elements, all assumed to be different from each other, can stand in a relation $r(v_0, v_1, \ldots, v_q)$. Here, the tuple $(v_0, v_1, \ldots, v_q)$ may be considered as ordered or unordered, depending on whether we want to let the relation depend on the order of elements or not. That is, in the unordered case, we have $r(v_0, v_1, \ldots, v_q) = r(v_{i_0}, \ldots, v_{i_q})$ for any permutation $(i_0, \ldots, i_q)$ of $(0, \ldots, q)$. Here, r takes its values in some set or space R which we leave unspecified for the moment. We write $r(v_0, \ldots, v_q) = o$ when this relation is trivial (whatever triviality might mean here), or perhaps with a better interpretation, absent; that is, we distinguish one particular member o of R that indicates the absence of a relation. We thus have a map

$$r : \bigcup_{q=0,1,\ldots} V^{q+1} \to R, \tag{3.1.1}$$

which in the unordered case reduces to a map from the collection of finite subsets of V to R. As a useful convention, one may then require that $r(\emptyset) = o$.

When we shall do cohomology theory below, we shall assume the following properties

(i)

$$r(v) \neq o, \tag{3.1.2}$$

that is, each element stands in a nontrivial relation with itself.

© Springer International Publishing Switzerland 2015
J. Jost, *Mathematical Concepts*, DOI 10.1007/978-3-319-20436-9_3

(ii)

$$\text{If } r(v_0, \ldots, v_q) \neq o, \text{ then also } r(v_{i_1}, \ldots, v_{i_p}) \neq o \quad (3.1.3)$$
$$\text{for any (different) } i_1, \ldots, i_p \in \{0, \ldots, q\},$$

that is, whenever some elements stand in a nontrivial relation, then this also holds for any nonempty subset of them.

When these conditions are satisfied and we associate a simplex to every collection of elements with $r(v_0, \ldots, v_q) \neq o$, we obtain a *simplicial complex* with vertex set V. In fact, we can take the preceding properties as the defining axioms for a simplicial complex.

Without these restrictions, we obtain a *hypergraph*.

Example Let $U_1, \ldots, U_m$ be a collection of nonempty subsets of some set M. We say that $U_{i_1}, \ldots, U_{i_p}$ stand in a relation, that is, $r(U_{i_1}, \ldots, U_{i_p}) \neq o$ iff

$$U_{i_1} \cap \cdots \cap U_{i_p} \neq \emptyset, \quad (3.1.4)$$

that is, if these sets have a nonempty intersection. The above conditions are then satisfied, as all the sets are nonempty, and hence $r(U_i) \neq o$, and if a collection of sets has a nonempty intersection, then this also holds for any subcollection. Thus, any such collection of sets defines a simplicial complex with the vertices corresponding to the sets $U_i, i = 1, \ldots, m$. This simplicial complex is called a *Čech complex*. This construction will be taken up in Sect. 4.6 below.

When the set M carries a measure μ (see Sect. 4.4), we can also put

$$r(U_{i_1}, \ldots, U_{i_p}) = \mu(U_{i_1} \cap \cdots \cap U_{i_p}). \quad (3.1.5)$$

We then obtained a *weighted simplicial complex*, that is, a complex where a real number is attached to every simplex.

When, as in the introductory Sect. 2.1.4, the relations are only defined for pairs of elements, that is, for $q = 1$, and if r can take only two possible values, which we may denote by o and 1, we obtain a loopless directed graph, also called a *digraph*. In the unordered case, this reduces to an undirected loopless graph, usually simply called a *graph*. In that case, the elements of V are called vertices or nodes. That is, in the ordered case, we put an edge $e := [v, w]$ from the vertex v to the vertex w when $r(v, w) = 1$. In the unordered case, we then simply have an edge between v and w. "Loopless" here means that we do not allow for edges from a vertex to itself. (The graph is loopless, because we assume in this chapter that only different elements can stand in a relation with each other. The selfrelation is not considered to be binary, but unary, $r(v) \neq o$, and therefore is represented by a vertex rather than an edge of the graph.) Also, the graphs are "simple", meaning that between any two vertices, there can be at most one edge.

When, instead of restricting the range of r to two values, we let r take values in $\mathbb{R}$ or $\mathbb{C}$ (and we identify o with 0), we obtained weighted graphs. The weight of the edge from v to w is then given by $r(v, w)$. In particular, we have the convention that the edge weight is 0 iff $r(v, w) = 0$, that is, in the absence of an edge from v to w.

3.2   Relations Specifying Elements

The alternative approach consists in starting with a set S of relations and a map

$$s : S \rightarrow \bigcup_{q=0,1,\ldots} V^{q+1} \qquad (3.2.6)$$

or to the set $\mathcal{P}(V)$ of subsets of V in the unordered case, possibly requiring that the range of s be confined to the finite subsets of V, in order to respect the convention set up in the preceding section.

For instance, to construct a graph, we take an edge set E and a vertex set V and a map

$$s : E \rightarrow V^2. \qquad (3.2.7)$$

For each edge $e \in E$, $s(e) = (v_1, v_2)$ then specifies its beginning v_1 and end v_2. When the graph is to be simple, we require that s be injective. When the graph is to be loopless, we require that the vertices $v_1 \neq v_2$. When we want to have an undirected graph, we let s take its values in the unordered pairs of vertices. Such a point of view is, for instance, useful in the study of random graphs, that is, where pairs of vertices are randomly connected by edges according to some probabilistic rule. In the simplest case, every pair of different vertices $v_1 \neq v_2$ is connected by an edge with probability $0 < p < 1$, without any dependencies among pairs. This construction was first proposed by Erdös and Renyi [34], and has become an important paradigm in graph theory. In this book, however, we do not enter the realm of stochastics, but may refer the curious reader to [63].

3.3   Isomorphisms

Whichever way we choose to specify and describe the category of relations between elements, the categorical description will include the notion of a morphism between two members of such a category. Since it is rather straightforward to work out the corresponding notion of morphism in each situation, we consider here only the simplest case of undirected graphs. Such a graph is specified by a vertex set V and an edge set E as explained in Sect. 3.1. A morphism

$$\gamma : (V_1, E_1) \rightarrow (V_2, E_2) \qquad (3.3.8)$$

is then given by a map $\eta : V_1 \rightarrow V_2$ between the vertex sets whose induced map on pairs of vertices, $(v, v') \mapsto (\eta(v), \eta(v'))$ maps edges in E_1 to edges in E_2. In other words, $\eta(v), \eta(v') \in V_2$ are connected by an edge iff $v, v' \in V_1$ are thus connected.

Two such graphs $\Gamma_1 = (V_1, E_1)$, $\Gamma_2 = (V_2, E_2)$ then are isomorphic if there exist morphisms $\gamma : \Gamma_1 \rightarrow \Gamma_2, \gamma' : \Gamma_2 \rightarrow \Gamma_1$ for which $\gamma' \circ \gamma$

and $\gamma \circ \gamma'$ are the identities of Γ_1 and Γ_2, resp. An automorphism of the graph $\Gamma = (V, E)$ is then an isomorphism from Γ to itself (recalling some general concepts of category theory).

In order to understand the automorphism group $A(\Gamma)$ of a graph $\Gamma = (V, E)$ with n vertices, we observe that the symmetric group $\mathfrak{S}_n$ operates by permutations on the vertex set V. The automorphism group of Γ then consists precisely of those elements of $\mathfrak{S}_n$ that map the edge set E to itself. For the trivial graph Γ_0, that is, the graph with an empty edge set, and the complete graph K_n, that is, the graph with an edge between any pair of the n vertices (see the figure below for K_4),

| Complete graph K_4 |

$A(\Gamma)$ is the entire symmetric group $\mathfrak{S}_n$. For any other graph, $A(\Gamma)$ is a proper subgroup of $\mathfrak{S}_n$, possibly the trivial group consisting only of the trivial permutation, that is, the permutation operating as the identity on V. The smallest graphs without any nontrivial isomorphism have 6 vertices; an example is shown in the figure below.

| Graph without symmetries |

While all smaller graphs do possess some nontrivial automorphisms, when the vertex sets gets larger, automorphisms become rarer, and in fact, a generic graph does not possess any nontrivial automorphisms. Also, checking whether two graphs (with the same number of vertices) are isomorphic is an NP-hard[1] problem, and therefore so is the determination of the automorphism group of a given graph. Here, however, we do not wish to enter into the corresponding details.

When some $s \in \mathfrak{S}_n$ does not yield an automorphism of a graph Γ, it maps Γ to another graph $s(\Gamma)$ that is isomorphic to Γ. That graph is thus obtained from Γ by a permutation of its vertices and the induced mapping of the edges. In turn, any graph isomorphic to Γ is easily seen to be obtained in this manner.

---

[1]A concept not explained here; for this and other notions of computational complexity theory, see for instance [92].

3.4   Moduli Spaces of Graphs

3.4.1   From the Vertices

When, instead of looking at a single relational structure, we would like to understand all possible such structures, we are led to so-called moduli spaces (the name being derived from Riemann's investigation of the various conformal structures (Riemann surfaces) that a given topological surface can be equipped with). Thus, let us look at the set of all binary relation structures between n elements. Each such structure is encoded by a directed graph (digraph) with a vertex set given by those n elements. In order to represent all these structures simultaneously, we consider each such structure, that is, each such digraph as an element (vertex) of another graph $D(n)$, and we put an oriented edge from the vertex v_Γ representing the digraph Γ to v_Δ representing Δ if Δ is obtained from Γ by addition of a directed edge.

Similarly, when we consider undirected graphs, we obtain a space $M(n)$ whose vertices stand for graphs Γ, Δ of n elements, and we again put an edge from v_Γ to v_Δ when the latter is obtained from the former by addition of an edge. Thus, $M(n)$ again is a directed graph, and if we wish, we can turn it into an undirected graph by ignoring the direction of its edges.

For weighted graphs, with edge weights in either $\mathbb{R}$ or $\mathbb{C}$, we have an additional structure, since we can add two weighted graphs on n elements (again with the convention that we put an edge of weight 0 in the absence of an edge between two vertices) by simpling adding the weights of corresponding edges. Therefore, the moduli spaces of weighted graphs with n vertices is a vector space (over $\mathbb{R}$ or $\mathbb{C}$ or whatever field the weights take their values in) of dimension equal to the number of ordered (for a directed graph) or unordered (for an undirected graph) pairs of vertices, that is, of dimension $n(n-1)$ or $\frac{n(n-1)}{2}$, resp.

All this is straightforward, but we have ignored one important aspect. In fact, a moduli space should represent the different objects in a category. Two isomorphic graphs, as defined in Sect. 3.3, should not be considered as different. To incorporate this into our formal framework, we consider the induced operation of the symmetric group $\mathfrak{S}_n$ on $M(n)$ (or $D(n)$, the two cases being completely analogous). The moduli space for graphs with n vertices is therefore

$$\mathfrak{M}(n) := M(n)/\mathfrak{S}_n, \tag{3.4.9}$$

the quotient of $M(n)$ by the operation of $\mathfrak{S}_n$.

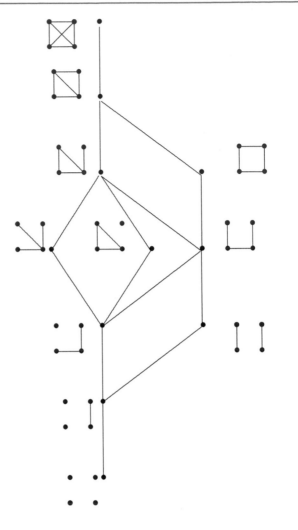

(3.4.10)

The figure shows the moduli space $\mathfrak{M}(4)$ in black, with the graph representing each vertex in $\mathfrak{M}(4)$ adjacent to it in color. Let us discuss the structure of $\mathfrak{M}(4)$ a bit further. $\mathfrak{M}(4)$ is a bipartite graph, one class representing graphs with an even number of edges (represented in red), the other class those with an odd number (represented in blue). Moreover, $\mathfrak{M}(4)$ possesses an automorphism obtained by exchanging each graph by its complementary graph. For instance, K_4 goes to the trivial graph while the 3-star and the triangle are interchanged and the 3-path is a fixed point of this automorphism.

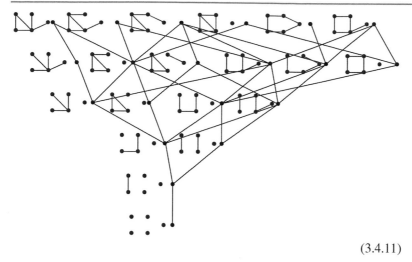

$$(3.4.11)$$

When we move on to graphs with 5 vertices, the diagram (3.4.11) already gets more complicated. Because of the symmetry that we have discovered for graphs with 4 vertices, between a graph and the complementary graph, here we have depicted only those graphs with ≤ 5 edges. We can also perform certain checks, for instance for the graphs with precisely 5 edges. We shall use some basic concepts from graph theory to evaluate our findings. A path is a sequence $v_1, \ldots, v_m$ of distinct vertices such that v_ν and $v_{\nu+1}$ are connected by an edge for $\nu = 1, \ldots, m - 1$. A graph is called connected if any two vertices can be joined by a path. A cycle is a closed path $v_0, v_1, \ldots, v_m = v_0$, that is, all the vertices $v_1, \ldots, v_m$ are distinct, but the initial vertex v_0 is the same as the terminal v_m. For $m = 3, 4, 5$, we speak of a triangle, quadrangle, or pentagon. A connected graph without cycles is called a tree. A tree with n vertices has $n - 1$ edges. A graph with more than $n - 1$ edges has to contain some cycle. In particular, a graph with 5 vertices and 5 edges must contain at least one cycle, that is, either a triangle, a quadrangle or a pentagon. When it contains a triangle, either the two remaining vertices are connected to different vertices of the triangle, to the same vertex of the triangle, or one of them is connected either with two vertices of the triangle, or with one vertex and the remaining node. Likewise, when we have a quadrangle, either there is a diagonal inside the quadrangle (which is the same as a vertex connected to two vertices of a triangle) or the remaining vertex is connected to one of the vertices of the quadrangle. Thus, we find 6 nonisomorphic graphs with 5 vertices and 5 edges.

As a different check, we can look at the sequences of vertex degrees. For the 6 graphs with 5 vertices and 5 edges, they are (in the order that they appear in the diagram)

$(1, 1, 2, 3, 3), (1, 1, 2, 2, 4), (1, 2, 2, 2, 3), (0, 2, 2, 3, 3), (2, 2, 2, 2, 2), (1, 2, 2, 2, 3).$

We note that the sum of the vertex degrees always has to be equal to twice the number of edges, that is, 10 for the current case. We also observe that there are two different graphs, the triangle with a two-path and the quadrangle connected to a single vertex, that have the same degree sequences. Thus, the degree sequence cannot always distinguish nonisomorphic graphs.

In any case, it is clear that when we move on to higher numbers of vertices, such an enumeration scheme of all isomorphism classes of graphs is no longer feasible, and therefore, we need to develop more abstract and general methods to investigate such families of graphs. One can ask many questions. For instance, how many connected nonisomorphic graphs are there with n vertices? Let A_n be that number. Then $A_2 = 1$, $A_3 = 2$, $A_4 = 6$, $A_5 = 21$. But what is the formula for A_n for arbitrary n? Not so easy to figure out.

3.4.2   From the Edges

The complete graph K_n on n vertices possesses $\frac{n(n-1)}{2}$ edges. We consider the simplicial complex Σ_N constituted by the $N := (\frac{n(n-1)}{2} - 1)$-dimensional simplex, that is, the one with $\frac{n(n-1)}{2}$ vertices $x_0, \ldots, x_N$, and all its subsimplices. Each such vertex x_j stands for an edge e_j of K_n. Thus, a subsimplex with vertices $x_{i_0}, \ldots, x_{i_k}$ represents the graph possessing the edges $e_{i_0}, \ldots, e_{i_k}$ corresponding to those vertices of Σ_N. Again, we have an induced operation of the symmetry group $\mathfrak{S}_n$ on Σ_N, and so our moduli space of graphs of n vertices in this setting is given by

$$\Sigma_N/\mathfrak{S}_n. \tag{3.4.12}$$

Ignoring the action of $\mathfrak{S}_n$ for the moment, the simplex Σ_N can also be seen as representing the complete graph K_n together with all its subgraphs. In the same manner, we could take any other graph Γ on n vertices and consider the simplicial complex representing Γ together with all its subgraphs.

3.4.3   From the Edge Weights

As we have already seen, the moduli space of graphs with (complex) edge weights is the vector space $\mathbb{C}^{n(n-1)/2}$, and we can add such weighted graphs by adding the weights of each edge. Again, the symmetry group $\mathfrak{S}_n$ permuting the vertices of a graph then operates on that vector space. Since this is the operation of a group on a vector space, we are therefore in the realm of representation theory. We shall now briefly present that theory and apply it to the example of weighted graphs with 4 vertices.

3.4.4   Representation Theory

Definition 3.4.1 A *representation* of the (finite) group G is a homomorphism

$$\rho : G \to \mathrm{Gl}(V) \tag{3.4.13}$$

into the automorphism group of a (finite-dimensional) complex vector space V. ρ is also said to give V the structure of a G-module.

One also usually simply says that V is a representation of G, assuming that ρ is clear from the context.

Here, $\mathrm{Gl}(V)$ is the group of invertible linear endomorphisms of the vector space V. When we identify V as $\mathbb{C}^n$ by choosing some vector space basis of V, then $\mathrm{Gl}(V)$ operates by invertible matrices, and so then does every element $g \in G$ via $\rho(g)$. We therefore often write gv in place of $\rho(g)v$ for $v \in V$.

A morphism between representations V and W of G is then given by a linear map $L : V \to W$ that commutes with the actions of G, that is the diagram

$$
\begin{array}{ccc}
V & \xrightarrow{\ L\ } & W \\
{\scriptstyle g}\downarrow & & \downarrow{\scriptstyle g} \\
V & \xrightarrow{\ L\ } & W
\end{array}
\tag{3.4.14}
$$

commutes for every $g \in G$.

We are then naturally interested in the representations of G up to isomorphism; we write $V \cong W$ for isomorphic representations.

The concept of a representation is compatible with natural operations on vector spaces. For representations V, W, the direct sum $V \oplus W$ and the tensor product $V \otimes W$ are also representations[2]; for instance

$$
g(v \otimes w) = gv \otimes gw. \tag{3.4.15}
$$

A representation ρ on V induces a representation ρ^* on the dual $V^* = \mathrm{Hom}(V, \mathbb{C})$ via the requirement that the pairing $\langle v^*, v \rangle := v^*(v)$ be preserved,

$$
\langle \rho^*(g)v^*, \rho(g)v \rangle = \langle v^*, v \rangle \text{ for all } v, v^*, \tag{3.4.16}
$$

namely

$$
\rho^*(g) = \rho(g^{-1})^t : V^* \to V^*. \tag{3.4.17}
$$

Therefore, when V, W are representations, so is

$$
\mathrm{Hom}(V, W) = V^* \otimes W. \tag{3.4.18}
$$

g then acts on $L \in \mathrm{Hom}(V, W)$ via

$$
L \mapsto g \circ L \circ g^{-1}, \text{ that is, } v \mapsto gL(g^{-1}v). \tag{3.4.19}
$$

L is said to be G-linear or G-equivariant if

$$
g \circ L \circ g^{-1} = L \text{ for all } g. \tag{3.4.20}
$$

In other words, the space of G-linear maps between the representations V and W is identified with the space

$$
\mathrm{Hom}^G(V, W) \tag{3.4.21}
$$

of elements of $\mathrm{Hom}(V, W)$ fixed under the action of G.

---

[2] The direct sum and the tensor product can be defined in categorical terms, see Sect. 8.2 for such constructions. For the present purposes, it is sufficient to consider the cases where $V = \mathbb{C}^n$ and $W = \mathbb{C}^m$, with bases $(e_1, \ldots, e_n)$ and $(f_1, \ldots, f_m)$. $\mathbb{C}^n \oplus \mathbb{C}^m$ is then the vector space $\mathbb{C}^{n+m}$ with basis $(e_1, \ldots, e_n, f_1, \ldots, f_m)$ and $\mathbb{C}^n \otimes \mathbb{C}^m$ is $\mathbb{C}^{nm}$ with basis vectors $(e_i \otimes f_j)$, $i = 1, \ldots, n$, $j = 1, \ldots, m$.

Definition 3.4.2 A subspace $V' \neq \{0\}$ of the representation V that is invariant under the action of G (i.e., $gv' \in V'$ for every $v' \in V'$, $g \in G$) is called a *subrepresentation*. When $V' \subsetneq V$, this subrepresentation is proper. The representation V is called *irreducible* if it does not contain a proper subrepresentation.

We have the following important

Lemma 3.4.1 *For each proper subrepresentation V' of V, there is a complementary subrepresentation, that is a complementary G-invariant subspace V'' with*

$$V = V' \oplus V''. \tag{3.4.22}$$

Proof Take any Hermitian product $\langle ., . \rangle$ on V and average it over G to obtain a G-invariant Hermitian product

$$\langle v, w \rangle_G := \sum_{g \in G} \langle gv, gw \rangle. \tag{3.4.23}$$

(Here, G-invariance means that $\langle hv, hw \rangle_G = \langle v, w \rangle_G$ for all $h \in G$; this G-invariance of (3.4.23) follows from the fact that when g runs through all the elements of G, so does gh for any $h \in G$.) We then take V'' as the subspace of V that is orthogonal to V' w.r.t. $\langle ., . \rangle_G$. $\square$

The method of the preceding proof, averaging a product over the G-action is simple, but very important.[3] In fact, for the representation theory of compact Lie groups, such an averaging via integration w.r.t. an invariant measure on the group is fundamental. For noncompact groups, however, it fails in general.

We immediately obtain

Lemma 3.4.2 (Schur) *If V, W are irreducible representations of G and $L \in \mathrm{Hom}^G(V, W)$, then either $L = 0$ or L is an isomorphism. If $V = W$, then $L = \lambda \, \mathrm{Id}$ for some $\lambda \in \mathbb{C}$. Thus,*

$$\dim \mathrm{Hom}^G(V, W) = \begin{cases} 1 & \text{if } V \cong W \\ 0 & \text{if } V \not\cong W \end{cases} \tag{3.4.24}$$

Proof The kernel and the image of L are invariant subspaces. Irreducibility implies that they are either 0 or the entire space V or W, resp. When $V = W$, L must have some eigenvalue λ, hence $L - \lambda \, \mathrm{Id}$ has nonzero kernel. By irreducibility again, this kernel then must be all of V, hence $L = \lambda \, \mathrm{Id}$. $\square$

Remark In Schur's lemma, it is important that the representations be complex, so as to ensure the existence of eigenvalues.

---

[3] The method of averaging has a more general scope than described here, as it can also be used for linear representations in vector spaces over fields other than $\mathbb{C}$ or $\mathbb{R}$. It breaks down, however, when working over a field whose characteristic divides the group order. Here, we refrain from explaining this in detail.

Corollary 3.4.1 *Any representation V of the finite group G can be uniquely decomposed as a direct sum of distinct unique irreducible representations V_i with unique multiplicities m_i*

$$V = V_1^{\oplus m_1} \oplus \cdots \oplus V_k^{\oplus m_k}. \tag{3.4.25}$$

(The uniqueness here of course holds up to permutation of the summands.)

Proof The decomposition follows from Lemma 3.4.1, and the uniqueness follows from Schur's Lemma 3.4.2. □

For a group G, we first of all have the trivial representation on $\mathbb{C}$, with

$$gv = v \text{ for all } v \in \mathbb{C}. \tag{3.4.26}$$

For the symmetric group $\mathfrak{S}_n$, we also have the permutation representation on the vector space $\mathbb{C}^n$ that permutes the coordinate axes. This representation is not irreducible because $\sum e_i$, where the e_i, $i = 1, \ldots, n$, are the standard basis vectors of $\mathbb{C}^n$, spans a 1-dimensional invariant subspace. On the complementary subspace

$$V := \{z \in \mathbb{C}^n : \sum z^i = 0\} \tag{3.4.27}$$

the representation is irreducible. This is called the standard representation.

Moreover, we have the alternating representation of $\mathfrak{S}_n$ on $\mathbb{C}$, defined by

$$v \mapsto \text{sgn}(g)v, \tag{3.4.28}$$

where the signum, $\text{sgn}(g)$, is 1 or -1, depending on whether g consists of an even or odd number of transpositions. (The reader may recall from Sect. 2.1.6 that any permutation can be written as a product of transpositions, that is, exchanges between two elements of the set on which it operates; in the sequel, we shall write the permutation exchanging i and j as (ij). The alternating group is then the kernel of this representation.)

In fact, any finite group G operates as a group of permutations, or more precisely, can be seen as a subgroup of a permutation group. In fact, it operates on itself by left translation, and this permutes the group elements. In general, when G operates by left translation on a finite set S, we have a homomorphism $G \to \text{Aut}(S)$ into the permutation group of S, and this then induces a representation of G on the vector space V_S with basis elements e_s corresponding to the elements $s \in S$, and the representation then is obviously

$$g \sum \lambda_s e_s = \sum \lambda_s e_{gs}. \tag{3.4.29}$$

Thus, the case where $S = G$ on which G operates by left translation is a special case, and the resulting representation is called the regular representation V_R of G. Note, however, that the permutation representation of $\mathfrak{S}_n$ is not the regular representation, because the former is induced by the operation of $\mathfrak{S}_n$ on a set of n elements whereas the latter comes from the action of $\mathfrak{S}_n$ on itself by left translation.

Returning to the above example of graphs with 4 vertices, $\mathfrak{S}_4$ operates on $M(4)$, or more precisely on $\mathbb{C}^6$, the space of weighted graphs with 4 vertices, by permutation of the edges. Since edges are given by pairs of vertices, this representation comes from the representation on $\mathrm{Sym}^2\mathbb{C}^4$, the symmetric 4×4 matrices with complex entries, induced by the permutation representation. Again, this representation is not irreducible. First of all, the subrepresentation on pairs of the form (j, j), $j = 1, 2, 3, 4$, is a permutation representation (which decomposes, as discussed, into the sum of the trivial and the standard representation). Since $\mathrm{Sym}^2\mathbb{C}^4$ is 10-dimensional, we are left with a 6-dimensional representation. Again, it contains a permutation representation as a subrepresentation operating on triples of edges sharing one vertex. For instance, when we permute the vertices 1 and 2, the edge triple e_{12}, e_{13}, e_{14} goes to the edge triple e_{12}, e_{23}, e_{24}. This permutation representation then splits, as always, into the sum of the trivial and the standard representation. The complementary representation is 2-dimensional. This representation, in fact, is a representation of the quotient group

$$\mathfrak{S}_3 = \mathfrak{S}_4/\{1, (21)(43), (31)(42), (41)(32)\}, \qquad (3.4.30)$$

where (ji) is the permutation of the vertices i and j. This representation permutes pairs of opposite edges. Thus, $\mathfrak{S}_3$ operates on the set of edge pairs $\{(e_{12}, e_{34}), (e_{13}, e_{24}), (e_{14}, e_{23})\}$ by permutation. It is only a 2-dimensional representation, because we have split off the trivial representation already. In other words, we have the standard representation of $\mathfrak{S}_3$.

We now proceed to a more systematic treatment of the representations of a finite group G. The group G operates on itself by conjugation

$$g \mapsto hgh^{-1} \text{ for } h \in G. \qquad (3.4.31)$$

The orbits of this action are called *conjugacy classes*; that is, the conjugacy class of g is

$$C(g) := \{hgh^{-1} : h \in G\}. \qquad (3.4.32)$$

Of course, when G is abelian, the conjugacy class of an element consists only of that element itself. Here, however, we are rather interested in non-abelian groups like the symmetric group $\mathfrak{S}_n$.

Since conjugacy is an equivalence relation on the elements of g (for instance, if $g_2 = hg_1h^{-1}$, then $g_1 = kg_2k^{-1}$ with $k = h^{-1}$, showing the symmetry), the group G is the disjoint union of its different conjugacy classes. This decomposition of the group G will be the key to understand its representations.

Group elements in the conjugacy class should be similar to each other, or more precisely, operate in essentially the same manner in representations. When g_1 and $g_2 = hg_1h^{-1}$ are conjugate, then the operation of g_2 is obtained from that of g_1 simply by shifting vectors v by the operation of h; in formulae, $g_2(hv) = hg_1h^{-1}hv = hg_1(v)$. Therefore, we shall now consider functions that are constant on the conjugacy classes of G, and we shall see that a representation produces a particular class function, the so-called character, which in turn determines this representation, see Corollary 3.4.3. In particular, we shall see that the number of irreducible representations of a finite group equals its number of conjugacy classes, see Corollary 3.4.2.

$\mathfrak{S}_3$, $\mathfrak{S}_4$

Definition 3.4.3 A *class function* on G is a complex valued function that is constant on the conjugacy classes of G. The vector space of class functions is denoted by $\mathbb{C}_{\text{class}}(G)$. On $\mathbb{C}_{\text{class}}(G)$, we have the Hermitian product

$$(\phi, \psi) := \frac{1}{|G|} \sum_g \overline{\phi(g)} \psi(g), \tag{3.4.33}$$

where, of course, the normalization factor $|G|$ denotes the number of elements of G.

The following lemma indicates that class functions are relevant for understanding group representations.

Lemma 3.4.3 *If $\phi : G \to \mathbb{C}$ is a class function and V a representation of G, then*

$$L_{\phi, V} := \sum_g \phi(g) g : V \to V \tag{3.4.34}$$

is G-equivariant.

Proof

$$L_{\phi, V}(hv) = \sum_g \phi(g) g(hv)$$

$$= \sum_g \phi(hgh^{-1}) hgh^{-1}(hv) \text{ since } hgh^{-1} \text{ runs through } G \text{ if } g \text{ does}$$

$$= h(\sum_g \phi(g) g(v)) \text{ since } \phi \text{ is a class function}$$

$$= h(L_{\phi, V}(v))$$

which is the condition for G-equivariance. $\qquad\square$

So, the point of the proof is that we can pull h from the right to the left of g even though h and g need not commute, because gh and hg are in the same conjugacy class and ϕ is constant on conjugacy classes.

(In fact, the converse also holds, that is, if $L_{\phi, V}$ is G-equivariant, then ϕ is a class function. The reader may prove that herself.)

We consider the class function $\phi(g) = \frac{1}{|G|}$ for all g and the associated G-equivariant

$$L := L_{1, V} = \frac{1}{|G|} \sum_g g : V \to V. \tag{3.4.35}$$

We have

Lemma 3.4.4 *L projects V onto the fixed point set*

$$V^G := \{v \in V : gv = v \text{ for all } g \in G\}. \tag{3.4.36}$$

Proof For any $h \in G$,

$$hLv = \frac{1}{|G|} \sum_g hgv$$

$$= \frac{1}{|G|} \sum_g gv \text{ by the familiar argument}$$

$$= Lv.$$

Thus, the image of L is contained in the fixed point set. Conversely, if v is fixed by all g, then it is also fixed by L, the average over all g. In particular, $L \circ L = L$. □

It will now turn out that representations can be understood in terms of particular class functions.

Definition 3.4.4 The *character* of a representation V of G is the class function

$$\chi_V(g) := \text{tr}(g_{|V}). \tag{3.4.37}$$

Of course, χ_V is a class function because the trace of a matrix is invariant under conjugation.

The main result of this section will be that the characters form an orthonormal basis w.r.t. to the Hermitian product (3.4.33) of $\mathbb{C}_{\text{class}}(G)$.

Let us start with some easy **examples:**

1. For the trivial representation, we have $\chi(g) = 1$ for every g.
2. For the alternating representation of $\mathfrak{S}_n$, we have $\chi(g) = \text{sgn}(g)$ for every g.
3. Let V_1 be a 1-dimensional representation of G. Then, identifying V_1 with $\mathbb{C}$, for every $g \in G$, $g1 = \lambda$ for some $\lambda \in \mathbb{C}$, and since $g^n = e$ for some $n \in \mathbb{N}$ as G is finite, we have $\lambda^n = 1$. Therefore, λ is a root of unity, and hence

$$|\chi_{V_1}(g)| = 1. \tag{3.4.38}$$

4. For the permutation representation of $\mathfrak{S}_n$, $\chi(g)$ equals the number of fixed points of g on the set of n elements on which $\mathfrak{S}_n$ operates. This is easily seen: The trace of a matrix is the sum of its diagonal elements, and when an element is fixed by g, we get a corresponding one on the diagonal of the matrix for the permutation representation of g, whereas when two elements are exchanged by g, the corresponding matrix contains a block of the form

$$\begin{pmatrix} 0 & 1 \\ 1 & 0 \end{pmatrix}, \tag{3.4.39}$$

that is, one with 0s on the diagonal.

5. The preceding then extends to any subgroup of the permutation group of a set S, in particular to the regular representation V_R of a group G induced by the left translations of G on itself. In particular, we have

$$\chi_{V_R}(e) = |G| \tag{3.4.40}$$

$$\chi_{V_R}(g) = 0 \text{ for } g \neq e. \tag{3.4.41}$$

$\boxed{\mathfrak{S}_n}$

6. Since the permutation representation splits as the direct sum of the trivial and the standard representation, we can deduce the characters for the latter from (3.4.42) below. $\chi(g)$ for the standard representation simply equals the number of fixed points of g minus 1.

We next compile some elementary general results about characters.

Lemma 3.4.5 *For representations V, W of G, the characters satisfy*

$$\chi_{V \oplus W} = \chi_V + \chi_W \tag{3.4.42}$$

$$\chi_{V \otimes W} = \chi_V \chi_W \tag{3.4.43}$$

$$\chi_{V^*} = \bar{\chi}_V \tag{3.4.44}$$

$$\chi_{\mathrm{Hom}(V,W)} = \bar{\chi}_V \chi_W \tag{3.4.45}$$

$$\chi_{\wedge^2 V} = \frac{1}{2}(\chi_V(g)^2 - \chi_V(g^2)) \tag{3.4.46}$$

$$\chi_{\mathrm{Sym}^2 V} = \frac{1}{2}(\chi_V(g)^2 + \chi_V(g^2)) \tag{3.4.47}$$

Proof All these formulae follow easily from the fact that the trace of a matrix is the sum of its eigenvalues. For instance, when the operation of g on V has eigenvalues λ_i, then the eigenvalues of the operation on $\wedge^2 V$ are $\lambda_i \lambda_j, i < j$, and

$$\sum_{i<j} \lambda_i \lambda_j = \frac{1}{2}((\sum \lambda_i)^2 - \sum \lambda_i^2), \tag{3.4.48}$$

and for $\mathrm{Sym}^2 V$, we can either use the eigenvalues $\lambda_i \lambda_j, i \leq j$, or the relation

$$V \otimes V = \mathrm{Sym}^2 V \oplus \bigwedge^2 V. \tag{3.4.49}$$

Also, (3.4.45) follows from (3.4.18), (3.4.43), (3.4.44).   □

While Lemma 3.4.5 is easy, it embodies an important principle: *Algebraic* relations between representations are converted into *arithmetic* relations between their characters.

We now come to the key observation

Lemma 3.4.6 *For any representation V of G,*

$$\dim V^G = \frac{1}{|G|} \sum_g \chi_V(g), \tag{3.4.50}$$

where V^G is the fixed point set of (3.4.36).

Proof By Lemma 3.4.4, for $L = \frac{1}{|G|} \sum_g g : V \to V$,

$$\dim V^G = \mathrm{tr} L = \frac{1}{|G|} \sum_g \mathrm{tr} g = \frac{1}{|G|} \sum_g \chi_V(g). \tag{3.4.51}$$

□

We can now bring everything together. We apply Lemma 3.4.6 to the representation $\mathrm{Hom}(V, W)$, where V, W are representations of G, recall from (3.4.45) that the character of this representation is $\overline{\chi}_V \chi_W$ and divide this by $|G|$ to obtain the Hermitian product (3.4.33) of the characters χ_V and χ_W and at the same time make (3.4.50) applicable and conclude that this Hermitian product is

$$\frac{1}{|G|} \sum_g \overline{\chi}_V(g) \chi_W(g) = \begin{cases} 1 & \text{if } V \cong W \\ 0 & \text{if } V \ncong W \end{cases} \tag{3.4.52}$$

Theorem 3.4.1 *The characters of the irreducible representations of G form an orthonormal basis of $\mathbb{C}_{\mathrm{class}}(G)$ w.r.t. the Hermitian product (3.4.33).*

Proof The orthonormality is contained in (3.4.52). In order to show that the characters span $\mathbb{C}_{\mathrm{class}}(G)$, we need to show that $\phi = 0$ is the only class function ϕ satisfying $(\phi, \chi_V) = 0$ for all characters of representations; in fact, since any representation is a sum of irreducible representations, it of course suffices to consider the characters of the irreducible representations. We consider the G-equivariant map $L_{\phi, V} = \sum_g \phi(g)g : V \to V$ of (3.4.34) for an irreducible representation V. By Schur's Lemma 3.4.2, $L_{\phi, V} = \lambda \mathrm{Id}$, and so

$$\lambda = \frac{1}{\dim V} \mathrm{tr} L_{\phi, V} = \frac{1}{\dim V} \sum_g \phi(g) \chi_V(g) = \frac{|G|}{\dim V} \overline{(\phi, \chi_{V^*})} = 0.$$

This means that

$$\sum_g \phi(g)g = 0 \text{ for all representations of } V. \tag{3.4.53}$$

But, for instance, for the regular representation, the operations of the gs are linearly independent. Therefore, (3.4.53) yields $\phi(g) = 0$ for all g, hence $\phi = 0$. □

This theorem has a number of immediate corollaries.

Corollary 3.4.2 *The number of irreducible representations of the finite group G is equal to the number of conjugacy classes of G.*

Proof The dimension of $\mathbb{C}_{\mathrm{class}}(G)$ is equal to the number of conjugacy classes of G. □

Corollary 3.4.3 *Representations are determined by their characters.*

Corollary 3.4.4 *For a representation V and an irreducible representation V_i, V_i occurs in V with multiplicity $m_i = (\chi_V, \chi_{V_i})$. In particular, V is irreducible iff*

$$(\chi_V, \chi_V) = 1. \tag{3.4.54}$$

Proof According to (3.4.25), we can decompose V as $V = V_1^{\oplus m_1} \oplus \cdots \oplus V_k^{\oplus m_k}$, and from (3.4.42), we get

$$(\chi_V, \chi_V) = \sum m_i^2, \tag{3.4.55}$$

and the results follow easily. $\qquad\square$

Corollary 3.4.5 *If V is an irreducible representation of G and V_1 is a 1-dimensional representation, then $V \otimes V_1$ is also irreducible.*

Proof By (3.4.43), $\chi_{V \otimes V_1} = \chi_V \chi_{V_1}$, and so by (3.4.38),

$$(\chi_V \chi_{V_1}, \chi_V \chi_{V_1}) = (\chi_V, \chi_V) = 1 \text{ since } V \text{ is irreducible,}$$

by the preceding Corollary, which in turn implies that $V \otimes V_1$ is irreducible. $\qquad\square$

Of course, this is trivial if V_1 is the trivial representation, but we may also apply it, for instance, to the alternating representation of the permutation group.

Corollary 3.4.6 *Unless G is trivial ($G = \{e\}$), the regular representation of G is not irreducible. In fact, it is the sum of all irreducible representations V_i, and each V_i appears in this sum with multiplicity $m_i = \dim V_i$; thus*

$$|G| = \dim V_R = \sum_{V_i \text{ irred. rep. of } G} (\dim V_i)^2, \tag{3.4.56}$$

and

$$0 = \sum \dim V_i \chi_{V_i}(g) \text{ for } g \neq e. \tag{3.4.57}$$

Proof It follows from (3.4.54), (3.4.40), (3.4.41) that for $|G| > 1$, the regular representation is not irreducible. We decompose it as a sum of the irreducible representations

$$V_R = V_1^{m_1} \oplus \cdots \oplus V_k^{m_k}, \tag{3.4.58}$$

hence

$$\dim V_R = \sum_i m_i \dim V_i. \tag{3.4.59}$$

From Theorem 3.4.1, (3.4.55), (3.4.40) and (3.4.41), we obtain

$$m_i = (\chi_{V_i}, \chi_{V_R}) = \frac{1}{|G|} \chi_{V_i}(e) |G| = \dim V_i, \tag{3.4.60}$$

and with (3.4.59), also (3.4.56) follows. Similarly, (3.4.57) is derived from (3.4.41). $\qquad\square$

As a test, we now apply the preceding theory to the symmetric group $\mathfrak{S}_4$. From the abstract discussion of representations and the action on the set of edges of graphs with 4 vertices, and the corresponding action on $M(4)$, we already know the following representations:

> Representations of $\mathfrak{S}_4$

1. The trivial representation V_0
2. The alternating representation V_1
3. The standard representation V
4. The representation V_3 of the quotient group $\mathfrak{S}_3$
5. The representation $V' := V \otimes V_1$, according to Corollary 3.4.5

We shall now verify from the characters of these representations and the preceding results that these are all the irreducible representations. As a preparation, we observe that $\mathfrak{S}_4$ possesses 5 conjugacy classes, represented by the elements $1, (21), (21)(43), (231)$, and (2341). These classes contain $1, 6, 3, 8$, and 6 elements, resp. We also recall that the permutation representation V_P is reducible as the sum of the standard and the trivial representation. We then have the following character table, first for V_P and then below the line for the irreducible representations

class # elements	1 1	(21) 6	(21)(43) 3	(231) 8	(2341) 6
V_P	4	2	0	1	0
V_0	1	−1	1	1	1
V_1	1	−1	1	1	−1
V	3	1	−1	0	−1
V_3	2	0	2	−1	0
V'	3	−1	−1	0	1

Character table for $\mathfrak{S}_4$

For instance, the character $\chi_{V_P}(g)$ is given by the number of fixed points of the permutation induced by g which directly gives the first row. The row for V_0 is obvious, and the row for V then follows from the relation $V_P = V \oplus V_0$, using (3.4.42). The row for V_1 is again obvious, and the row for V' then follows from $V' = V \otimes V_1$, using (3.4.43). Finally, the characters for V_3 can be either determined directly or deduced from the others with the help of (3.4.56) and (3.4.57).

In fact, we may also observe that as the rows of the table of the characters are orthogonal, so are the columns, and we have

$$\sum_{V_i \text{ irred. rep. of } G} \overline{\chi_{V_i}(g)} \chi_{V_i}(h) = 0 \quad \text{if } g \text{ and } h \text{ are not conjugate.}$$

(3.4.61)

We may also check this relation in the preceding character table for $\mathfrak{S}_4$.

Here, we have used the operation of $\mathfrak{S}_4$ on the vertex and edge sets of graphs with 4 elements to derive its character table. As an alternative, we can also use its operation on the 3-dimensional cube by permutations of its 4 spatial diagonals. It is an instructive exercise for the reader to carry this through.

More importantly, the reader is invited to investigate how the preceding analysis generalizes to the symmetric group $\mathfrak{S}_d$ for arbitrary $d \in \mathbb{N}$.

A reference for this section is [40] which I have followed for many of the details. The material is also presented in [17, 75, 111], for instance.

The representation theory of finite groups was originally developed by Frobenius and I.Schur, see [99]. Similar methods apply to representations of compact Lie groups, as shown by Peter and Weyl [93]. More generally, the Cartan-Weyl theory of representations of compact Lie groups decomposes irreducible representations under abelian subgroups, see for instance [54, 69, 116].

Spaces

4

In the history of science, there have been different conceptualizations of the relation between physical objects or bodies and space. One view, which underlies Aristotelian and Cartesian physics, is that space simply surrounds bodies. In that perspective, bodies have the ontological priority. Without bodies, there is no space. In the other view, which emerged in the natural philosophy of the 16th century and which is basic for Galilean and Newtonian physics, bodies fill a preexisting space. It is this latter view that also underlies the topological concepts to be developed in this chapter although its traces might be somewhat difficult to discern in the abstract definitions. Therefore, let me illustrate this aspect first in informal terms. Topology is concerned with a set and its subsets and it equips that set with the structure of a space by singling out a particular collection of subsets as open. The complements of open sets are then called closed. Closed sets contain their boundary, whereas open sets don't. For the standard topology of the real line, intervals of the form $\{x \in \mathbb{R} : a < x < b\}$ are open, those of the form $\{x \in \mathbb{R} : a \leq x \leq b\}$ closed, when $a < b \in \mathbb{R}$ are fixed. The boundary of such an interval consists of the points a and b. Similarly, in $\mathbb{R}^d$, open sets are generated by strict inqualities, for instance $\{x \in \mathbb{R}^d : a < f(x) < b\}$ for some continuous function f, and closed sets by weak such inequalities. Under some nondegeneracy assumption which we don't spell out here, the boundary would consist of the hypersurfaces $\{x : f(x) = a\}$ and $\{x : f(x) = b\}$.

At this level of abstraction, physical bodies could be considered as closed subsets of the space. Now let us explore the implications of this in terms of naive physics. The boundary might represent the skin of a body. A natural intuition is that two bodies could touch each other along their skins, like when you are shaking hands with another person or get in some other way in physical contact. This, however, is not compatible with the above topology. In that topology, two sets can touch only along some interface, and that interface between two disjoint sets can only belong to one of them. Or putting it differently, if two closed sets touched each other, they would have to share a single skin. For instance, if $B_1 = \{x : a \leq f(x) \leq b\}$ and $B_2 = \{x : b \leq f(x) \leq c\}$, they share the interface $\{f(x) = b\}$. In that sense, the topology that we are going to develop is somewhat counterintuitive, or at least not compatible with an Aristotelian view of physics. Instead of having two bodies touch each other along their individual skins, one should rather

© Springer International Publishing Switzerland 2015
J. Jost, *Mathematical Concepts*, DOI 10.1007/978-3-319-20436-9_4

think of two chambers separated by a common wall. For instance, the open chambers $C_1 = \{x : a < f(x) < b\}$ and $C_2 = \{x : b < f(x) < c\}$ share the wall $\{f(x) = b\}$ which belongs to neither of them. (Ignore the fact that this wall is infinitely thin.)

Of course, for a real physical understanding of what happens when two bodies touch each other, we should go to atomic physics, and to explore the relation between physical objects and physical space, we should look into the theory of general relativity. But this is not our concern here, and in the sequel, I shall develop the mathematical concept of a topological space. This concept, which is essentially due to Hausdorff [48], has a far wider significance than abstractly modelling physics in three-dimensional space, and in fact, it constitutes one of the basic tools of modern mathematics.

Actually, in the preceding informal discussion, we have used the ordering $<$ of the real line in order to illustrate its topology. The abstract concept of a topological space that we are going develop will, however, not need any such ordering. It simply imposes some structure upon the power set of a given set. This should also serve as an important example that we need to disentangle concepts in order to understand their full scope. That will also guide our considerations in Chap. 5.

4.1   Pretopological and Topological Spaces

For a set X, let $\mathcal{P}(X)$ be its power set, that is, the set of all its subsets. We recall from (2.1.88)–(2.1.91) that $\mathcal{P}(X)$ has an algebraic structure with the following operations

$$\text{Complement: } A \mapsto X \setminus A \tag{4.1.1}$$

$$\text{Union: } (A, B) \mapsto A \cup B \tag{4.1.2}$$

$$\text{Intersection: } (A, B) \mapsto A \cap B := X \setminus (X \setminus A \cup X \setminus B) \tag{4.1.3}$$

$$\text{Implication: } (A, B) \mapsto A \Rightarrow B := (X \setminus A) \cup B, \tag{4.1.4}$$

and we also have the relations

$$A \cup (X \setminus A) = X \tag{4.1.5}$$

$$A \cap (X \setminus A) = \emptyset \tag{4.1.6}$$

for all $A \in \mathcal{P}(X)$.

The preceding relations turn $\mathcal{P}(X)$ into a Boolean algebra (Definition 2.1.11). Here, the intersection $\cap$ corresponds to the logical and $\wedge$, the union $\cup$ to the logical or $\vee$, and the complement $\setminus$ to the negation $\neg$. In fact,

$$x \in A \cap B \Leftrightarrow (x \in A) \wedge (x \in B) \tag{4.1.7}$$

$$x \in A \cup B \Leftrightarrow (x \in A) \vee (x \in B) \tag{4.1.8}$$

$$x \in X \setminus A \Leftrightarrow \neg(x \in A) \quad (\Leftrightarrow x \notin A). \tag{4.1.9}$$

For the statement of the relations $\cap, \cup, \setminus$, we do not need to invoke the points of X. In fact, essentially all what follows can be developed on the basis of this structure on $\mathcal{P}(X)$ without involving those points. The systematic exploration of this aspect is called pointless topology (note the pun!).

In this section, we shall introduce and study the notion of a topological space which is defined in terms of classes of subsets of $\mathcal{P}(X)$, called open, that are characterized by properties involving the union $\cup$ and the intersection $\cap$ operators, but do not need the complement $\backslash$, see Theorem 4.1.1 below. In Sect. 4.2, we shall introduce another notion, that of a measurable space, which is defined in terms of classes of subsets of $\mathcal{P}(X)$, called measurable, whose characterization will also involve the complement.

We start with a more general concept, which, as we shall argue, is also of independent interest.

Definition 4.1.1 X is called a *pretopological space* if it possesses an operator $^\circ:\mathcal{P}(X) \to \mathcal{P}(X)$ with the following properties

(i) $X^\circ = X$.
(ii) $A^\circ \subset A$ for all $A \in \mathcal{P}(X)$.
(iii) $A^\circ \cap B^\circ = (A \cap B)^\circ$ for all $A, B \in \mathcal{P}(X)$.

A° is called the *interior* of A. A is called *open* if $A = A^\circ$.
   $\mathcal{O}(X) \subset \mathcal{P}(X)$ is the collection of open subsets of X.
   X is called a *topological space* if the operator $^\circ$ in addition satisfies

(iv) $A^{\circ\circ} = A^\circ$ for all $A \in \mathcal{P}(X)$.

Thus, in a topological space, the interior of a set is always open.

Lemma 4.1.1 *In a pretopological space X,*

$$A \subset B \text{ implies } A^\circ \subset B^\circ. \qquad (4.1.10)$$

Proof Since $A \cap B = A$ when $A \subset B$, we have

$$A^\circ = (A \cap B)^\circ = A^\circ \cap B^\circ \subset B^\circ. \qquad \square$$

In particular, we conclude that

$$(\emptyset)^\circ = \emptyset. \qquad (4.1.11)$$

Thus, on $\emptyset$, the only pretopology is that with (4.1.11). On a set with a single element, $X = \{1\}$, we also have only a single pretopology, because we must also have $X^\circ = X$. On a set with two elements, $X = \{1, 2\}$, we can put $\{1\}^\circ$ as either $\{1\}$ or $\emptyset$, and we have the same options for the other single element set, $\{2\}$. Thus, altogether, we have four different pretopologies on this set. Each such pretopology would also be a topology. On a three element set, $X = \{1, 2, 3\}$, we could, for instance, put $\{1, 2\}^\circ = \{1\}$ and $\{1\}^\circ = \emptyset$, in which case the resulting pretopology would not be a topology.

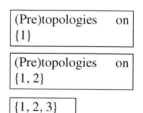

(Pre)topologies on $\{1\}$

(Pre)topologies on $\{1, 2\}$

$\{1, 2, 3\}$

Lemma 4.1.2 *In a topological space, $A°$ is the largest open subset of A.*

Proof If A' is an open subset of A, that is, $(A')° = A' \subset A$, then by Lemma 4.1.1, $(A')° \subset A°$. Thus, any other open subset of A is contained in $A°$.   □

Theorem 4.1.1 *A topological space X is defined by a collection $\mathcal{O}(X) \subset \mathcal{P}(X)$ whose members are called the open subsets of X, satisfying*

  (i) $X \in \mathcal{O}(X)$.
  (ii) $\emptyset \in \mathcal{O}(X)$.
  (iii) If $A, B \in \mathcal{O}(X)$, then also $A \cap B \in \mathcal{O}(X)$.
  (iv) For any collection $(A_i)_{i \in I} \subset \mathcal{O}(X)$, also $\bigcup_{i \in I} A_i \in \mathcal{O}(X)$.

Remark If one employs the conventions $\bigcup_{i \in \emptyset} A_i = \emptyset$, $\bigcap_{i \in \emptyset} A_i = X$ and requires condition (iii) for any finite intersection, then the first two conditions become superfluous.

Proof Given an operator $°$ as in Definition 4.1.1, we need to show that the sets with $A° = A$ satisfy the conditions of the theorem. The first three conditions are clear from the axioms of the definition and (4.1.11). Now suppose that the sets A_i satisfy $A_i^{\circ} = A_i$. Then for $B = \bigcup_{i \in I} A_i$, since $A_i \subset B$ for all i, we then also have $A_i^{\circ} \subset B°$ by (4.1.10), hence $B° \supset \bigcup_{i \in I} A_i^{\circ} = \bigcup_{i \in I} A_i = B$, hence $B° = B$. This verifies condition (iv) of the theorem.

Conversely, when we have a collection $\mathcal{O}(X)$ as in the theorem, we define A' as the largest member of $\mathcal{O}(X)$ contained in A. For condition (iii) of Definition 4.1.1, it is clear that $A' \cap B' \subset (A \cap B)'$ because $A' \cap B'$ is an open subset of $A \cap B$. Conversely, when $(A \cap B)'$ is not contained in $A' \cap B'$, then it is either not contained in A' or not contained in B'. In the latter case, $(A' \cap B') \cup B'$ is an open subset of B larger than B' which is in contradiction with the definition of B' as the largest open subset of B.

At this point, the reader may wonder why we have not yet used condition (iv) of Definition 4.1.1. But it still remains to show that the collection $\mathcal{O}(X)$ that we have defined with the help of $°$ yields the original topology that we have started with. For this, we need to show that $A' = A°$ for all $A \in \mathcal{P}(X)$. Since $\mathcal{O}(X)$ is closed under taking unions, A' is the union of all members of $\mathcal{O}(X)$ contained in A. Now by condition (iv) of Definition 4.1.1, $A° \in \mathcal{O}(X)$, hence $A° \subset A'$. Conversely, since $A' \subset A$ and by definition of $\mathcal{O}(X)$, $(A')° = A'$, we get from Lemma 4.1.1 that $A' \subset A°$. Thus, $A' = A°$.   □

From the proof of Theorem 4.1.1, we see that any pretopology defines a topology whose open sets are those with $A° = A$. Unless the original pretopology is a topology, however, not every set of the form $A°$ is open.

The collection $\mathcal{O}(X)$ of open sets of a topological space in general does not form a Boolean algebra, because complements of open sets need not be open. However, it is a Heyting algebra which, as we recall from (2.1.41),

(2.1.42), (2.1.43), means the following. First, a lattice with 0 and 1 is a partially ordered set with two binary associative and commutative operations $\wedge, \vee$ and two distinguished elements 0, 1, satisfying

$$a \wedge a = a, \quad a \vee a = a \tag{4.1.12}$$

$$1 \wedge a = a, \quad 0 \vee a = a \tag{4.1.13}$$

$$a \wedge (b \vee a) = a = (a \wedge b) \vee a \tag{4.1.14}$$

for any element a.

A lattice with 0 and 1 is a Heyting algebra if for any elements a, b, there exists an exponential b^a, that is, according to (2.1.80), an element $a \Rightarrow b$ characterized by

$$c \leq (a \Rightarrow b) \text{ iff } c \wedge a \leq b. \tag{4.1.15}$$

$\mathcal{O}(X)$ with the inclusion $\subset$ as the partial order $\leq$, with $\emptyset, X$ as 0, 1 and with $\cap, \cup$ as $\wedge, \vee$ is a lattice. When we put

$$U \Rightarrow V := \bigcup W \text{ for all open } W \text{ with } W \cap U \subset V, \tag{4.1.16}$$

it becomes a Heyting algebra. The key point here is that the complement of a set is replaced by the interior of its complement, in order to stay inside the category $\mathcal{O}(X)$. We remark that Heyting algebras also replace Boolean algebras when one goes from classical logic to intuitionistic propositional calculus where one abandons the law of the excluded middle, that is, the exclusive complementarity between a statement and its complement. This issue will be taken up in Sect. 9.3.

Lemma 4.1.3 *In a topological space X, for any collection of subsets $(A_i)_{i \in I}$,*

$$\left(\bigcup_{i \in I} A_i\right)^\circ = \bigcup_{i \in I} A_i^\circ. \tag{4.1.17}$$

Proof This follows from Theorem 4.1.1 which implies that in a topological space, the union of open sets is open. □

Example 1. For a set X, $\mathcal{O} = \{\emptyset, X\}$ is the smallest $\mathcal{O}$ satisfying all the axioms for a topological space. This topology is called the indiscrete topology.

2. For a set X, $\mathcal{O} = \mathcal{P}(X)$ is the largest $\mathcal{O}$ satisfying all the axioms for a topological space. This topology is called the discrete topology.

3. Let (X, d) be a metric space. The metric induces a topology on X whose open sets are obtained from unions of balls $U(x, r) := \{y \in X : d(x, y) < r\}$ for some $x \in X, r \geq 0$. ($U(x, 0) = \emptyset$, $\bigcup_{x \in X} U(x, r) = X$ for any $r > 0$.)

(a) When $X = \mathbb{R}^d$ and d is the Euclidean metric, this yields the standard topology of Euclidean space. Note that this topology can be generated by many different metrics. For instance, any metrics $d(.,.)$ that are generated by some norm $\|.\|$, that is, $d(x, y) = \|x - y\|$, are equivalent to each other in the sense that they generate the same topology on $\mathbb{R}^d$.

Indiscrete topology
$\mathcal{O} = \{\emptyset, X\}$

Discrete topology
$\mathcal{O} = \mathcal{P}(X)$

Euclidean metric

Trivial metric

Discrete topology

(b) From the trivial metric (2.1.51), we obtain the discrete topology, because in that case any ball $U(x, r)$ for $0 < r < 1$ reduces to the singleton set $\{x\}$. Therefore, any such singleton set is open, and since any set U is the union of the singleton sets of its points, $U = \bigcup_{x \in U}\{x\}$, therefore every U is open.

4. $X = \mathbb{R}$. Let $\mathcal{O}$ consist of $\emptyset$, X and all intervals (ξ, ∞), $\xi \in \mathbb{R}$.
5. Let X contain infinitely many points. We let $\mathcal{O}$ consist of $\emptyset$ and all complements of finite sets. This topology is called the cofinite topology.
6. For a vector space V, let $\mathcal{O}$ consist of all complements of finitely many affine linear subspaces of V.

The preceding examples can be cast into a slightly more systematic form with the help of

Definition 4.1.2 $\mathcal{B} \subset \mathcal{O}(X)$ is called a *basis*[1] for the topology $\mathcal{O}(X)$ if every $U \in \mathcal{O}(X)$ can be written as a union of elements of $\mathcal{B}$.

In the examples, we can then identify the following bases: $\{X\}$ for 1, $\{\{x\}, x \in X\}$ in 2, the balls $U(x, r)$ with $x \in X$ and rational r in 3 (use the triangle inequality to check that the intersection of finitely many balls can also be represented as the union of balls centered at elements of this intersection), the intervals (ξ, ∞) with rational ξ in 4.

With this notion, we also naturally have

7. Let $(X, \mathcal{O}(X))$, $(Y, \mathcal{O}(Y))$ be topological spaces. We can then equip the product $X \times Y = \{(x, y) : x \in X, y \in Y\}$ with the topology $\mathcal{O}(X \times Y)$ that has as basis sets $U \times V$ with $U \in \mathcal{O}(X)$, $V \in \mathcal{O}(Y)$. This topology on $X \times Y$ is called the product topology.

We can also start with a $\mathcal{B} \subset \mathcal{P}(X)$ that satisfies

(i)
$$\bigcup_{B \in \mathcal{B}} B = X, \tag{4.1.18}$$

(ii)

If $B_1, B_2 \in \mathcal{B}, x \in B_1 \cap B_2$, there exists $B_0 \in \mathcal{B}$ with $x \in B_0 \subset B_1 \cap B_2$ (4.1.19)

and then let $\mathcal{O}$ consist of all unions of elements of $\mathcal{B}$. This defines a topology on X with basis $\mathcal{B}$.

When $\mathcal{B} \subset \mathcal{P}(X)$ does not satisfy (4.1.19), it need not be the basis of any topology on X. For instance, take $X = \{-1, 0, 1\}$, and let $B_1 = \{-1, 0\}$, $B_2 = \{0, 1\}$. Then $\{B_1, B_2\}$ cannot be the basis of any topology because we cannot recover the set $\{0\} = B_1 \cap B_2$ by taking unions.

To address this deficiency, we formulate

---

[1] In [66], this is called a *base* instead of a basis.

Definition 4.1.3 $\mathcal{S} \subset \mathcal{O}(X)$ is called a *subbasis* for the topology $\mathcal{O}(X)$ if every $U \in \mathcal{O}(X)$ can be written as a union of finite intersections of elements of $\mathcal{S}$.

In contrast to the previous example, we can now take any $\mathcal{S} \subset \mathcal{P}(X)$. Then $\mathcal{O}$, defined as the collection of all finite intersections of unions of members of $\mathcal{S}$, is a topology on X with subbasis $\mathcal{S}$. We say that $\mathcal{S}$ *generates* the topology $\mathcal{O}$.

We also have

Lemma 4.1.4 $\mathcal{B} \subset \mathcal{O}(X)$ *is a basis for the topology* $\mathcal{O}(X)$ *iff for every* $U \in \mathcal{O}(X)$ *and every* $x \in U$, *there exists some* $V \in \mathcal{B}$ *with* $x \in V \subset U$.

Proof Let $\mathcal{B}$ be a basis. If $x \in U \subset \mathcal{O}(X)$, then since U is the union of members of $\mathcal{B}$, there must exist some $V \in \mathcal{B}$ with $x \in V \subset U$. For the converse, let $U \in \mathcal{O}(X)$ and let W be the union of all $V \in \mathcal{B}$ with $V \subset U$. Thus $W \subset U$. But when $x \in U$, by assumption there exists some $V \in \mathcal{B}$ with $x \in V \subset W$. Hence also $U \subset W$, and altogether $U = W$. Thus, every open set is the union of members of $\mathcal{B}$. □

Definition 4.1.4 An open set containing an element x is called an *(open) neighborhood* of x.

Definition 4.1.5 $\mathcal{U} \subset \mathcal{O}(X) \setminus \{\emptyset\}$ is called an *open covering* of the topological space X if $\bigcup_{U \in \mathcal{U}} U = X$.

The following is one of the most important concepts of mathematical analysis, but it will not be explored in this book.

Definition 4.1.6 A subset K of the topological space $(X, \mathcal{O}(X))$ is called *compact* if every open covering $\mathcal{U}$ of K, i.e., $K \subset \bigcup_{U \in \mathcal{U}} U$, possesses a finite subcovering. That means that there exist finitely many $U_1, \ldots, U_m \in \mathcal{U}$ such that

$$K \subset \bigcup_{j=1,\ldots,m} U_j. \tag{4.1.20}$$

The topological space $(X, \mathcal{O}(X))$ is called *locally compact* if every $x \in X$ is contained in some compact subset $K \subset X$ that contains some open neighborhood U of x, i.e., $x \in U \subset K$.

We next describe how a topology can be transferred from a set to its subsets.

Definition 4.1.7 Let $(X, \mathcal{O}(X)$ be a topological space. The topology on $Y \subset X$ defined by $\mathcal{O}(Y) := \{U \cap Y, U \in \mathcal{O}(X)\}$ is called the *induced topology* on Y.

One might also consider a topological structure as a device for localizing constructions in a set, or better, as the framework that defines such localizations. In that sense, the question naturally emerges whether such a concept of localization is compatible with the notion of a set as composed of elements. An open set is conceived as surrounding the points contained in it. Such a compatibility is then expressed by the following

Definition 4.1.8 A topological space $(X, \mathcal{O}(X))$ is called a *Hausdorff space* if for every two elements $x_1 \neq x_2 \in X$ there exist disjoint open sets, i.e., $U_1 \cap U_2 = \emptyset$ with $x_1 \in U_1, x_2 \in U_2$.

Thus in a Hausdorff space any two distinct points possess disjoint open neighborhoods.

We note, however, that the topologies of examples 4, 5 and 6 do not satisfy the Hausdorff property.

Instead of defining a topology in terms of open sets, we can also define it in terms of complements of open sets, the so-called *closed sets*. Likewise, a pretopology can be defined by a *closure operator* instead of an interior operator, as we shall now describe.

Definition 4.1.9 The *closure* $\overline{A}$ of a subset A of a pretopological space X is $\overline{A} = X \backslash (X \backslash A)^\circ$, the complement of the interior of its complement. A is called *closed* if $\overline{A} = A$, or equivalently, if $X \backslash A$ is open.

The examples 5 and 6 can now be more easily described by saying that the finite or the affine linear subsets are the closed ones.

On the basis of Definition 4.1.9, an equivalent definition of a (pre)topological space can be given in terms of the Kuratowski closure operator.

Theorem 4.1.2 *X is a pretopological space if it possesses a closure operator⁻ with the following properties*

(i) $\overline{\emptyset} = \emptyset$.
(ii) $A \subset \overline{A}$ *for all* $A \in \mathcal{P}(X)$.
(iii) $\overline{A \cup B} = \overline{A} \cup \overline{B}$ *for all* $A, B \in \mathcal{P}(X)$.

It is a topological space if the closure operator in addition satisfies

(iv) $\overline{\overline{A}} = \overline{A}$ *for all* $A \in \mathcal{P}(X)$.

In some interpretation, $\overline{A}$ is that part of X that you can reach from A by applying some operation. The first property then says that you cannot reach anything from nothing. The second property says that you can reach all your starting points, that is, nothing is lost by an operation. The third property says that from a union of starting sets you cannot reach more than the combination of what you can reach from each single set. Finally, condition (iv) says that whatever you can reach, you can already reach in a single step. In other words, when you can reach more and more by applying more and more steps, then the closure operation does not come from a topological structure. For instance, in a graph Γ, one can define the closure of a set of vertices as the union of this set with the set of all neighbors of these vertices. Then, except when every vertex can be reached from each other vertex in a single step, that is, except for a complete graph, the closure operator does not satisfy (iv). The same construction is also possible for directed graphs where the closure operator reaches the forward neighbors of a vertex set, again in addition to the members of the set itself. Thus, let Γ be a directed graph. y is called a (forward) neighbor of $x \in \Gamma$ if there is a directed edge from x to y. For each $x \in \Gamma$, we let $\bar{x}$ (abbreviated for $\overline{\{x\}}$) be the set containing x and its neighbors, and for $A \subset \Gamma$, we then put

$$\bar{A} := \bigcup_{x \in A} \bar{x}. \tag{4.1.21}$$

This then defines a closure operator, hence a pretopology. Conversely, given a pretopological space with a closure operator as above, we can construct a directed graph by connecting each x with all the elements of $\overline{\{x\}}$.

Another important example of a closure operator arises from a dynamical system. For instance, we may consider

$$\dot{x}(t) = F(x(t)) \text{ for } x \in \mathbb{R}^d, t > 0 \tag{4.1.22}$$
$$x(0) = x_0 \tag{4.1.23}$$

for some sufficiently regular F, say uniformly Lipschitz continuous[2] so that the Picard-Lindelöf theorem implies the existence of a unique solution of (4.1.22) with initial values (4.1.23) for all $t \geq 0$. Thus, (4.1.22) constitutes a special case of Definition 2.5.1. For $A \subset \mathbb{R}^d$, we may then put

$$\bar{A}^T := \{x(t), 0 \leq t \leq T\} \text{ where } x(t) \text{ is a solution of (4.1.22) with } x(0) \in A. \tag{4.1.24}$$

For each $T > 0$, this then defines a closure operator. The closed sets then are the forward invariant sets of the dynamical system (4.1.22), that is, those $B \subset \mathbb{R}^d$ with $x(t) \in B$ for all $t \geq 0$ if $x(0) \in B$. Note that this does not depend on the choice of $T > 0$ as follows readily from the semigroup property

$$x(t_1+t_2) = y(t_2) \text{ where } y(t) \text{ is the solution of (4.1.22) with } y(0) = x(t_1). \tag{4.1.25}$$

We also note that a set that is closed w.r.t. this closure operator need not be closed w.r.t. the standard topology of $\mathbb{R}^d$. For instance, for $F(x) = -x$, any open ball $U(0, r)$ w.r.t. the Euclidean metric on $\mathbb{R}^d$ is closed w.r.t. the dynamical system.

We may also define an infinitesimal closure operator

$$\bar{A} := \bigcap_{T > 0} \bar{A}^T. \tag{4.1.26}$$

In any case, in view of these examples, the name "pretopological space" does not seem so fortunate. The notion of a topology is a static one, whereas

---

[2]This means that there exists some constant $K < \infty$ with the property that $|F(x) - F(y)| \leq K|x - y|$ for all $x, y \in \mathbb{R}^d$.

the notion of a pretopology has a dynamic or operational content, and its thrust is therefore principally different from that of a topology.

If we wish to consider pretopological or topological spaces as objects of a category, we need to identify the corresponding morphisms. That is what we shall now do. (Again, it might be useful to remind ourselves that we have categories at different levels. A topological space is a category by itself, whose objects are the open sets and whose morphisms are the inclusions. We are now interested in the higher level category whose objects are (pre)topological spaces.)

Definition 4.1.10 A map $f : X \to Y$ between pretopological spaces is called *continuous* if

$$f^{-1}(A^\circ) \subset (f^{-1}(A))^\circ \qquad (4.1.27)$$

for any subset of Y. (Here, the interior operator for both spaces X and Y is denoted by the same symbol $^\circ$.)

It is clear that the identity map 1_X of a pretopological space is continuous, and that the composition of continuous maps is again continuous. Therefore, we have the categories **Pretop** and **Top** of pretopological and topological spaces with continuous maps as morphisms.

Lemma 4.1.5 *A map* $f : X \to Y$ *between pretopological spaces is continuous iff*

$$f(\overline{B}) \subset \overline{f(B)} \qquad (4.1.28)$$

for any subset of X.

Proof This follows directly from the definitions as the reader will readily check.   □

Lemma 4.1.6 *A map* $f : X \to Y$ *between topological spaces is continuous iff the preimage* $f^{-1}(A)$ *of any open subset of* Y *is an open subset of* X.

Proof We need to show that $f^{-1}(A)$ is open when $A \subset Y$ is open. In that case, $A^\circ = A$, hence $f^{-1}(A^\circ) = f^{-1}(A)$. If f is continuous, then $f^{-1}(A) \subset f^{-1}(A)^\circ$ which implies that these two sets are equal. Thus $f^{-1}(A)$ is indeed open.   □

Example (we use the same numbering as for the above examples of topological spaces)

1. For the indiscrete topology on a set X, any map $f : Y \to X$ is continuous where $(Y, \mathcal{O}')$ is any topological space. In fact, this fact characterizes the indiscrete topology. Conversely, for the indiscrete topology, if $g :$

$X \to Y$ is a continuous map into any topological space that has the Hausdorff property, then g has to be constant.

Indiscrete topology

2. For the discrete topology on X, any map $g : X \to Y$ into any topological space $(Y, \mathcal{O}')$ is continuous, and this characterizes the discrete topology.

Discrete topology

3. Continuous maps $f : (X, d) \to (Y, d')$ between metric spaces are characterized by the property that for every $x \in X$ and $\epsilon > 0$, there exists some $\delta > 0$ with the property that $f(U(x, \delta)) \subset U(f(x), \epsilon)$. (This example explains how the continuity concept is related to the ϵ-δ-criterion that you have encountered in calculus.)

4. A function $f : (X, \mathcal{O}) \to \mathbb{R}$ is called lower semicontinuous if for every $x \in X$ and $\epsilon > 0$, there exists some $U \in \mathcal{O}$ with $x \in U$ with the property that

$$f(y) > f(x) - \epsilon \text{ if } y \in U. \tag{4.1.29}$$

These lower semicontinuous functions are precisely those functions on $(X, \mathcal{O})$ that are continuous for the topology on $\mathbb{R}$ for which the open sets are $\emptyset$, X and all intervals (ξ, ∞), $\xi \in \mathbb{R}$.

5. A map between cofinite topological spaces is continuous if the preimage of every point is finite.

6. For a vector spaces V, W and topologies consisting of all complements of affine linear subspaces, the continuous maps are the affine linear ones.

Definition 4.1.11 A bijective map $f : X \to Y$ between topological spaces for which both f and its inverse f^{-1} are continuous is called a homeomorphism.

We now consider the continuous functions $C(X)$ from a topological space $(X, \mathcal{O}(X))$ to $\mathbb{R}$ (with its standard topology, but the subsequent construction does not depend on the choice of the topology on $\mathbb{R}$). $\mathbb{R}$ has a field structure, with the operations $+$ and $\cdot$. Therefore, we can also add and multiply $\mathbb{R}$-valued functions and obtain a ring structure on $C(X)$. Why, then, is $C(X)$ only a ring, but not a field? The reason is that we cannot divide by any function that assumes the value 0 at some point. In a field, however, the only element by which one cannot divide is the 0 element; in $C(X)$, this element is, of course, the function $\equiv 0$. In fact, $C(X)$ is even an algebra, because we can multiply functions by scalars $\lambda \in \mathbb{R}$. This constructions then obviously applies to functions with values in any field, and not only for $\mathbb{R}$. In particular, we can take the complex scalars $\mathbb{C}$ in place of $\mathbb{R}$.

When we now have a continuous map $f : X \to Y$ between topological spaces, it induces a contravariant ring (or algebra) homomorphism

$$f^* : C(Y) \to C(X)$$
$$\phi \mapsto \phi \circ f. \tag{4.1.30}$$

Established references for topological spaces are [66, 95]. The connections with analysis are well explained in [2]. For pretopological spaces, we refer to [23], one of the founding texts of the field. For recent contributions to pretopological spaces, see [105, 106, 107] and the literature therein. We now describe some applications developed in those papers.

We consider the set of all binary strings of length n, that is, objects of the form $(x_1, \ldots, x_n)$ with $x_i \in \{0, 1\}$. When we connect two strings that differ in precisely one position, that is, that can be obtained from each other by a single point mutation, or equivalently, that have Hamming distance 1 from each other (see (2.1.53)), we obtain the n-dimensional hypercube W_n as the graph representing this structure. For each $x \in W_n$, we let $\overline{x}$ be the set of its neighbors in W_n. Of course, we can also apply this to other sets of strings, where the entries are chosen from some arbitrary alphabet L, in place of the binary one $\{0, 1\}$. For instance, we may consider genetic sequences composed of four letters, denoted by A, T, C, G. In this manner, we obtain a pretopology representing the biological concept of a point mutation.

Letters A, T, C, G of genetic sequences

Another important biological operation is recombination by genetic cross-over. Again, we could work with an arbitrary alphabet instead of the binary one, without any substantial change of mathematical content. We consider equal cross-over, meaning that for two binary strings $x = (x_1, \ldots, x_n)$, $y = (y_1, \ldots, y_n)$ of length n, their possible recombinations $R(x, y)$ are the strings

Genetic recombination

$$(x_1, \ldots, x_k, y_{k+1}, \ldots, y_n) \text{ and } (y_1, \ldots, y_\ell, x_{\ell+1}, \ldots x_n) \qquad (4.1.31)$$

for $k, \ell = 1, \ldots, n$. We thus have

1.
$$R(x, x) = \{x\}$$

2.
$$R(x, y) = R(y, x)$$

3.
$$\{x, y\} \subset R(x, y)$$

In general, however, it is not true that

$$R(x, z) \subset R(x, y) \text{ if } z \in R(x, y).$$

For instance, for $x = (0000)$, $y = (1111)$, we have $z = (0011) \in R(x, y)$ and $w = (0010) \in R(x, z)$, but $w \notin R(x, y)$. Thus, if, following Gitchoff-Wagner [41], we defined the closure of A as

$$A^1 := \bigcup_{x, y \in A} R(x, y), \qquad (4.1.32)$$

condition (iv) for a topological space would not be satisfied. This deficit can be remedied, however, by considering the union of all iterations of recombinations starting from a given set of strings. Then by such iterations, starting from two strings x, y as before, we can generate all strings $(u_1, \ldots, u_n)$ where each u_i independently can be either x_i or y_i. Thus, for instance, by iterated recombinations, we can generate any string of length 4 from $x = (0000)$ and $y = (1111)$. Formally, we put

$$A^n := \bigcup_{x, y \in A^{n-1}} R(x, y) \text{ for } n \geq 2, \qquad \text{and}$$

$$\overline{A} := \bigcup_{n \in \mathbb{N}} A^n. \qquad (4.1.33)$$

This then satisfies condition (iv). Condition (iii) for a Kuratowski closure operator, however, is neither satisfied by (4.1.32) nor by (4.1.33). In contrast to property (iv), this deficit cannot be remedied by looking at the union of iterations of recombination. Thus, recombination leads to a structure on binary sequences that is more general than a pretopology. The failure of condition (iii) points to the advantage of genetic diversity in a population with sexual recombination. From the union of two gene pools, in general, one can generate more genomes by recombination than from the two gene pools in isolation.

4.2 σ-Algebras

We now introduce a class of set systems whose defining properties include complementation. There are several motivations for this. When we shall introduce measures in Sect. 4.4, we wish to have the property that the measure of a subset $A \subset X$ and that of its complement $X \backslash A$ add up to the measure of X. In particular, whenever we are able to measure A, we also want to measure its complement. More conceptually, we may want to think in terms of alternatives, for instance when we make observations or perform measurements. For example, when we make a scalar observation, that is, evaluate a function $f : X \to \mathbb{R}$ at some point $x \in X$, we may ask the question "Is $f(x) > a$?" for some specified value $a \in \mathbb{R}$. We then have the two complementary subsets $\{x \in X : f(x) > a\}$ and $\{x \in X : f(x) \leq a\}$ that represent the two possible answers, yes or no, to our question. For such reasons, we wish to work with classes of subsets that are closed under complementation and also satisfy certain other properties which are also related to the issues just discussed. The relevant definition is

Definition 4.2.1 Let X be a set. A σ-algebra of subsets of X is a subset $\mathcal{B}$ of $\mathcal{P}(X)$ satisfying:

(i) $X \in \mathcal{B}$.
(ii) If $B \in \mathcal{B}$, then so is $X \backslash B$.
(iii) If $B_n \in \mathcal{B}$ for all $n \in \mathbb{N}$, then so is $\bigcup_{n \in \mathbb{N}} B_n$.

$(X, \mathcal{B})$ is then called a *measurable space*.

The preceding properties imply:

(iv) $\emptyset \in \mathcal{B}$.
(v) If $B_1, \ldots, B_m \in \mathcal{B}$, then so is $\bigcap_{j=1}^{m} B_j$.

Thus, on $X = \{0, 1\}$, we have two different σ-algebras, one with $\mathcal{B} = \{\emptyset, X\}$, and the other with $\mathcal{B} = \{\emptyset, \{0\}, \{1\}, X\}$.

$\boxed{\sigma\text{-algebras on } \{0, 1\}}$

In order to obtain a σ-algebra, we can start with any collection of subsets of X and close it up under complements and countable unions. If X is a topological space, we can therefore take the smallest σ-algebra containing all open subsets of X. The sets in this σ-algebra are called Borel sets.

So far, however, one might be naturally inclined to simply use the entire Boolean algebra $\mathcal{P}(X)$ as the σ-algebra. One important reason why one works with smaller σ-algebras than $\mathcal{P}(X)$ is that when the σ-algebra is too large, it becomes too restrictive to satisfy the properties required in the Definition 4.4.1 of a measure in Sect. 4.4. Another reason, that σ-algebras are tools to account for the distinctions that can be made on the basis of observations, will be analyzed in detail below.

Having thus defined the category of measurable spaces, we can easily identify the morphisms.

Definition 4.2.2 A map $T : (X, \mathcal{B}(X)) \to (Y, \mathcal{B}(Y))$ between measurable spaces is called *measurable* if for all $A \in \mathcal{B}(Y)$ also $T^{-1}(A) \in \mathcal{B}(X)$.

A continuous map between topological spaces is always measurable for any Borel measure because the preimages of open sets are open, and so the preimages of Borel sets are Borel.

In turn, when we have a measurable space $(Y, \mathcal{B})$ and a map $f : X \to Y$, we can equip X with the σ-algebra $f^\star\mathcal{B}$ consisting of all sets $f^{-1}(A)$ with $A \in \mathcal{B}$. Such a σ-algebra may be quite small. In particular, we cannot distinguish between points with the same value of f.

Let us elaborate on this issue and dwell a little on the interpretation of σ-algebras in terms of observables. In the most basic case, we check the elements of X for some property p, and we put $p(x) = 1$ when x possesses that property and $p(x) = 0$ if not. We then obtain a σ-algebra $\mathcal{B}$ on X from the two sets

$$A := p^{-1}(1) = \{x \in X; p(x) = 1\} \text{ and } X \setminus A = p^{-1}(0) = \{x \in X; p(x) = 0\}. \tag{4.2.1}$$

It could happen that either all or none of the elements of X possess the property p in which case $\mathcal{B}$ is the trivial σ-algebra consisting only of X itself and $\emptyset$. Otherwise, we would have four members,

$$\mathcal{B} = \{A, X \setminus A, X, \emptyset\}. \tag{4.2.2}$$

This σ-algebra $\mathcal{B}$ then expresses all the distinctions that we can make on the basis of our observations, that is, the trivial observation whether x is in the set X and the observation whether it possesses the property p.

When we then have several such properties $p_1, \ldots, p_n$ that we can observe, we get the corresponding sets $A_j := p_j^{-1}(1)$ and their complements for $j = 1, \ldots, n$, and by taking intersections and unions, we can generate the corresponding σ-algebra $\mathcal{B}_n$. This simply expresses the fact that we can now make distinctions by combining our observations. For instance $A_1 \cap X \setminus A_2$ is the set of those elements that satisfy property p_1, but not p_2, and $A_2 \cup A_3$ contains those points that satisfy p_2 or p_3. In particular, if the set X is otherwise opaque for us and we can only perform observations that check properties p_j, then the σ-algebra $\mathcal{B}_n$ reflects our knowledge and our ability to discriminate between the points of X on the basis of our observations regarding the properties $p_1, \ldots, p_n$. When we then make an additional observation, the σ-algebra gets larger as our knowledge or our discrimination ability increase. More precisely, we can define that a new property p_{n+1} is independent of the properties $p_1, \ldots, p_n$ if the σ-algebra $\mathcal{B}_{n+1}$ is strictly larger than $\mathcal{B}_n$.

When we can make scalar measurements, that is, if we have a function $f : X \to \mathbb{R}$, then we obtain the σ-algebra $\mathcal{F}$ generated by the sets

$$f^{-1}\{f > a\} = \{x \in X : f(x) > a\} \text{ for } a \in \mathbb{R}. \qquad (4.2.3)$$

For every $a \in \mathbb{R}$, we can thus make the binary distinction whether $f(x) > a$ or not. Alternatively, $\mathcal{F}$ is also generated by the sets

$$f^{-1}\{f = a\} = \{x \in X : f(x) = a\} \text{ for } a \in \mathbb{R}. \qquad (4.2.4)$$

Thus, we can discriminate between points x_1, x_2 with $f(x_1) \neq f(x_2)$, but we cannot distinguish between points with the same value of f.

The σ-algebra $\mathcal{B}$ also defines an equivalence relation where x_1, x_2 are equivalent iff $x_1 \in A \Leftrightarrow x_2 \in A$ for all $A \in \mathcal{B}$.

Again, when we can perform further such measurements, say $f_1, \ldots, f_n$, we can make further distinctions as the resulting σ-algebra gets larger, or equivalently, the equivalence classes defined by this σ-algebra get smaller. It may happen that we can uniquely identify each point $x \in X$ by the values of m such suitably chosen measurements. In fact, one could try to define the dimension of a topological space X as the smallest number m with the property that each $x \in X$ possesses an open neighborhood U with continuous functions $f_1, \ldots, f_m : U \to \mathbb{R}$ such that the resulting σ-algebra on U contains the singleton sets $\{y\}$ for all $y \in U$. Here, the functions could depend on the neighborhood U, but not on the point x itself. If there is no finite such m, we could declare the dimension of x to be infinite. While this may seem quite natural, there are technical difficulties involved with this approach. Although these can eventually be overcome, we do not want to go into further details here. See, for instance, [36]. Below, in Sect. 5.3, we shall consider a more constrained class of spaces, the differentiable manifolds, for which the definition of the dimension is easier.

4.3 Set Systems

In this section, we develop a more abstract and unifying perspective on topologies and σ-algebras with the concepts of category theory. Both a Heyting algebra $\mathcal{O}$ of open sets and a σ-algebra $\mathcal{B}$ are defined in terms of the operations of the union $\cup$, the intersection $\cap$ and the complement $\setminus$ of subsets of some set. These operations do not behave naturally under maps $f : X \to Y$. In fact,

$$f(A \cap B) \subset f(A) \cap f(B), \qquad (4.3.1)$$

where inequality can occur when f is not injective. Similarly, in general

$$f(X\backslash A) \neq f(X)\backslash f(A) \subset Y\backslash f(A), \tag{4.3.2}$$

where the right hand side is larger when f is not surjective, but where also the left-hand side could be larger for a non-injective f. The phenomena should be clear from this diagram:

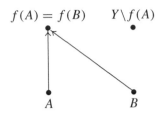

Remarkably, however, when applying f^{-1}, we obtain a homomorphism of the Boolean algebra $\mathcal{P}$. In fact, we have

$$f^{-1}(V \cap W) = f^{-1}(V) \cap f^{-1}(W) \tag{4.3.3}$$

and

$$f^{-1}(V \cup W) = f^{-1}(V) \cup f^{-1}(W), \tag{4.3.4}$$

as well as

$$f^{-1}(W\backslash V) = f^{-1}(W)\backslash f^{-1}(V). \tag{4.3.5}$$

In fact, we also have the more general relations

$$f^{-1}(\bigcap_{i \in I} A_i) = \bigcap_{i \in I} f^{-1}(A_i), \quad f^{-1}(\bigcup_{i \in I} A_i) = \bigcup_{i \in I} f^{-1}(A_i) \tag{4.3.6}$$

for any collection $(A_i)_{i \in I} \subset \mathcal{P}(Y)$. All these relations are easily checked from the basic fact that

$$x \in f^{-1}(A) \quad \Leftrightarrow \quad f(x) \in A, \tag{4.3.7}$$

and the relations (4.1.7)–(4.1.9). We happily leave this to the reader as a simple exercise to keep him alert.

We therefore define the functor

$$f^*(V) := f^{-1}(V) \text{ for } V \subset Y, \tag{4.3.8}$$

to express this contravariant behavior.

This morphism $f^* : \mathcal{P}(Y) \to \mathcal{P}(X)$ also preserves the partial ordering structure of inclusion $\subset$, as it should. This simply means that

$$V \subset W \quad \Rightarrow \quad f^*(V) \subset f^*(W). \tag{4.3.9}$$

In fact, in turn any such morphism $F : \mathcal{P}(Y) \to \mathcal{P}(X)$ yields a map $f : X \to Y$. We simply put

$$f(x) := y \quad \text{for all } x \in F(\{y\}). \tag{4.3.10}$$

Since if $y_1 \neq y_2$, then $\{y_1\} \cap \{y_2\} = \emptyset$ and hence also $F(\{y_1\}) \cap F(\{y_2\}) = \emptyset$, the definition (4.3.10) causes no ambiguity.

We now want to check to what extent the preceding extends to subfamilies of $\mathcal{P}$, such as a Heyting algebra $\mathcal{O}$ of the open sets of some topology or a σ-algebra $\mathcal{B}$.

Definition 4.3.1 A *set system* $\mathcal{S}$ on the set X is a subset of $\mathcal{P}(X)$ that is closed under certain operations involving only the union, intersection, and complement of (possibly infinitely many) sets.

$\mathcal{G} \subset \mathcal{P}(X)$ is called a *subbasis* of the set system $\mathcal{S}$, or is said to generate $\mathcal{S}$, if every element of $\mathcal{S}$ can be obtained from elements of $\mathcal{G}$ by the operations defining $\mathcal{S}$.

For instance, the set system $\mathcal{O}(X)$ is defined in terms of the operations $\cup$ and $\cap$, see Theorem 4.1.1. The definition of a σ-algebra also involves the complement $\setminus$, but in fact, two of the three operations would suffice in any case because, for instance, $\cap$ can be expressed in terms of $\cup$ and $\setminus$, see (4.1.3).

The preceding considerations, or more precisely the relations (4.3.3)–(4.3.6), imply

Lemma 4.3.1 *Let $\mathcal{S}(Y)$ be a set system on Y, and $f : X \to Y$ a map. Then $f^* $ yields a morphism of set systems. This means that $f^*\mathcal{S}(Y) = f^{-1}\mathcal{S}(Y)$ is a set system on X, of the same type as $\mathcal{S}(Y)$, that is, defined by the same operations as $\mathcal{S}(Y)$.*

In particular, when $\mathcal{O}(Y)$ defines a topology on Y, then $f^*\mathcal{O}(Y)$ defines a topology on X, and if $\mathcal{B}(Y)$ is a σ-algebra on Y, then $f^*\mathcal{B}(Y)$ yields a σ-algebra on X.

We may then also ask whether a morphism $F : \mathcal{S}(Y) \to \mathcal{S}(X)$ defines a map $f : X \to Y$, as was the case for $\mathcal{P}$. In fact, we need to require that $\mathcal{S}(Y)$ be Hausdorff in the sense that it separates the points of Y, as otherwise such a morphism need not define a map, that is, a relation between points in X and Y such that every point in X is related to precisely one point in Y. For instance, if Y has the indiscrete topology, we then only have the relations $F(Y) = X$ and $F(\emptyset) = \emptyset$ which is not sufficient to yield a map. That $\mathcal{S}(Y)$ is Hausdorff means that for $p \neq q \in Y$, there exist $U, V \in \mathcal{S}(Y)$ with $U \cap V = \emptyset, p \in U, q \in V$. Then also $F(U) \cap F(V) = \emptyset$. In this sense, when we put

$$F(p) := \lim_{\overrightarrow{x \in U}} F(U) \qquad (4.3.11)$$

as a colimit (as defined in Sect. 8.2), we can assign distinct preimages to all those points in Y where this colimit is not empty. On the other hand, since the morphism F satisfies $F(Y) = X$, every point in X occurs as some such preimage. Thus, F defines a map $f : X \to Y$.

Lemma 4.3.2 *This map f preserves the structure of the set system $\mathcal{S}$. For instance, if $\mathcal{S} = \mathcal{O}$ defines a topology, then f is continuous, and if $\mathcal{S} = \mathcal{B}$ is a σ-algebra, then f is measurable.*

Here, we refer to the characterization of continuity established in Lemma 4.1.6.

We can also express this in a diagram

$$
\begin{array}{ccc}
(X, f^*\mathcal{S}(Y)) & \xrightarrow{\ \ f\ \ } & (Y, \mathcal{S}(Y)) \\
\downarrow & & \downarrow \\
X & \xrightarrow{\ \ f\ \ } & Y
\end{array}
\tag{4.3.12}
$$

The upper f then indicates that f is a morphism for the category of set systems of type $\mathcal{S}$, according to Lemma 4.3.2. In fact, it is a pullback[3] in the sense that whenever we have a morphism $h : (W, \mathcal{S}(W)) \to (Y, \mathcal{S}(Y))$ which, at the level of maps between sets, i.e., downstairs in (4.3.13), factors as $h = f \circ g$, then it also induces an $\mathcal{S}$-morphism $(W, \mathcal{S}(W)) \to (X, \mathcal{S}(X))$. That is, there is a dashed arrow making the following diagram commutative.

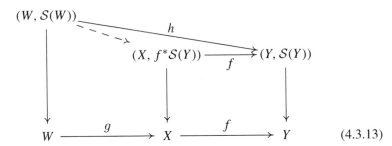

$$\tag{4.3.13}$$

To see this, we observe that since $h = f \circ g$, for any $A \in \mathcal{S}(Y), h^{-1}(A) = g^{-1} f^{-1}(A) = g^{-1}(B)$ for $B = f^{-1}(A)$. And since h is assumed to be an $\mathcal{S}$-morphism, we have $h^*\mathcal{S}(Y) \subset \mathcal{S}(W)$. Thus, for any $B \in f^*\mathcal{S}(Y)$, we find a preimage in $\mathcal{S}(W)$. This yields the $\mathcal{S}$-morphism $\mathcal{S}(W) \to f^*\mathcal{S}(Y)$, that is, the dashed arrow in the diagram. Compare also the general discussion of pullbacks around (8.2.34).

There is another characterization of $f^*\mathcal{S}(Y)$.

Lemma 4.3.3 *Suppose that the set system $\mathcal{S}(Z)$ on Z satisfies the Hausdorff property. A mapping $g : (X, f^*\mathcal{S}(Y)) \to (Z, \mathcal{S}(Z))$ is an $\mathcal{S}$-morphism iff there exists an $\mathcal{S}$-morphism $h : (Y, \mathcal{S}(Y)) \to (Z, \mathcal{S}(Z))$ with $g = h \circ f$.*

Proof For the map h to exist, we need to show that whenever $f(x_1) = f(x_2)$ for $x_1, x_2 \in X$, then also $g(x_1) = g(x_2)$. Now if $f(x_1) = f(x_2)$, then there is no $U \in f^*\mathcal{S}(Y)$ that separates x_1, x_2, that is, contains one of these two points, but not the other. If now $g(x_1) \neq g(x_2)$, then by the Hausdorff property of $\mathcal{S}(Z)$, there exists some $V \in \mathcal{S}(Z)$ containing one of them, but not the other. But then $g^{-1}(V)$ would separate x_1 and x_2. But this is not compatible with the fact that since g is an $\mathcal{S}$-morphism, we must have

---

[3] See (8.2.34) for a general definition of pullbacks.

$g^{-1}(V) \in f^*\mathcal{S}(Y)$. This contradiction shows that there exists such a map h, and it is then readily verified that it is also an $\mathcal{S}$-morphism. This establishes the forward direction of the lemma. The reverse direction is trivial.

We can also represent the content of Lemma 4.3.3 by a commuting diagram

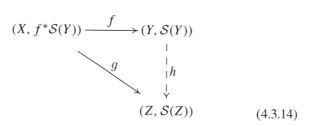

$$(4.3.14)$$

We now come to an important generalization of Lemma 4.3.1.

Lemma 4.3.4 *Let $Y_a, a \in A$ be a collection of sets indexed by some index set A. Let $\mathcal{S}(Y_a)$ be a set system on Y_a (where all these set systems are supposed to be defined through the same operations; for instance, they could all be topologies). Let X be a set, and let $f_a : X \to Y_a$ be a map for every $a \in A$. We then obtain a set system $\mathcal{S}(X)$ with subbasis $\bigcup_{a \in A} f^\star(\mathcal{S}(Y_a))$, that is, generated by the sets $f^{-1}(U_a)$ with $U_a \in \mathcal{S}(Y_a)$.*

With respect to this set system, each f_a is then an $\mathcal{S}$-morphism. For instance, if the set systems are topologies, then each f_a is continuous.

The *proof* is clear, because the statement is essentially a definition. □

We shall now use this Lemma to define topologies (or, by the same scheme, other set systems) on products when we have one on each factor. This will generalize the Example 7 of Sect. 4.1 (after Definition 4.1.2).

Definition 4.3.2 Let $Y_a, a \in A$ again be a collection of sets indexed by some index set A. The *Cartesian product*

$$\underset{a \in A}{\mathsf{X}} Y_a \qquad\qquad (4.3.15)$$

is then defined as the set of all mappings y defined on A with $y(a) \in Y_a$ for every $a \in A$. We also have the projection $p_b : \mathsf{X}_{a \in A} Y_a \to Y_b$ with $p_b(y) = y(b)$ for $b \in A$.

Suppose that each Y_a carries a topology $\mathcal{O}(Y_a)$. Then the product topology on $\mathsf{X}_{a \in A} Y_a$ is generated by the sets $p_b^{-1}(U_b)$ for $U_b \in \mathcal{O}(Y_b)$.

We may recall (2.1.1) where we have identified the elements of a set Y via maps $f : \{1\} \to Y$. The notion of a product extends this to index sets other than the trivial $\{1\}$.

We should point out that even when the index set A is infinite, only *finite* intersections of sets of the form $p_b^{-1}(U_b)$ need to be open.

Importantly, not only can we pull back set systems, but we can also push them forward. We can then construct quotients instead of products, as we shall now explain.

Definition 4.3.3 Let $f : X \to Y$ be a map, and let $\mathcal{S}(X)$ be a set system on X. We then define the set system $\mathcal{S}(Y)$ as the collection of all subsets B of Y for which $f^{-1}(B) \in \mathcal{S}(X)$. This set system is called the *quotient set system* on Y induced by the map f. In particular, when $\mathcal{S}(X)$ defines a topology on X, then the resulting topology on Y is called the *quotient topology*.

It is easy to check that $\mathcal{S}(Y)$ is a set system of the desired type, indeed. In particular, $f : (X, \mathcal{S}(X)) \to (Y, \mathcal{S}(Y))$ then becomes an $\mathcal{S}$-morphism.

In fact, $\mathcal{S}(Y)$ is the largest set system on Y for which f is an $\mathcal{S}$-morphism. In contrast, in the situation of Lemma 4.3.1, $f^{\star}\mathcal{S}(Y)$ is the smallest set system on X for which f is an $\mathcal{S}$-morphism.

We leave it to the reader as an exercise to prove the following

Lemma 4.3.5 *Let* $f : (X, \mathcal{O}(X)) \to (Y, \mathcal{O}(Y))$ *be a continuous map between topological spaces where* $\mathcal{O}(Y)$ *is the quotient topology. Let* $(Z, \mathcal{O}(Z))$ *be another topological space. Then a map* $g : (Y, \mathcal{O}(Y)) \to (Z, \mathcal{O}(Z))$ *is continuous iff* $g \circ f : (X, \mathcal{O}(X)) \to (Z, \mathcal{O}(Z))$ *is continuous.*

4.4   Measures

Topology is concerned with qualitative properties whereas geometry is based on measurements, that is, quantitative aspects. In this section, we shall be concerned with the question of how to measure the size of subsets of a set, and the conceptual requisites to make this possible.

The idea of a measure on a set X is to assign to subsets of X nonnegative real numbers, or possibly also the value ∞, that express the sizes of those subsets. Therefore, the measure should be additive, that is, the measure of a disjoint union of subsets should equal the sum of the measures of those subsets. This, however, leads to the difficulty that, while one can take unions over arbitrary index sets, one can take sums of real numbers only for countable sequences. In order to overcome this difficulty, some constructions are necessary. They amount to restricting the class of subsets to which one can assign a measure, but this restriction should be such that the "important" or "natural" subsets of X should still have a measure. In particular, when X carries a topology as given by an algebra $\mathcal{O}(X)$ of open subsets, then those open sets should be measurable. However, as the additive requirement should include that the measure of a subset of X and the measure of its complement add up to the measure of X itself, we are faced with the difficulty that in general the complement of an open set need not be open itself. Therefore, we should like to require that complements of open sets, that is, closed sets also be measurable. In particular, this indicates that the concept of a σ-algebra as introduced in Sect. 4.2 should be useful.

Here, the σ-algebra should not be too small because we might like to measure as many subsets of X as possible. More precisely, we want to make sure that certain particular classes of subsets of X be measurable, for instance the open sets if X happens to carry a topology. As already explained, in order to obtain a σ-algebra, we can start with any collection of subsets of X and close it up under complements and countable unions. In particular, when X is a topological space, we can take the Borel σ-algebra, that is, the smallest σ-algebra containing all open subsets of X.

On the other hand, the σ-algebra should not be too large, as otherwise it becomes too restrictive to satisfy the properties in the next definition. Thus, the Borel σ-algebra is typically a good choice in measure theory.

Definition 4.4.1 A *measure* on $(X, \mathcal{B})$ is a function

$$\mu : \mathcal{B} \to \mathbb{R}^+ \cup \infty$$

satisfying:

(i) $\mu(\bigcup_{n \in \mathbb{N}} B_n) = \sum_{n \in \mathbb{N}} \mu(B_n)$, if $B_i \cap B_j = \emptyset$ for all $i \neq j$, i.e., if the sets B_n are pairwise disjoint,

(ii) $\mu(X) \neq 0$.

A triple $(X, \mathcal{B}, \mu)$ with the preceding properties is called a *measure space*. When $\mathcal{B}$ is the Borel σ-algebra, we speak of a *Borel measure*. A particularly important class of Borel measures are the *Radon measures*, i.e., those that take finite values on compact sets and for which the measure of any Borel set is given by the supremum of the measures of its compact subsets.

(i) implies

(iii) $\mu(\emptyset) = 0$ and

(iv) $\mu(B_1) \leq \mu(B_2)$ if $B_1 \subset B_2$.

Remark The preceding properties are modeled after the Lebesgue measure which plays a fundamental role in analysis, because it leads to *complete* function spaces, like the Hilbert space L^2. For that purpose, it is important to require the additivity (i) for *countable* families of disjoint measurable sets, and not only for finite ones. But this is not our topic, and we refer to [59]. We shall utilize measures for putting weights on the simplices of a simplicial complex, and for sketching certain applications in biology. Nevertheless, the concept of a measure is not really central for this book, and some readers may wish to skip it.

The *null sets* of a measure μ are those subsets A of X with $\mu(A) = 0$. Thus, the null sets of a measure are negligible w.r.t. that measure. Two measures are called *compatible* if they have the same null sets. If μ_1, μ_2 are compatible Radon measures, they are absolutely continuous with respect to each other in the sense that there exists a function $\phi : X \to \mathbb{R}^+ \cup \infty$ that is integrable (see e.g. [59]) with respect to either of them, such that

$$\mu_2 = \phi \mu_1, \quad \text{or, equivalently,} \quad \mu_1 = \phi^{-1} \mu_2. \qquad (4.4.1)$$

ϕ is called the *Radon-Nikodym derivative* of μ_2 with respect to μ_1.

We observe that, being integrable, ϕ is finite almost everywhere, i.e., the set of points where it takes the value ∞ is a null set (with respect to both μ_1 and μ_2) on X, and since the situation is symmetric between ϕ and ϕ^{-1}, ϕ is also positive almost everywhere.

When $(X, \mathcal{O}(X))$ is a locally compact Hausdorff space, then the Riesz representation theorem says that the Radon measures are precisely the positive continuous linear functionals on $C^0(X)$, the space of continuous functions $f : X \to \mathbb{R}$. Here, a functional

$$L : C^0(X) \to \mathbb{R} \qquad (4.4.2)$$

is *positive* whenever $L(f) \geq 0$ for all $f \geq 0$ (that is, $f(x) \geq 0$ for all $x \in X$). In order to define when L be continuous, we equip $C^0(X)$ with the following metric

$$d(f, g) := \sup_{x \in X} |f(x) - g(x)|. \qquad (4.4.3)$$

In this way, according to the general scheme whereby a metric induces a topology, we define a basis $U(f, \epsilon)$ of open sets ($f : X \to \mathbb{R}$ continuous, $\epsilon \geq 0$) on $C^0(X)$ via

$$g \in U(f, \epsilon) \qquad \Leftrightarrow \qquad (g - f)^{-1}(U(0, \epsilon)) = X \qquad (4.4.4)$$

where $U(0, \epsilon)$ of course is an open interval in $\mathbb{R}$.

Definition 4.4.2 A measure μ satisfying

(iv) $\mu(X) = 1$

is called a *probability measure*.

We let $\mathcal{M}(X)$ be the set of all probability measures on $(X, \mathcal{B})$. $\mathcal{M}(X)$ contains the Dirac measures supported at the points of X. That is, for any

| Dirac measure |

$x \in X$, we have the measure δ_x defined by

$$\delta_x(A) := \begin{cases} 1 & \text{if } x \in A \\ 0 & \text{if } x \notin A \end{cases} \qquad (4.4.5)$$

For a Dirac measure, of course, the requirements of Definition 4.4.1 are easily satisfied even when we take $\mathcal{P}(X)$ as our σ-algebra. For other measures, like the Lebesgue measure on $\mathbb{R}$, this is not so, and one needs to restrict to the Borel σ-algebra. We do not go into those details here, however.

Also, when we have a random walk process operating on X with time $t \in \mathbb{R}^+$ (continuous time) or $t \in \mathbb{N}$ (discrete time), we have the transition probabilities $p(A, B, t)$, denoting the probability that a particle starting in the set A at time 0 is in the set B at time t. Here, the sets $A, B, \ldots$ are assumed to be contained in some appropriate σ-algebra $\mathcal{B}$, and $p(A, ., t)$ is a probability measure on $(X, \mathcal{B})$. It therefore satisfies

$$p(A, B, t) + p(A, X \setminus B, t) = 1 \text{ for all } B \in \mathcal{B}, t \geq 0. \qquad (4.4.6)$$

We also have

$$p(A, A, 0) = 1 \text{ for all } A \in \mathcal{B}. \qquad (4.4.7)$$

The underlying process here could be the mutational drift in a population of genomes, that is, in genotype space.

Other important examples of measures arise in applications from distributions of objects. For instance, in molecular biology, for any protein represented in a cell, we can use the density ϕ of that protein to define some (Radon) measure $\phi\mu$ for some background measure μ, say the Lebesgue measure. In Sect. 4.7 below we shall address the question of how to compare different such measures representing the distribution of proteins or whatever else is of interest in some application.

> Protein colocalization

Definition 4.4.3 A map $T : (X, \mathcal{B}(X), \mu_X) \rightarrow (Y, \mathcal{B}(Y), \mu_Y)$ between measure spaces is called *measure preserving* when $\mu_X(T^{-1}(A)) = \mu_Y(A)$ for all $A \in \mathcal{B}(Y)$.

In turn, when we have a measure space $(X, \mathcal{B}, \mu)$ and a map $f : X \rightarrow Y$, we can equip Y with the σ-algebra $f_\star\mathcal{B}$ generated by the sets $f(A)$ with $A \in \mathcal{B}$ and the measure $f_\star\mu$ defined by the property

$$f_\star\mu(B) := \mu(f^{-1}(B)) \tag{4.4.8}$$

for $B \in \mathcal{B}$. (Note that in general $\phi^\star\mu$ with $\phi^\star\mu(C) := \mu(\phi(C))$ for a map $\phi : Z \rightarrow X$ does not define a measure as the additivity property is violated when, for instance, there are two disjoint sets C_1, C_2 with $\phi(C_1) = \phi(C_2)$.)

For instance, we can look at the genotype-phenotype map f that associates to a genome $x \in X$ (a string in the alphabet with letters A, C, G, T, but for the mathematical structure behind this, we could also take our binary alphabet as above) the phenotype $y \in Y$ (whatever that is–here we do not specify the structure of the set or topological space Y) produced by it (or, perhaps better, the collection of possible phenotypes that can be produced by it, depending on external influences and circumstances, but we suppress this important aspect for the sake of the present discussion). In general, many different genotypes can lead to the same phenotype, that is, the map f is not injective. The set of genotypes leading to one and the same phenotype is called a neutral basin. Genotype space carries a natural metric given by the Hamming distance (2.1.53) between genomes, that is, the number of mutations required to move from one genome to another one.

When we then have a measure on genotype space, for instance, given by the distribution of genomes in a population, we then obtain an induced measure on phenotype space. Also the above family of measures $p(A, , t)$ corresponding to mutational drift in genotype space then induces a corresponding family of probability measures in phenotype space. Thus, when we have two phenotypes K, L, we can look at the genotypes giving rise to them, $A := f^{-1}K, B := f^{-1}L$ and then have the transition probability

$$\pi(y, z, t) := (f_\star p)(y, z.t) = p(A, B, t). \tag{4.4.9}$$

Thus, the transition probabilities between phenotypes are simply given by those between the underlying genotypes. This assumes that the underlying genotype is not known in detail, that is, that all genotypes that can produce the given phenotype are possible. When one instead knows which of the possible genotypes is underlying the genotype in question, then one may wish to replace A in (4.4.9) by the set containing only that particular genotype.

We may call a phenotype set $K \subset Y$ almost absorbing if

$$\pi(K, Y \setminus K, t) \ll 1 \text{ for all } t < T \text{ for some } T \gg 0. \tag{4.4.10}$$

That is, one would have to wait for a very long time before a different phenotype will emerge from those in the set K, because there is strong selective pressure to maintain the phenotypic characteristics encoded in K. For some other phenotype set $L \subset Y$, we may then ask for the time accumulated transition probability

$$\sum_{t=1}^{T} \pi(L, K, t) \tag{4.4.11}$$

or for the minimal waiting time

$$\inf_{t>0} \pi(L, K, t) > \delta \text{ for some given } \delta > 0 \tag{4.4.12}$$

to reach that almost absorbing phenotype with some nonnegligible probability from the original phenotype L.

We also observe that for two phenotypes y, z, the reachabilities

$$r(y, z) := r_T(y, z) := \sum_{t=1}^{T} \pi(y, z, t) \tag{4.4.13}$$

for some given $T > 0$ in general are different in the two directions, that is

$$r(y, z) \neq r(z, y), \tag{4.4.14}$$

as are the waiting times

$$\inf_{t>0} \pi(y, z, t) > \delta \tag{4.4.15}$$

before one is reached from the other with some minimal probability $\delta > 0$.

4.5   Sheaves

In this section, we shall elaborate upon the concept of a presheaf as already defined in Sect. 2.4 and to be further developed below in Sect. 8.4, but we shall concentrate on the case where the category in question is that of open subsets of some topological space. We shall also try to make this largely independent of Chap. 8.

In fact, the constructions here are analogous to those in Sect. 2.4, with the difference that now we do not consider the category $\mathcal{P}(S)$ of subsets of some set S, but rather the category $\mathcal{O}(X)$ of open subsets of some topological space X.

In this sense, we define a bundle over the topological space X as a surjective continuous map $p : Y \to X$ from some other topological space. Analogously to Sect. 2.4, for $x \in X$, the preimage $p^{-1}(x)$ is called the fiber over x. X is called the base space, and Y is the total space. A section of such a bundle is a continuous map $s : X \to Y$ with $p \circ s = 1_X$.

We say that the fiber is modeled after the topological space F if for every $x \in X$, there exists some open U with $x \in U$ and a homeomorphism, that is, a map that is bijective and continuous in both directions (see Definition 4.1.11),

$$\phi : p^{-1}(U) \to U \times F. \tag{4.5.1}$$

When the model F has some additional structure, such as that of a vector space, and if

$$\phi_x : p^{-1}(x) \to \{x\} \times F, \tag{4.5.2}$$

is an isomorphism with respect to that structure for every x (where we either assume that each fiber possesses this structure already, or that we can use the ϕ_x to endow the fibers with such a structure so that they become isomorphisms), we say that the bundle has such a structure. For instance, we speak of a vector bundle when F carries the structure of a vector space. Thus, such a structured bundle locally is a product of the base space and the fiber. Globally, however, that need not be the case, that is, in general a fiber bundle need not be a product $X \times F$. In algebraic topology, invariants are constructed that express the twisting of a bundle, that is, its deviation from a global product structure.

Such structured bundles then also form corresponding categories. The morphisms have to map the fiber over a point x to the fiber over its image and induce a morphism in the category constituted by the fibers. For instance, in the category of vector bundles, a morphism between two such bundles then induces linear maps, that is, vector space morphisms, between fibers.

After this preparation, we shall now recall the definition of a presheaf and then that of a sheaf. First we shall again work in the general context of a category $\textbf{Sets}^{\textbf{C}^{\text{op}}}$ for some fixed small category $\textbf{C}$. This time, the category $\textbf{C}$ that we have in mind is $\mathcal{O}(X)$, the objects of which are the open subsets U of some topological space X and the morphisms the inclusions $V \subset U$. $\mathcal{O}(X)$ has the structure of a poset, with the ordering relation $\leq$ given by the inclusion $\subset$. This poset has a largest element, X, and a smallest one, $\emptyset$. (In fact, $\mathcal{O}(X)$ is a Heyting algebra as explained above, in Sect. 4.1, and we also have the operations of intersection and union.)

From Sects. 2.4 and 4.5, we now recall the basic

Definition 4.5.1 An element P of $\textbf{Sets}^{\textbf{C}^{\text{op}}}$ is called a *presheaf* on $\textbf{C}$.

For an arrow $f : V \to U$ in $\textbf{C}$, and $x \in PU$, the value $Pf(x)$, where $Pf : PU \to PV$ is the image of f under P, is called the *restriction* of x along f.

Thus, a presheaf expresses that we can restrict collections of objects from an open set to its open subsets. This then allows us to concentrate on the opposite question, that is, when we can extend such collections of objects from open sets to larger open sets containing them. The obstructions to such extensions can be determined in terms of the nonvanishing of certain algebraic invariants that come from cohomology theory as will be developed below in Sect. 4.6.

We now return to the setting of topological spaces, where, as already indicated, the category of interest is $\mathbf{C} = \mathcal{O}(X)$, the category of open subsets of a topological space X. For a presheaf $P : \mathcal{O}(X)^{\mathrm{op}} \to \mathbf{Sets}$, we then have the restriction maps

$$p_{VU} : PV \to PU \text{ for } U \subset V \tag{4.5.3}$$

that satisfy

$$p_{UU} = 1_{PU} \tag{4.5.4}$$

and

$$p_{WU} = p_{VU} \circ p_{WV} \text{ whenever } U \subset V \subset W. \tag{4.5.5}$$

Analogously to Sect. 2.4, we have

Definition 4.5.2 The presheaf $P : \mathcal{O}(X)^{\mathrm{op}} \to \mathbf{Sets}$ is called a *sheaf* if it satisfies the following condition. If $U = \bigcup_{i \in I} U_i$ for some family $(U_i)_{i \in I} \subset \mathcal{O}(X)$ and $\pi_i \in PU_i$ satisfies $p_{U_i, U_i \cap U_j} \pi_i = p_{U_j, U_i \cap U_j} \pi_j$ for all $i, j \in I$, then there exists a unique $\pi \in PU$ with $p_{UU_i} \pi = \pi_i$ for all i.

Thus, the π_i that are compatible in the sense that the restrictions of π_i and π_j to $U_i \cap U_j$ always agree, can be patched together to an element π of PU that restricts to π_i on PU_i.

The sheaves that occur in geometry take their values not in a category of arbitrary sets, but in the category of Abelian groups or category of modules over some commutative ring R with 1. That is, to each open subset U of X, we associate some group GU or some R-module MU. The presheaf condition implies that whenever $V \subset U$, we obtain a homomorphism $g_{VU} : GU \to GV$ (or $m_{VU} : MU \to MV$), with $g_{UU} = 1_{GU}$ and $g_{WU} = g_{WV} \circ g_{VU}$ for $U \subset V \subset W$. The sheaf condition says that whenever we have elements $a_i \in GU_i$ with $g_{U_i, U_i \cap U_j} a_i = g_{U_j, U_i \cap U_j} a_j$ for all i, j, then there exists a unique $a \in GU$ with $g_{UU_i} a = a_i$ for all i. The uniqueness means that when $a_i = 0$ for all i, then also $a = 0$ whenever $g_{UU_i} a = a_i$ for all i.

We also note that functoriality implies

$$P\emptyset = 0, \text{ the trivial group or module,} \tag{4.5.6}$$

for any such presheaf P.

As a simple example, we can take the constant sheaf that assigns to each nonempty U the same group G (and, as always, to the empty set the trivial group 0). Essentially the same type of sheaf arises, when, on a topological space X, we take the locally constant functions with values in some set K. Thus, for each open $U \subset X$, PU is the set of locally constant functions from U to K.

The typical examples of sheaves are given by spaces of functions. For instance, we can take as MU the real-valued continuous functions on the open set U. This is an $\mathbb{R}$-module. Given a continuous function f_i on U_i with $f_{i|U_i \cap U_j} = f_{j|U_i \cap U_j}$, then there exists a unique continuous function f on $U = \bigcup U_i$ with $f_{|U_i} = f_i$ for all i.

| Constant sheaf |

| Continuous functions C^0 |

When X is a complex manifold, we can also assign to an open set U the space of holomorphic functions on U. The fundamental difference between continuous and holomorphic functions rests on the fact that a holomorphic function is globally determined by its value on some open set, or even its power series expansion at some point. In that sense, sheaves of holomorphic functions are similar to sheaves of locally constant functions, rather than to sheaves of continuous functions.

When V is a vector bundle over the topological space X, the sheaf of continuous sections of V is a sheaf of modules over the sheaf of rings of continuous functions on X. (It is straightforward to define what a sheaf of modules over a sheaf of rings is.)

We now wish to construct a topology on a presheaf $P : \mathcal{O}(X)^{\mathrm{op}} \to \mathbf{Sets}$. Let $x \in U, V \in \mathcal{O}(X)$, that is, two open neighborhoods of the point x. We say that $f \in PU, g \in PV$ have the same germ at x if there exists some open $W \subset U \cap V$ with $x \in W$ and with $f_{|W} = g_{|W}$, that is, if they coincide on some neighborhood of x. The important point is that this neighborhood may depend on f and g. The example that one should have in mind here is that of the sheaf of continuous functions. For analytic functions, we could simply take the power series expansion at the point x and take this as the infinitesimal representation of a function at a point x. Since continuous functions are not determined by infinitesimal properties, we rather check whether they agree locally in the vicinity of the point x, in order to identify them when we want to consider some kind of infinitesimal neighborhood of x only. In any case, we have defined an equivalence relation, and the equivalence class of such an f is called the germ of f at x and denoted by $\mathrm{germ}_x f$. The set of all such equivalence classes, that is, the set of all germs at x is denoted by P_x and called the stalk of P at x. For each open neighborhood U of x, we therefore have a function

$$\mathrm{germ}_x : PU \to P_x, \tag{4.5.7}$$

and in fact, we have

$$P_x = \varinjlim_{x \in U} PU \tag{4.5.8}$$

as a colimit. We consider the disjoint union of all stalks,

$$\Lambda_P := \coprod_{x \in X} P_x, \tag{4.5.9}$$

and the projection

$$p : \Lambda_P \to X$$
$$(\phi \in P_x) \mapsto x, \tag{4.5.10}$$

that is, we map each germ_x to its base point x. For $x \in U$ and $f \in PU$, we then have a function

$$\ell(f) : U \to \Lambda_P \tag{4.5.11}$$
$$x \mapsto \mathrm{germ}_x f. \tag{4.5.12}$$

We now take the topology on Λ_P that is generated by the basis consisting of the images $\ell(f)(U)$ for all such $U \in \mathcal{O}(X), f \in PU$. Thus, an open set is a union of such images. For this topology, each $\ell(f)$ is a homeomorphism onto its image, that is, continuous in both directions. For the continuity of $\ell(f)$, we use the fact that the image $\ell(f)(U)$, a basic open set, consists of all germs $\text{germ}_x f$ for $x \in U$. Now an open set containing such a germ is of the form $\ell(g)(V)$ where $g \in PV$ satisfies $\text{germ}_x g = \text{germ}_x f$. By the definition of a germ, there then exists some open $W \subset U \cap V$ with $f_{|W} = g_{|W}$. Thus, for every element in every open set of ΛP, the preimage under $\ell(f)$ contains an open subset of U that is mapped to that open set in ΛP. Thus, $\ell(f)$ is continuous. Since it maps open sets U to open sets $\ell(f)(U)$ and is injective, it is indeed a homeomorphism onto its image.

Also, the projection p from (4.5.10) is a local homeomorphism, as each point in $\text{germ}_x f$ has the open neighborhood $\ell(f)(U)$, and $p \circ \ell(f) = 1_U, \ell(f) \circ p = 1_{\ell(f)(U)}$. To each such presheaf P, we can then assign the sheaf of continuous sections of ΛP. One verifies that if P is already a sheaf, then this assignment is an isomorphism. In particular, every sheaf is such a sheaf of continuous sections of a bundle.

4.6   Cohomology

In this section, we present some generalization of Čech cohomology for presheaves. A reference for cohomology theory is [104].

We recall the basic definition of Sect. 3.1 and start with a set V whose elements are denoted by v or $v_0, v_1, v_2, \ldots$. We assume that finite collections of different elements can stand in a relation $r(v_0, v_1, \ldots, v_q)$. We write $r(v_0, \ldots, v_q) = o$ when this relation is trivial (whatever triviality might mean here). We assume the following properties

(i)
$$r(v) \neq o, \tag{4.6.1}$$

that is, each element stands in a nontrivial relation with itself.

(ii)
$$\text{If } r(v_0, \ldots, v_q) \neq o, \text{ then also } r(v_{i_1}, \ldots, v_{i_p}) \neq o$$
$$\text{for any (different) } i_1, \ldots, i_p \in \{0, \ldots, q\}, \tag{4.6.2}$$

that is, whenever some elements stand in a nontrivial relation, then this also holds for any nonempty subset of them.

When we associate a simplex to every collection of elements with $r(v_0, \ldots, v_q) \neq o$, we obtain a simplicial complex with vertex set V. In fact, we can take the preceding properties as the defining axioms for a simplicial complex.

An example arises for an open covering $\mathcal{U}$ of a topological space $(X, \mathcal{O}(X))$. Here, for $U_0, \ldots, U_q \in \mathcal{U}, r(U_0, \ldots, U_q) \neq o$ iff $U_0 \cap \cdots \cap U_q \neq \emptyset$. In fact, we observe here that each open covering $\mathcal{U}$ of a topological space defines a simplicial complex with vertex set $V = \mathcal{U}$. This simplicial complex is called the nerve of the covering.

We can also consider the relations $r(v_0, \ldots, v_q)$ as the objects of a category $\mathbf{R}$, with a morphism $r(v_{i_1}, \ldots, v_{i_q}) \rightarrow r(v_0, \ldots, v_q)$ whenever $i_1, \ldots, i_p \in \{0, \ldots, q\}$. This category possesses coproducts,[4] as $r(v_0, \ldots, v_q)$ is the coproduct of $r(v_0), \ldots, r(v_q)$. In fact, in the sequel, for $\mathbf{R}$ we can take any category with coproducts and define $v_{0\ldots q}$ as the coproduct of the objects $v_0, \ldots, v_q$, replacing the more cumbersome $r(v_0, \ldots, v_q)$ by $v_{0\ldots q}$.

We now assume that we have a functor G from the category $\mathbf{R}$ to the category $\mathbf{Ab}$ of Abelian groups. We denote the group operation in an Abelian group by $+$ and the inverse by $-$, and, as customary, we write $g_1 - g_2$ for $g_1 + (-g_2)$. The trivial group with a single element is denoted by 0. Thus, for every morphism $r_1 \rightarrow r_2$ in $\mathbf{R}$, we obtain a group homomorphism $G(r_1 r_2) : G(r_1) \rightarrow G(r_2)$ with

$$G(rr) = 1_{G(r)} \tag{4.6.3}$$

and

$$G(r_1 r_3) = G(r_2 r_3) G(r_1 r_2) \text{ for } r_1 \rightarrow r_2 \rightarrow r_3. \tag{4.6.4}$$

We also have

$$G(o) = 0 \tag{4.6.5}$$

since o is the terminal object (in the sense of Definition 8.2.1 below) in $\mathbf{R}$ (unless all elements of V together stand in a nontrivial relation (in which case the above simplicial complex consists of a single simplex)), and 0 is the terminal object in $\mathbf{Ab}$.

Definition 4.6.1 For $q = 0, 1, \ldots$, we let $C^q(\mathbf{R}, G)$ be the group of mappings γ that associates to each ordered $(q + 1)$-tuple $(v_0, v_1, \ldots, v_q)$ of elements of V an element $\gamma(v_0, \ldots, v_q) \in G(r(v_0, \ldots, v_q))$. For subsequent convenience, we put $C^q = 0$ whenever $q \in \mathbb{Z}$ is negative or larger than the number of elements of V. We define the *coboundary operator*

$$\delta = \delta^q : C^q(\mathbf{R}, G) \rightarrow C^{q+1}(\mathbf{R}, G)$$

$$(\delta\gamma)(v_0, \ldots v_{q+1}) := \sum_{i=0}^{q+1} (-1)^i \gamma(v_0, \ldots, \widehat{v_i}, \ldots, v_{q+1}) \tag{4.6.6}$$

where $\widehat{v_i}$ means omission of the argument v_i.

We easily check

Lemma 4.6.1

$$\delta^q \circ \delta^{q-1} = 0 \tag{4.6.7}$$

for all q, or leaving out the superscript q,

$$\delta \circ \delta = 0. \tag{4.6.8}$$

Lemma 4.6.1 implies

---

[4]See (8.2.47) for the general definition of a coproduct in a category.

Corollary 4.6.1

$$\mathrm{im}\,\delta^{q-1} \subset \mathrm{ker}\,\delta^q, \tag{4.6.9}$$

that is, the image of δ^{q-1} is contained in the kernel of δ^q.

We therefore define

Definition 4.6.2 The qth *cohomology group* of the category $\mathbf{R}$ with values in G is the quotient group

$$H^q(\mathbf{R}, G) := \mathrm{ker}\,\delta^q / \mathrm{im}\,\delta^{q-1}. \tag{4.6.10}$$

Here, it is important that we are working with Abelian groups, because that implies that the subgroup $\mathrm{im}\,\delta^{q-1}$ of $\mathrm{ker}\,\delta^q$ is automatically a normal subgroup so that the quotient group is well defined, see Lemma 2.1.22.

Thus, cohomology groups identify those elements of C^q whose coboundary vanishes up to what is considered trivial, that is, what is in the image of the coboundary operator from C^{q-1}, which we know to be in the kernel anyway from (4.6.7). Thus, the cohomology group H^q reflects what is new in dimension or degree q in the sense that it is not derived from what is there in degree $q - 1$. For instance when we have three sets with a nonempty intersection $U_1 \cap U_2 \cap U_3 \neq \emptyset$, then of course each pairwise intersection between them also has to be nonzero. However, when we have $U_1 \cap U_2 \neq \emptyset$, $U_2 \cap U_3 \neq \emptyset$, $U_1 \cap U_3 \neq \emptyset$, but with $U_1 \cap U_2 \cap U_3 = \emptyset$, then there is something nontrivial that is reflected by a nontrivial cohomology class.

The cohomology theory developed so far applies to any collection $A_1, \ldots, A_n$ of subsets of some set S, and it reflects the intersection patterns between those sets. In particular, when X is a topological space, we can apply it to any collection $\mathcal{U}$ of open subsets $U_1, \ldots, U_n$. When we are interested in constructing invariants of X itself, the question arises to what extent this depends on the choice of those open subsets. We shall now describe the corresponding result. Another collection $\mathcal{U}'$ of open sets $U_1', \ldots, U_m'$ is called a refinement of $\mathcal{U}$ if for every μ there exists some $\nu(\mu)$ with $U_\mu' \subset U_{\nu(\mu)}$. In this situation, we obtain an induced map

$$\rho_\nu : C^q(\mathcal{U}, G) \to C^q(\mathcal{U}', G) \tag{4.6.11}$$

by restricting maps to G from the $U_{\nu(\mu)}$ or from intersections of such sets to the U_μ and their intersections. We have

$$\delta \circ \rho = \rho \circ \delta \tag{4.6.12}$$

for the coboundary operator. Therefore, we obtain an induced homomorphism of cohomology groups

$$\rho : H^q(\mathcal{U}, G) \to H^q(\mathcal{U}', G) \tag{4.6.13}$$

When we choose another refinement function λ with $U_\mu' \subset U_{\lambda(\mu)}$ for all μ, then we check that this leads to same ρ in (4.6.13). Thus, the homomorphism of cohomology groups does not depend on the choice of the refinement map ν.

We then define the cohomology groups of X as the colimits (see Sect. 8.2) of those of its open coverings when these coverings get finer and finer,

$$H^q(X, G) := \lim_{\overrightarrow{\mathcal{U}}} H^q(\mathcal{U}, G). \tag{4.6.14}$$

In practice, the computation of these cohomology groups uses the following theorem of Leray that tells us that we do not need to pass to arbitrary refinements, but simply need to check that the intersections in the covering are cohomologically trivial.

Theorem 4.6.1 *Whenever the covering $\mathcal{U}$ of the topological space $X, \mathcal{O}$ is acyclic in the sense that*

$$H^q(U_{i_1} \cap \ldots U_{i_p}, G) = 0 \text{ for all } q > 0, i_1, \ldots, i_p, \tag{4.6.15}$$

then

$$H^q(X, G) = H^q(\mathcal{U}, G) \text{ for all } q. \tag{4.6.16}$$

Thus, when there is no local cohomology in the covering $\mathcal{U}$, it captures all the cohomology of X. In this sense, cohomology expresses global qualitative properties of a topological space and a presheaf of Abelian groups on it. This is fundamental for algebraic topology. In some applications sketched below, however, we are interested in the cohomology of specific coverings that need not be acyclic in the sense of (4.6.15).

4.7 Spectra

We now assume that we have some nonnegative Radon measure μ on the measurable topological space $(X, \mathcal{B}, \mathcal{O})$. We assume that we have a collection $\phi_0\mu, \phi_1\mu, \ldots$ of Radon measures compatible with μ. We take these $\nu_i := \phi_i\mu$ as the ν_i of our vertex set V, and put, for any subcollection $\nu_{i_1}, \ldots, \nu_{i_p}$,

$$(\nu_{i_1}, \ldots, \nu_{i_p}) := \int \prod_{j=1}^p (\phi_{i_j})^{1/p} \mu := \left(\prod_{j=1}^p (\phi_{i_j})^{1/p} \mu \right)(X) \in \mathbb{R}. \tag{4.7.1}$$

In particular, this allows us to define relations

$$r(\nu_{i_1}, \ldots, \nu_{i_p}) := \begin{cases} 1 & \text{if } (\nu_{i_1}, \ldots, \nu_{i_p}) \neq 0 \\ o & \text{else.} \end{cases} \tag{4.7.2}$$

In this way, we obtain a weighted simplicial complex. The simplices are given by those subcollections $\nu_{i_1}, \ldots, \nu_{i_p}$ with $r(\nu_{i_1}, \ldots, \nu_{i_p}) = 1$, and to each such simplex, we associate the weight $(\nu_{i_1}, \ldots, \nu_{i_p})$. The simplicial structure expresses the topology of the overlap patterns, and the weights reflect the geometry, that is, the amounts of overlap between subcollections of our Radon measures.

We start with the topological aspects and then refine them by geometric aspects. As in Sect. 4.6, we thus obtain a category $\mathbf{R}$, and we now consider an $\mathbb{R}$-module functor M, that is, we associate to each object r of $\mathbf{R}$ a module $M(r)$ over the real numbers, i.e., a real vector space, to get the groups $C^q(\mathbf{R}, M)$, the coboundary operators δ^q and the cohomology groups $H^q(\mathbf{R}, M)$.

We also assume that each $M(r)$ carries a scalar product $\langle ., .\rangle$. This product is assumed to be positive, that is, $\langle v, v\rangle > 0$ for $v \neq 0$. In particular, $\langle v, v\rangle = 0$ implies $v = 0$, that is, the product is nondegenerate.

In the simple case when $M(r) = \mathbb{R}$ for $r \neq o$, we can simply take the product in $\mathbb{R}$. When $M(r)$ is a space of (integrable) functions, we may put

$$\langle \gamma_1(\nu_0, \ldots, \nu_q), \gamma_2(\nu_0, \ldots, \nu_q)\rangle := \int \gamma_1(\nu_0, \ldots, \nu_q)\gamma_2(\nu_0, \ldots, \nu_q) \prod_{i=0}^{q} (\phi_i)^{1/(q+1)} \mu,$$
(4.7.3)

that is, use the measures $\nu_0, \ldots, \nu_q$ to integrate functions.

Given two elements of $C^q(\mathbf{R}, M)$, that is, mappings γ_1, γ_2 that associate to each ordered $(q + 1)$-tuple $(\nu_0, \nu_1, \ldots, \nu_q)$ elements $\gamma_\alpha(\nu_0, \ldots, \nu_q) \in M(r(\nu_0, \ldots, \nu_q))$ $(\alpha = 1, 2)$, we can define their product

$$(\gamma_1, \gamma_2)_q := \sum_{\nu_0,\ldots,\nu_q} \langle \gamma_1(\nu_0, \ldots, \nu_q), \gamma_2(\nu_0, \ldots, \nu_q)\rangle \int \prod_{i=0}^{q} (\phi_i)^{1/(q+1)} \mu.$$
(4.7.4)

With these products, we can define adjoints $\delta^{*q} : C^q(\mathbf{R}, M) \to C^{q-1}(\mathbf{R}, M)$ of the coboundary operators by

$$(\delta^{*q}\gamma, \eta)_{q-1} = (\gamma, \delta^{q-1}\eta)_q \text{ for } \gamma \in C^q(\mathbf{R}, M), \eta \in C^{q-1}(\mathbf{R}, M).$$
(4.7.5)

Because of (4.6.7), we also have

$$\delta^{*q-1} \circ \delta^{*q} = 0.$$
(4.7.6)

We then obtain the generalized Laplacian

$$\Delta^q := \delta^{q-1} \circ \delta^{*q} + \delta^{*q+1} \circ \delta^q : C^q(\mathbf{R}, M) \to C^q(\mathbf{R}, M).$$
(4.7.7)

From the definition, we see that the Δ^q are selfadjoint in the sense that

$$(\Delta^q\gamma_1, \gamma_2) = (\gamma_1, \Delta^q\gamma_2) \text{ for all } \gamma_1, \gamma_2.$$
(4.7.8)

Therefore, all eigenvalues of Δ^q are real. Here, λ is an eigenvalue of Δ^q if there exists some $g \neq 0$ with

$$\Delta^q g = \lambda g.$$
(4.7.9)

Such a g is then called an eigenfunction for the eigenvalue λ. Also, the collection of the eigenvalues of an operator is called its spectrum.

Since

$$(\Delta^q g, g) = (\delta^q g, \delta^q g) + (\delta^{*q} g, \delta^{*q} g)$$
(4.7.10)

we see that

$$\Delta^q g = 0 \text{ iff } \delta^q g = 0 \text{ and } \delta^{*q} g = 0.$$
(4.7.11)

Thus, the eigenfunctions for the eigenvalue 0 correspond to cohomology classes, and the multiplicity of the eigenvalue 0 is therefore a topological invariant. In fact, one can show that each cohomology class, that is, each element of $H^q(\mathbf{R}, M)$ is represented by a unique eigenfunction of Δ^q. Here, the nondegeneracy of the scalar product of course plays a crucial role.

We can utilize these constructions to associate numerical invariants, the eigenvalues of the Laplacians Δ^q, to collections of densities of objects. For instance, when we have different proteins in a cell, the distribution of each protein i yields a density or Radon measure ν_i, and we can then try to characterize the joint distribution of different proteins,[5] their colocalization, by the topological invariants given by the cohomology groups $C^q(\mathbf{R}, \mathbb{Z})$ or $C^q(\mathbf{R}, \mathbb{R})$ or for some other ring, and by the geometric invariants given by the spectra of the Laplacians Δ^q. Those topological invariants encode the qualitative properties of the overlap, that is, which collections of proteins occur together, and which ones don't, whereas the geometric invariants encode also the quantitative aspects as reflected by the amount of overlap.

Of course, the foregoing can also be applied to many other—biological or non-biological—examples. For instance, we can study the distributions of members of different species in an ecosystem.

Returning to the formalism, what we have done here amounts to the following. To a collection of compatible Radon measures on a topological space, we have associated a weighted simplicial complex. The simplices correspond to the intersections of support sets of the measures, and the weights are then given by the induced measures on those intersections. We have then utilized the weights to define scalar products of cochains. With the help of those products, we have then defined adjoints δ^* of the coboundary operators δ, and Laplace operators Δ. The spectra of those Laplace operators then are supposed to yield geometric invariants of the collection of Radon measures, in addition to the topological invariants obtained from the Čech cohomology of the intersection patterns of the support sets.

In fact, for the analysis of the spectra, it might be simpler to consider the operators

$$\Delta_d^q := \delta^{q-1} \circ \delta^{*q} \text{ and } \Delta_u^q := \delta^{*q+1} \circ \delta^q \qquad (4.7.12)$$

(with d for "down" and u for "up") separately. Of course, these operators satisfy

$$\Delta^q = \Delta_d^q + \Delta_u^q. \qquad (4.7.13)$$

The reason is the following general fact about eigenvalues of products of operators A, B. Whenever v is an eigenfunction of AB with eigenvalue $\lambda \neq 0$, that is,

$$ABv = \lambda v \neq 0, \qquad (4.7.14)$$

then $Bv \neq 0$ is an eigenfunction of BA with the same eigenvalue, as

$$BA(Bv) = \lambda Bv. \qquad (4.7.15)$$

Therefore, Δ_d^q and Δ_u^{q-1} have the same nonvanishing eigenvalues. Thus, their spectra differ at most by the multiplicity of the eigenvalue 0.

Also, if v is an eigenfunction of Δ_d^q with eigenvalue $\lambda \neq 0$, that is, $\delta^{q-1} \circ \delta^{*q} v = \lambda v$, then

$$\delta^{*q+1} \circ \delta^q \lambda v = \delta^{*q+1} \circ \delta^q \circ \delta^{q-1} \circ \delta^{*q} v = 0$$

---

[5]For a technology that can generate such protein colocalization data, see [98].

by (4.6.7), hence $\Delta_u^q v = 0$. Similarly, by (4.7.6), when w is an eigenfunction of Δ_u^q with eigenvalue $\lambda \neq 0$, then $\Delta_d^q w = 0$. Therefore, the spectrum of Δ^q is the union of the nonvanishing parts of the spectra of Δ_d^q and Δ_u^q and the eigenvalue 0 with its multiplicity. Combined with the above fact that Δ_d^q and Δ_u^{q-1} have the same nonvanishing eigenvalues, we see that for the spectral analysis, the appropriate operators to investigate are Δ_d^q and Δ_u^q, and there exist relations between the spectra for the different possible values of q. More precisely, for each q, some part of the nonzero spectrum is shared with the spectrum at $q-1$, and the other part is shared with $q+1$. See [52].

What Is Space?

5

5.1 A Conceptual Deconstruction and a Historical Perspective

In order to approach this question, we might first ask "What is the space that we are living in?". And a first answer might be "Three-dimensional Euclidean space". But what is that three-dimensional Euclidean space, and how can it be described? Perhaps we think that we know the answer: Each point x in three-dimensional Euclidean space is described uniquely by its Cartesian coordinates x^1, x^2, x^3, that is, by a triple of real numbers. That is, we suppose that we have three mutually orthogonal coordinate axes that emanate from a common origin "0", and that the point x is characterized by the lengths x^1, x^2, x^3 of its projections onto those three coordinate axes. Furthermore, we have a Euclidean scalar product between points x and y that can be computed from their coordinate representations

$$\langle x, y \rangle := x^1 y^1 + x^2 y^2 + x^3 y^3. \tag{5.1.1}$$

From this Euclidean scalar product, we obtain the norm

$$\|x\| := \sqrt{\langle x, x \rangle} \tag{5.1.2}$$

of x and the angle α

$$\langle x, y \rangle =: \cos \alpha \, \|x\| \|y\| \tag{5.1.3}$$

between x and y. In particular, when $\langle x, y \rangle = 0$, we say that x and y are orthogonal to each other.

Furthermore, we have the structure of a vector space, that is, we can add two points

$$x + y := (x^1 + y^1, x^2 + y^2, x^3 + y^3) \tag{5.1.4}$$

and multiply points by real numbers α,

$$\alpha x := (\alpha x^1, \alpha x^2, \alpha x^3). \tag{5.1.5}$$

In other words, algebraic operations on coordinates induce the corresponding operations on the points of Euclidean space.

Euclidean space also supports vector calculus, that is, processes of differentiation and integration. In particular, we can compute derivatives (tangent

Euclidean 3-space

© Springer International Publishing Switzerland 2015
J. Jost, *Mathematical Concepts*, DOI 10.1007/978-3-319-20436-9_5

vectors) of smooth curves, and measure the lengths of curves, the areas of curved surfaces, and the volumes of bodies.

Thus, we have a lot of structure. Perhaps too much for the beginning. In particular, some part of the structures are simply arbitrary conventions. For instance, what determines the position of the origin 0 of the Cartesian structure in physical space? Clearly, this is an arbitrary convention. The Euclidean structure also depends on the assumption or convention that the coordinate axes are orthogonal to each other, and the lengths of vectors depend on the choices of units on the coordinate axes. Where do they come from?

The Cartesian representation of Euclidean space may also obscure some of its geometric properties. We think that Euclidean space is homogeneous, that is, its geometric properties are the same at every point, and isotropic, that is, its geometric properties are the same in every direction. However, the Cartesian structure distinguishes one particular point, the origin 0, as well as the three directions of the coordinate axes.

Thus, in particular, we have seen that our naive conception of space confounds two structures, the Cartesian or vector space structure with its coordinate representations and the Euclidean metric structure. And it remains unclear what either of them has to do with physical space.

So, perhaps we should start differently and ask "What are the essential or characteristic properties of space?". Many answers to this question have been proposed in the history of philosophy, mathematics, and physics, and it might be insightful to examine some of them.

To that end, we shall now give a caricature of the history of concepts of space, with little regard for historical accuracy, but rather for the purpose of identifying some important conceptual aspects.

Aristotle (384–322) who lived before Euclid (ca.325–265) had a rather different concept of space than Euclid. He argued that since for instance a stone falls to the ground, the location of the stone in the air is different from its location on the ground. In other words, there are differences between "up" and "down", and space is neither homogeneous nor isotropic. For Aristotle, space was the collection of "natural" locations of physical objects. Aristotle thought of a categorical difference between matter and movement, and therefore between rest and motion. The natural state of matter was rest, and it moved only in order to attain its natural position of rest. This became a conceptual obstacle that physicists needed a long time to overcome.

In contrast to this, the painters of the Renaissance, led by Filippo Brunelleschi (1377–1446), saw space as the medium in which light rays propagate according to the laws of Euclidean geometry and developed the constructions of geometrical perspective for their two-dimensional representations. It is somewhat remarkable, and perhaps ironic, that they postulated an objective spatial reality in order to create subjective illusions, that is, a perception of spatial depth in planar paintings.

Galileo Galilei (1564–1642) considered space as the idealized medium for the kinetics and dynamics of physical bodies and emphasized the invariance properties of the laws of physics that depend on the homogeneity and isotropy of (Euclidean) space. This was a key step for eventually overcoming the Aristotelian distinction between movement and rest. For Isaac

Newton (1643–1717), space was an invariant container in which bodies could exercise forces upon each other which then led to their physical dynamics. For that purpose, Newton conceived space as absolute, and he considered centrifugal forces as evidence of movement against such an absolute reference frame. Notwithstanding his fundamental concept of force that overcame the Cartesian misconception of identifying matter and extension and that turned out to be basic for the subsequent progress of physics, his notion of absolute space constituted a step backwards from the insights of Galilei. In contrast, for Gottfried Wilhelm Leibniz (1646–1716), space simply encoded relations between objects, instead of existing independently of the objects contained in it. This idea both anticipated some aspect of relativistic physics and also contributed to more abstract notions of space of modern mathematics. There, we speak of a space when we wish to express any kind of qualitative or quantitative relation between objects that need not be of a physical nature. For instance, a graph codifies binary relations between discrete objects. But in fact, even for the description of physical systems, it is often useful to employ a more abstract notion of space. For instance, the state space of a collection of N particles in three-dimensional Euclidean space represents the positions of these particles by $3N$ coordinates, and their phase space employs $6N$ coordinates for their positions and momenta or velocities. More generally, state or phase spaces for dynamical systems describe their possible degrees of freedom. Actually, before Newton and Leibniz, Descartes (1596–1650), who had translated his name as Cartesius into Latin, had already introduced coordinates for the geometric description and analytic treatment of algebraic relations, such as $z = w^2$ or $z^2 = w^3$. This leads to two-dimensional Cartesian space, and we can naturally extend this to construct N-dimensional Cartesian space. When we base this on the concept of real numbers $\mathbb{R}$ (which was only clarified in the 19th century by Richard Dedekind (1831–1916) and others, much later than Descartes), then N-dimensional Cartesian space is $\mathbb{R}^N$. Furthermore, the concept of a vector space, which underlies this construction, was only introduced later by Hermann Grassmann (1809–1877). The vector space $\mathbb{R}^N$ is not yet equipped with the Euclidean metric structure.

$\boxed{\mathbb{R}}$

An entirely new approach was introduced by Bernhard Riemann (1826–1866) [96]. In modern terminology, he conceived of space as a differentiable manifold in which one could perform (infinitesimal) measurements of lengths and angles, that is, what one calls a Riemannian manifold in his honor. Riemann was the key person behind the conceptual analysis and clarification of the notion of space in mathematical terms. In particular, Riemann was the first to clearly distinguish the topological and the metric structure of space.

Partly building upon Riemann and partly criticizing him, the physicist Hermann von Helmholtz (1821–1894) argued that the characteristic empirical property of physical space is that one can freely move bodies around in it without deforming them. In modern terminology, this led him to the conclusion that physical space must be a three-dimensional Riemannian manifold of constant curvature, that is, either a Euclidean, spherical, or hyperbolic space form. Sophus Lie (1842–1899), however, identified some gaps in the mathematical reasoning of Helmholtz and developed the abstract concept of geometric invariance transformation. The Lie groups yielded a fundamental concept for the subsequent development of the quantum theory of physics, and Lie groups and Riemannian geometry constitute the basic language of modern quantum field theory.

For Albert Einstein (1879–1955), however, in contrast to Helmholtz, bodies were not moving invariantly in an invariant space, but rather the properties of space and the movements of bodies were intertwined; again, the concept of a Riemannian manifold turned out to be the appropriate notion for the physics of general relativity.

The work of Riemann also inspired the axiomatic approach to geometry, that is, the program of characterizing space, or classes of possible spaces, in terms of abstract properties. Important contributions came from Georg Cantor (1845–1918) and David Hilbert (1862–1943). On the basis of Cantor's set theory, the notion of a topological space could be developed, and Hilbert started a systematic axiomatic program for geometry (and other areas of mathematics). For a more detailed treatment of the conceptual history of geometry, I refer to my commentary in [96].

In algebraic geometry, one is interested in the solution sets of algebraic equations or relations. Again, Riemann, by conceiving Riemann surfaces as geometric objects expressing the multivalued character of such solutions, had an important impact, and eventually, the concept of an algebraic variety emerged. In contrast to a manifold which has the same local structure throughout, an algebraic variety may possess singularities, that is, points or subsets with more complicated local surroundings than others. This makes coordinate descriptions less convenient, and the alternative of characterizing points as zero sets of suitable collections of local functions shaped the development of modern algebraic geometry. In the hands of Alexander Grothendieck (1928–2014), this led to the general notion of a scheme that unifies algebraic geometry and arithmetic.

On the basis of this rather cursory narrative of the history of geometry, we can now identify several feasible approaches to the notion and description of spaces.

1. *Manifold*: Description by coordinate charts, that is, by locally relating to a model space
2. *Scheme*: Description through local functions, with points corresponding to maximal ideals in function spaces
3. Description of a given space through homomorphisms, that is, structure preserving maps, from variable other spaces: $\mathrm{Hom}(-, S)$
4. Characterization in terms of invariants, as in cohomology theory
5. Assembling a space from local pieces (simplices, cells) so that the properties of the space are reduced to the combinatorics of this assembly pattern
6. Characterization of a space in terms of the group of its structure preserving transformations. In the basic situation, those structure preserving transformations (homeomorphisms, diffeomorphisms, isometries, . . .) operate transitively on the space. More complicated spaces are then assembled from or decomposed into such homogeneous pieces or strata. Perhaps the simplest instance is a manifold with boundary.

7. *Metric space*: Defining a space through distance relations between its elements

Each of these conceptualizations contains a drive towards generalization and abstraction. For instance, when we describe a space by its algebra of continuous functions, then we can simply turn things around and consider every algebra as defining some space. For instance, while the algebra of continuous functions is commutative—when $f, g : M \to \mathbb{R}$ are functions on some space M, then $f(x)g(x) = g(x)f(x)$ for every $x \in M$—, one can then also consider noncommutative algebras (perhaps with suitable additional properties) and declare them to define a noncommutative space, see [24]. We shall see the power of such an approach in Sect. 5.4, albeit there we shall still work within the commutative setting.

5.2 Axioms of Space

On the basis of the mathematical structures developed in Chap. 4, in particular in Sect. 4.1, we shall now develop a more abstract and axiomatic theory of space.

We now define a *basic* space through a series of axioms and then add an axiom that will allow us to introduce additional structures. In short, in a space we shall be able to identify points, but not will not be able to distinguish between them.

Let us try to start with the following property.

Axiom 1 *A space consists of points.*

Axiom 1 is motivated by set theory. It is, however, problematic in the sense that it concerns an ontological aspect, as opposed to an operational one. We therefore try to replace this axiom by the following construction.

The space under consideration will be denoted by X. When we think of X as consisting of points, then X is a set, and we can then also consider subsets of X. The subsets of X constitute the Boolean algebra $\mathcal{P}(X)$ with the operations of complement, union, intersection, and the inclusion relation, as described in Sect. 4.1. As pointed out there, we can also perform pointless topology, that is, consider $\mathcal{P}(X)$ as an abstract formal category, without having to take recourse to any points. We now develop this aspect in the way that we start from that category $\mathcal{P}(X)$ and construct the points from appropriate systems of members of $\mathcal{P}(X)$. That is, the points of X, instead of being fundamental or constitutive for X, become derived objects.

Within the framework of category theory, there exists a natural approach to this issue, using the (ultra)filters introduced in Sect. 2.1.5 in the proof of Theorems 2.1.2, see (2.1.98)–(2.1.101). Let us briefly recall those definitions here. Let X be a set. A filter $\mathcal{F}$ on $\mathcal{P}(X)$ is then a subset of $\mathcal{P}(X)$ satisfying

$$\emptyset \notin \mathcal{F}, \qquad X \in \mathcal{F}, \qquad (5.2.1)$$

$$\text{if } U \in \mathcal{F}, \ U \subset U', \qquad \text{then also } U' \in \mathcal{F}, \qquad (5.2.2)$$

$$\text{if } U_1, \ldots, U_n \in \mathcal{F}, \text{ then also } U_1 \cap \cdots \cap U_n \in \mathcal{F}. \qquad (5.2.3)$$

Heuristically, one could say that a filter specifies a class of sets that share some property. For instance, if $A \subset X$, then $\mathcal{F}(A) := \{V \subset X : A \subset V\}$, the collection of all subsets of X containing A, is a filter. Thus, if $x \in X$ is a point, it would be captured by the filter $\mathcal{F}(\{x\})$. This filter then enjoys an additional property: it is what we have called an ultrafilter or a maximal filter, that is, whenever for some filter $\mathcal{G}$

$$\mathcal{F} \subset \mathcal{G}, \text{ then } \mathcal{F} = \mathcal{G}. \qquad (5.2.4)$$

Equivalently, $\mathcal{F}$ is an ultrafilter iff

$$\text{for all } U \in \mathcal{P}(X), \text{ either } U \in \mathcal{F} \text{ or } X \backslash U \in \mathcal{F}. \qquad (5.2.5)$$

In particular, for the ultrafilter $\mathcal{F}(\{x\})$, $S \in \mathcal{F}(\{x\})$ iff $x \in S$. Conversely, we could then use ultrafilters to define or identify the points x of X. Here, two different situations could arise. We could either have

$$\bigcap_{U \in \mathcal{F}} U \neq \emptyset, \qquad (5.2.6)$$

in which case we might consider this intersection as a point of X, or

$$\bigcap_{U \in \mathcal{F}} U = \emptyset, \qquad (5.2.7)$$

in which case there would be no point in X corresponding to $\mathcal{F}$. In the latter case, we might wish to add an "ideal" point to X. For instance, when X is the open interval $(0, 1) = \{t \in \mathbb{R} : 0 < t < 1\}$, the filter consisting of all subsets of $(0, 1)$ that contain an open interval $(0, \epsilon)$ for some $\epsilon > 0$ would have an empty intersection, but would naturally define the point 0 that we might then add to $(0, 1)$ to close it off at its left side. Similarly, we could write down a filter defining the point 1. Here, however, we do not explore the important issue of completing a space.

So far, the concept of a filter has been developed for a *set*, using its power set. When we have a *topological space* X, we might wish to replace the power set $\mathcal{P}(X)$ by the set $\mathcal{O}(X)$ of its open subsets and carry out the analogous construction there. In fact, the example of the open interval $(0, 1)$ just discussed finds its natural place in the context of topological spaces. And the issue of completion briefly alluded to in that example is meaningful only for topological spaces, but not for general sets.

Moreover, filters or ultrafilter do not behave nicely under mappings, reflecting the fact that maps need neither be injective, i.e., could map several points to the same one, nor surjective, i.e., could leave out some points in the target. We shall address this issue from a somewhat different perspective in Sect. 5.4 below.

We now step somewhat outside of category theory and try to investigate what we really need from an abstract perspective to define or identify points. For that purpose, we assume that we have $\mathcal{Q}(X) \subset \mathcal{P}(X)$ satisfying:

(i)

$$\emptyset, X \in \mathcal{Q} \qquad (5.2.8)$$

(ii) Whenever $(Q_i)_{i \in I} \subset Q$ for some index set I, then also

$$\bigcup_{i \in I} Q_i \in Q \tag{5.2.9}$$

Thus, the requirements for Q are somewhat weaker than those imposed upon the collection of open sets $\mathcal{O}$ in Sect. 4.1, but for most examples, one may simply take that collection $\mathcal{O}$ for Q.

A *focus* $\mathcal{U}$ is then a subset of $Q(X)$ with the property that for all $U, V \in \mathcal{U}$, there exists some $W \in \mathcal{U}$ with

$$W \subset U \cap V \tag{5.2.10}$$

and

$$\bigcap_{U \in \mathcal{U}} U \neq \emptyset. \tag{5.2.11}$$

We note that in (5.2.10), we do not require that the intersection of two members of a focus be in the focus. In fact, that intersection need not even be a member of Q. We might, however, take the union of all members of Q that are contained in that intersection. That union is again in Q, and in fact, it is the largest member with that property. We do not require, however, that it also be a member of the focus $\mathcal{U}$.

The focus $\mathcal{V}$ is a refinement of $\mathcal{U}$ if for every $V \in \mathcal{V}$, there exists some $U \in \mathcal{U}$ with

$$V \subset U. \tag{5.2.12}$$

A focus $\mathcal{U}$ is called *pointlike* if

$$\bigcap_{U \in \mathcal{U}} U \subset \bigcap_{V \in \mathcal{V}} V \text{ for all refinements } \mathcal{V}, \tag{5.2.13}$$

that is, one cannot obtain a smaller asymptotic intersection by making the elements U of the focus smaller. We may then say that *the pointlike focus $\mathcal{U}$ defines the point* $\bigcap_{U \in \mathcal{U}} U$. In other words, in place of points, we can consider pointlike foci. The notion of a point is then no longer a primitive notion, but rather becomes a derived one, in fact derived by the asymptotics of the operation of taking smaller and smaller members of foci.

We note that when $Q(X)$ is a collection $\mathcal{O}(X)$ of open sets as in the definition of a topological space, see Sect. 4.1, then a focus becomes a net in the terminology of topology. We may think in terms of such examples, but should keep in mind that, so far, we have not requested that Q be closed under finite intersections. We have only required that the intersection of two members of a focus contains another member of that focus.

An easy example is $\mathbb{R}$ with its standard topology. $x \in \mathbb{R}$ is then obtained from the net $(x - \frac{1}{n}, x + \frac{1}{n})$ for $n \in \mathbb{N}$. As another example, let us consider the cofinite topology on $\mathbb{R}$, introduced in Sect. 4.1, that is, the topology whose open sets are the complements of finitely many points. For $x \in \mathbb{R}$, we consider the focus consisting of all open sets of the form $\mathbb{R} \setminus \{x_1, \ldots, x_n\}$, where $x \neq x_1, \ldots, x_n$. These open sets then constitute a focus whose intersection is the point x.

Axiom 1 can then be replaced by

Axiom 2 *The points can be identified by foci. This identification should satisfy the Hausdorff property.*

By the Hausdorff property, we mean that for any two foci $\mathcal{U}_1, \mathcal{U}_2$ identifying different points, there exist disjoint members $U_i \in \mathcal{U}_i$, that is,

$$U_1 \cap U_2 = \emptyset. \tag{5.2.14}$$

Axiom 3 *The space is coherent in the sense that when Q is an element of $\mathcal{Q}(X)$, different from $\emptyset$, X, then its complement $X \setminus Q$ is not contained in $\mathcal{Q}$.*

In particular, when a focus identifies a point, we cannot use the complements of the members of the focus to identify another point. We can also interpret the preceding as an *exclusion principle*, by saying that $Q \in \mathcal{Q}$ *includes* the point $x \in Q$, and that it *excludes* the points y in its complement $X \setminus Q$. Axiom 3 then means that the inclusion Q of a point x, that is, $Q \in \mathcal{Q}$ with $x \in Q$, cannot simultaneously define the inclusion of another point y excluded by Q. In the context and terminology of set theoretic topology, that simply means that the set $X \setminus U$ cannot be open at the same time as U when both of them are nonempty. In other words, the space X is assumed to be connected as a topological space.

The Axioms 2, 3 exclude many types of topological spaces. For instance, the indiscrete topology where $\mathcal{O} = \{X, \emptyset\}$ does not satisfy the Hausdorff property whereas the discrete topology where $\mathcal{O} = \mathcal{P}$ violates Axiom 3. In that sense, Axiom 2 stipulates that $\mathcal{Q}$ has to be rich enough, but Axiom 3, in contrast, requires that it not be too rich.

So far, we have discussed the identification of points. We now come to another, in a certain sense, opposite aspect, their distinction. In a set without additional structure, all elements are equivalent in the sense that each element can be transformed into each other one by a permutation of the set, that is, by an automorphism of the set structure. Expressed differently, its transformation group operates transitively on the set. Here, we are employing a general concept from category theory. When A is a member of a category, an invertible morphism of A, that is, a map from A to itself that is structure preserving in both directions, is called a transformation of A. Since, being morphisms, transformations can be composed, and since by definition, they are invertible, they form a group, the transformation group of A.

Of course, on a set without a topological or some other structure, we cannot identify points. However, even when we can identify points as required in the preceding axioms, we may still be able to transform points into each other in a manner that preserves the structure. In fact, this is what we are going to reqire now for a *basic space*.

Axiom 4 *The points cannot be distinguished from each other. They are equivalent under transformations of the space.*

Informally stated, a basic space looks the same at each of its points, so that we can transform one into the other without constraints. Thus, a

| Indiscrete topology |
| Discrete topology |

basic space X has to possess a *transitive group of transformations*. When this requirement is not satisfied, we might speak of a *composite space*. Examples are manifolds with boundary (composed of the interior and the boundary), stratified spaces (composed of several basic strata), or algebraic varieties with singularities (which are examples of stratified spaces, the largest stratum consisting of the non-singular points).

We might also take the transformation group T as the primary object in place of the space. From that perspective, Axiom 4 says that the space X is a homogeneous space for T. In particular, it yields a representation of T. This aspect is fundamental in elementary particle physics where its symmetries, that is, a Lie group T, characterize a particle, and its manifestation in experimental data arises from a representation of T.

Axiom 4 provides an operationalization of the *equivalence principle*. When we have a structure $\mathcal{Q} \subset \mathcal{P}$ as above, we may consider transformations as invertible morphisms of $\mathcal{Q}$. That is, they have to preserve the operations of union and intersection and the relation of inclusion. Thus, foci are transformed into foci, and the Hausdorff property ensures that points correspond to points. For a topological space $(X, \mathcal{O})$, we have described this in Sect. 4.1. We recall that, in the terminology of points for simplicity, a *homeomorphism* of $(X, \mathcal{O})$ is a bijective map $h : X \to X$ with the property that the image and the preimage of every open set is open. That is, a homeomorphism preserves the topological structure. Homeomorphisms can be inverted and composed, that is, if h, g are homeomorphisms, then h^{-1} and $h \circ g$ are homeomorphisms as well. The homeomorphisms therefore constitute a group of transformations of the topological space X. Axiom 4 then means that for any two points $x, y \in X$, there exists some homeomorphism h with

$$h(x) = y. \tag{5.2.15}$$

This does not hold for all topological spaces. It is valid, however, for the perhaps most important class, for manifolds, to be introduced in Sect. 5.3 below.

Axiom 5 *A structure on the space consists of constraints on its transformations.*

We shall now describe some examples of this that involve some more advanced concepts that will be defined only subsequently, in Sect. 5.3 and its Sects. 5.3.1 and 5.3.2, or not at all in this text. For instance, a differentiable manifold is characterized by constraining the transformations to be diffeomorphisms. Its diffeomorphism group still operates transitively on a differentiable manifold. This is no longer valid for richer structures that lead to stronger constraints on the transformation, for instance for the structure of a complex manifold whose transformations would be biholomorphic maps, that is, invertible holomorphic maps. In such a case, we should rather consider local transformations. In the case of a complex manifold, this would be biholomorphic maps between open sets, e.g. small neighborhoods of points. A complex manifold is then locally homogeneous in the sense that for any two points z, w, there exist neighborhoods U of z and V of w and a biholomorphic map $h : U \to V$ with $h(z) = w$. We should formalize this property as the concept of *local space*.

For a semisimple Lie group G of noncompact type, a homogeneous space is given by the quotient G/K of G by a maximal compact subgroup K. G/K is also a special case of a Riemannian manifold (on which G operates by isometries), but a general Riemannian manifold need not be homogeneous, and in fact not even in the local sense just described. It, however, can be characterized in terms of infinitesimal isometries as we shall see below. We should formalize this property as the concept of *infinitesimal space*. More generally, the isometries of a metric space constitute the transformation group for a metric structure. Again, in general, a metric space need not possess any global or local isometries.

5.3   Manifolds

In the preceding, we have employed certain notions, those of a manifold, of a differentiable and of a Riemannian manifold, that we have not yet defined. We therefore have to proceed to the corresponding definitions. For the technical aspects, a reference is [58]. We first describe the principle before proceeding to a technical definition.

Axiom 6 *A manifold is a principle relating a basic space to the model space* $\mathbb{R}^d$ *(considered as a topological space) by local descriptions in such a manner that the identity of points can be traced across the different descriptions. Those descriptions can be arranged in such a manner that locally always finitely many of them suffice.*

We shall now explain the terms employed in this axiom in a more formal manner. For this purpose, we shall have to introduce some technical terms, like paracompactness, that shall not be further explored or discussed in the sequel. Some readers may therefore wish to skip this and proceed directly to Definition 5.3.1.

A *covering* $(U_\alpha)_{\alpha \in A}$ (A an arbitrary index set) is a collection of subsets of our space M whose union equals M. When M is a topological space we can speak of an *open covering*, that is, one where all the U_α are open sets. A covering is called *locally finite* if each $p \in M$ has a neighborhood that intersects only finitely many U_α. The topological space M is called *paracompact* if any open covering possesses a locally finite refinement. This means that for any open covering $(U_\alpha)_{\alpha \in A}$ there exists a locally finite open covering $(V_\beta)_{\beta \in B}$ with

$$\forall \beta \in B \ \exists \alpha \in A : V_\beta \subset U_\alpha. \tag{5.3.1}$$

Paracompactness is a technical condition that ensures the existence of partitions of unity. These constitute a useful analytical tool, and we briefly state the corresponding result now, even though we shall not utilize it here.

Lemma 5.3.1 *Let X be a paracompact Hausdorff space, $(U_\alpha)_{\alpha \in A}$ an open covering. Then there exist a locally finite refinement $(V_\beta)_{\beta \in B}$ of (U_α) and continuous functions $\varphi_\beta : X \to \mathbb{R}$ with*

(i) $\{x \in X : \varphi_\beta(x) \neq 0\} \subset V_\beta$ for all $\beta \in B$,
(ii) $0 \leq \varphi_\beta(x) \leq 1$ for all $x \in X$, $\beta \in B$,
(iii) $\sum_{\beta \in B} \varphi_\beta(x) = 1$ for all $x \in X$.

In (iii), there are only finitely many nonvanishing summands at each point since only finitely many φ_β are nonzero at any given point because the covering (V_β) is locally finite.

The collection of functions $\varphi_\beta : X \to \mathbb{R}$ as in the Lemma is then called a partition of unity subordinate to (U_α). A *proof* can be found in [95], for instance.

We recall from Sect. 4.1 that a map between topological spaces is called *continuous* if the preimage of any open set is again open. A bijective map which is continuous in both directions is called a *homeomorphism*.

Definition 5.3.1 A *manifold M* of *dimension d* is a connected paracompact Hausdorff space together with the collection of all homeomorphisms from open sets $U \subset M$ to open sets $\Omega \subset \mathbb{R}^d$, such that every $x \in M$ is contained in some such open U that is homeomorphic to an open $\Omega \subset \mathbb{R}^d$.

Such a homeomorphism

$$x : U \to \Omega$$

is called a *(coordinate) chart*.

It is customary to write the Euclidean coordinates of $\mathbb{R}^d$, $\Omega \subset \mathbb{R}^d$ open, as

$$x = (x^1, \ldots, x^d), \tag{5.3.2}$$

and these are then considered as local coordinates on our manifold M when $x : U \to \Omega$ is a chart.

In order to verify that a manifold from Definition 5.3.1 satisfies Axiom 4, we state

Theorem 5.3.1 *Let M be a manifold, and $p, q \in M$. Then there exists a homeomorphism*

$$h : M \to M \text{ with } h(p) = q. \tag{5.3.3}$$

We sketch the

Proof 1. A manifold M is path connected in the sense that for every two points $p \neq q$, there exists a continuous and injective map $c : [0, 1] \to M$ with $c(0) = p$, $c(1) = q$. (This sounds intuitively plausible, but the proof of this fact is not trivial.)

2. There exist a neighborhood U of $c([0, 1])$ and a chart

$$x : U \to \Omega = \{x \in \mathbb{R}^d : |x - p| < 1$$
$$\text{for some } p = (\lambda, 0, \ldots, 0), 0 \le \lambda \le 1\} \quad (5.3.4)$$

that maps $c([0, 1])$ to $\{(\lambda, 0, \ldots, 0), 0 \le \lambda \le 1\}$, with $p = c(0)$ corresponding to $(0, 0, \ldots, 0)$.

3. There exists a homeomorphism $\eta : \Omega \to \Omega$ that is the identity in a neighborhood of the boundary of Ω in $\mathbb{R}^d$ and maps $(0, 0, \ldots, 0)$ to $(1, 0, \ldots, 0)$, that is, the point corresponding to $c(0) = p$ to the point corresponding to $c(1) = q$.

4. Extending $x^{-1} \circ \eta$ to all of M by letting it be the identity outside U yields a homeomorphism of M with $h(p) = q$. $\qquad\square$

A point $p \in U_\alpha$ is determined by $x_\alpha(p)$; hence it is often identified with $x_\alpha(p)$. Often, the index α is also omitted, and the components of $x(p) \in \mathbb{R}^d$ are called *local coordinates* of p.

Thus, putting the preceding more abstractly, a manifold is a principle for identifying each point across different local representations (coordinate charts), as required in Axiom 6. That is, when

$$x_i : U_i \to \Omega_i \subset \mathbb{R}^d, \ i = 1, 2, \quad (5.3.5)$$

are two charts with $p \in U_1 \cap U_2$, then the representation $x_1(p)$ by the chart x_1 is identified with the representation $x_2(p)$ by the chart x_2.

Now, according to the concept of a manifold, one cannot distinguish between two homeomorphic manifolds. That is, when $h : M_1 \to M_2$ is a homeomorphism, then the point $p \in M_1$ has to be identified with the point $h(p) \in M_2$. In terms of coordinate charts, the charts $x : U \to \Omega$ of M_1 correspond to the charts $x \circ h^{-1} : h(U) \to \Omega$ of M_2. Here, it is important to realize that when $h : M \to M$ is a homeomorphism of the manifold M to itself, we then have two different, but equivalent representations of M, in the sense that the point p in M in the first representation corresponds to the point $h(p)$ in M in the second representation, but no longer to the point p in that representation.

The second important aspect is that charts allow us to transfer structures from $\mathbb{R}^d$ to a manifold, by imposing a compatibility condition on chart transitions. Again, we state an axiom before a more formal and concrete definition.

Axiom 7 *A structure on a manifold is a compatibility condition between local descriptions.*

Definition 5.3.2 An *atlas* is a family $\{U_\alpha, x_\alpha\}$ of charts for which the U_α constitute an open covering of M. We say that an atlas $\{U_\alpha, x_\alpha\}$ on a manifold is of structural type $\mathbf{C}$ where $\mathbf{C}$ is some category if all chart transitions

$$x_\beta \circ x_\alpha^{-1} : x_\alpha(U_\alpha \cap U_\beta) \to x_\beta(U_\alpha \cap U_\beta)$$

are isomorphisms for the category $\mathbf{C}$ (in case $U_\alpha \cap U_\beta \ne \emptyset$).

Any atlas consisting of charts that are compatible in this sense is contained in a maximal one, namely the one consisting of all charts compatible with the original one. A maximal atlas of charts that are compatible in this sense is called a **C**-structure, and a manifold of type **C** of dimension d is a manifold of dimension d with such an atlas.

The following is the most basic example of such a structure.

Definition 5.3.3 An atlas $\{U_\alpha, x_\alpha\}$ on a manifold is called *differentiable* if all chart transitions

$$x_\beta \circ x_\alpha^{-1} : x_\alpha(U_\alpha \cap U_\beta) \to x_\beta(U_\alpha \cap U_\beta)$$

are differentiable of class C^∞ (in case $U_\alpha \cap U_\beta \neq \emptyset$). A maximal differentiable atlas is called a differentiable structure, and a *differentiable manifold* of dimension d is a manifold of dimension d with a differentiable structure.

Remark

1. We are requiring here the differentiability class C^∞ ("smoothness"), that is, we request the functions to be infinitely often differentiable, but one can, of course, also work with other differentiability classes. Working with C^∞ is the laziest choice, as it obviates the need to always specify the precise order of differentiability required in a given situation.
2. Since the inverse of $x_\beta \circ x_\alpha^{-1}$ is $x_\alpha \circ x_\beta^{-1}$, chart transitions are differentiable in both directions, i.e. *diffeomorphisms*. Thus, they are isomorphisms for the corresponding category, as required.
3. It is easy to show that the dimension of a differentiable manifold is uniquely determined. For a general manifold that is not necessarily differentiable, this is much harder.
4. Since any differentiable atlas is contained in a maximal differentiable one, it suffices to exhibit some differentiable atlas if one wants to construct a differentiable manifold.

Definition 5.3.4 A map $h : M \to M'$ between differentiable manifolds M and M' with charts $\{U_\alpha, x_\alpha\}$ and $\{U'_\alpha, x'_\alpha\}$ is called *differentiable* if all maps $x'_\beta \circ h \circ x_\alpha^{-1}$ are differentiable (of class C^∞, as always) where defined. Such a map is called a *diffeomorphism* if it is bijective and differentiable in both directions.

As before, one cannot distinguish between two diffeomorphic differentiable manifolds.

For purposes of differentiation, a differentiable manifold locally has the structure of the vector space $\mathbb{R}^d$. That is, now, in addition to the topological structure of $\mathbb{R}^d$, we also use its vector space structure. As will become clear below, the vector space $\mathbb{R}^d$, however, is an infinitesimal rather than a local model for a differentiable manifold.

According to the concept of a differentiable manifold, the differentiability of a map can be tested in local coordinates. The diffeomorphism requirement for the chart transitions then guarantees that differentiability defined in this manner is a consistent notion, i.e. independent of the choice of a chart. It is important to realize, however, that the values of derivatives depend on the local charts employed. Therefore, they are not intrinsically defined. This issue will be taken up in Sect. 5.3.1. Whether the Jacobian of a map is zero or not, however, does not dependent on the local coordinates, and this therefore is an intrinsic property. This will be utilized in the proof of Theorem 5.3.2.

In any case, we can also formulate the preceding as

Axiom 8 *A differentiable manifold is a principle for identifying points and vectors across different local representations.*
This aspect will form the content of Sect. 5.3.1.

A differentiable manifold satisfies the analogue of Theorem 5.3.1

Theorem 5.3.2 *Let M be a differentiable manifold, and $p, q \in M$. Then there exists a diffeomorphism*

$$h : M \to M \text{ with } h(p) = q. \tag{5.3.6}$$

For the

Proof We observe that the constructions of the proof of Theorem 5.3.1 can be carried out in the differentiable category. In particular, on a differentiable manifold, we may find a *differentiable* curve $c : [0, 1] \to M$ with nowhere vanishing derivative with $c(0) = p, c(1) = q$, and the homeomorphism $\eta : \Omega \to \Omega$ of that proof can be constructed as a diffeomorphism.   □

Example

1. The simplest example of a d-dimensional manifold is $\mathbb{R}^d$ itself.

2. The *sphere*

$$S^n := \{(u^1, \ldots, u^{n+1}) \in \mathbb{R}^{n+1} : \sum_{i=1}^{n+1} (u^i)^2 = 1\} \tag{5.3.7}$$

is a differentiable manifold of dimension n. This manifold is covered by the following two charts. On $U_1 := S^n \setminus \{(0, \ldots, 0, 1)\}$ (the sphere without the north pole), we put

$$x_1(u^1, \ldots, u^{n+1}) := (x_1^1(u^1, \ldots, u^{n+1}), \ldots, x_1^n(u^1, \ldots, u^{n+1}))$$

$$:= \left(\frac{u^1}{1 - u^{n+1}}, \ldots, \frac{u^n}{1 - u^{n+1}}\right)$$

and on $U_2 := S^n \setminus \{(0, \ldots, 0, -1)\}$ (the sphere without the south pole),

$$x_2(u^1, \ldots, u^{n+1}) := (x_2^1(u^1, \ldots, u^{n+1}), \ldots, x_2^n(u^1, \ldots, u^{n+1}))$$

$$:= \left(\frac{u^1}{1 + u^{n+1}}, \ldots, \frac{u^n}{1 + u^{n+1}}\right).$$

$\boxed{\mathbb{R}^d}$

$\boxed{\text{Sphere}}$

3. Similarly, n-dimensional hyperbolic space, represented as the hyperboloid of revolution

$$H^n := \{u \in \mathbb{R}^{n+1} : (u^1)^2 + \ldots (u^n)^2 - (u^{n+1})^2 = -1, u^{n+1} > 0\},$$
(5.3.8)

is a manifold.

> Hyperbolic space

4. Let $w_1, w_2, \ldots, w_n \in \mathbb{R}^n$ be linearly independent and define $u_1, u_2 \in \mathbb{R}^n$ as equivalent if there are $m_1, m_2, \ldots, m_n \in \mathbb{Z}$ with

$$u_1 - u_2 = \sum_{i=1}^{n} m_i w_i.$$

Let π be the projection mapping $u \in \mathbb{R}^n$ to its equivalence class. The *torus* $T^n := \pi(\mathbb{R}^n)$ can then be made a differentiable manifold (of dimension n) as follows: Suppose Δ_α is open and does not contain any pair of equivalent points. We put

$$U_\alpha := \pi(\Delta_\alpha),$$
$$x_\alpha := (\pi_{|\Delta_\alpha})^{-1}.$$

> Torus

5. In general, any open subset of a (differentiable) manifold is again a (differentiable) manifold.
6. The Cartesian product $M \times N$ of two differentiable manifolds M, N also naturally carries the structure of a differentiable manifold. In fact, if $\{U_\alpha, x_\alpha\}_{\alpha \in A}$ and $\{V_\beta, y_\beta\}_{\beta \in B}$ are atlases for M and N, resp., then $\{U_\alpha \times V_\beta, (x_\alpha, y_\beta)\}_{(\alpha,\beta) \in A \times B}$ is an atlas for $M \times N$ with differentiable chart transitions.

According to Definition 5.3.2, one can put any type of restriction on the chart transitions, for example, require them to be affine, algebraic, real analytic, conformal, Euclidean volume preserving, . . . , and thereby define a class of manifolds with that particular structure. Perhaps the most important example is the notion of a complex manifold.

Definition 5.3.5 A *complex manifold* of complex dimension d ($\dim_\mathbb{C} M = d$) is a differentiable manifold of (real) dimension $2d$ ($\dim_\mathbb{R} M = 2d$) whose charts take values in open subsets of $\mathbb{C}^d$ with *holomorphic*[1] chart transitions.

Analogous to the case of a differentiable manifold, since chart transitions operate in both directions, they are in fact biholomorphic, that is, bijective with holomorphic inverses.

---

[1] Let us recall the definition of a holomorphic map. On $\mathbb{C}^d$, we use complex coordinates $z^1 = x^1 + iy^1, \ldots, z^d = x^d + iy^d$. A complex function $h : U \to \mathbb{C}$ where U is an open subset of $\mathbb{C}^d$, is called holomorphic if $\frac{\partial h}{\partial \bar{z}^k} := \frac{1}{2}(\frac{\partial h}{\partial x^k} + i \frac{\partial h}{\partial y^k}) = 0$ for $k = 1, \ldots, d$, and a map $H : U \to \mathbb{C}^m$ is then holomorphic if all its components are. We do not want to go into the details of complex analysis here, and we refer the reader to [58] for more details about complex manifolds.

For a complex manifold, the analogue of Theorems 5.3.1, 5.3.2 in general no longer holds. We only have

Theorem 5.3.3 *Let M be a complex manifold, $p, q \in M$. Then there exist open neighborhoods U, V of p and q, resp., and a biholomorphic map $h : U \to V$ with $h(p) = q$.*

Proof We choose two local charts x, y whose domains contain p and q, resp., with $x(p) = 0 = y(q)$. By multiplying with an appropriate scaling factor, if necessary, we may assume that the images of x and y both contain the unit ball $U(0, 1) := \{z \in \mathbb{C}^d : |z| < 1\}$. We may then take $U := x^{-1}(U(0, 1))$, $V := y^{-1}(U(0, 1))$ and the biholomorphic map $h := y^{-1} \circ x : U \to V$, that is, the biholomorphic map that is represented by the identity in our local charts.   □

5.3.1 Differential Geometry

We now explore the concept of a differentiable manifold in more detail, according to **three principles**

1. Through local coordinate representations and the compatibility requirement between such local descriptions, differential calculus becomes available on a differentiable manifold. The objects of that calculus, like tangent vectors, however, are described differently in different local descriptions, and they therefore need to be related by transformations. The resulting transformation rules constitute the subject of **tensor calculus**.
2. According to the concept of a manifold, the geometric content should not depend on the local descriptions. This leads to Riemann's idea of **invariants**.
3. Along with the differential calculus, we can also utilize the Euclidean structure on $\mathbb{R}^d$, in order to introduce metric relations on a differentiable manifold. This leads to the profound concept of a **Riemannian manifold**. The basic invariants of a Riemannian manifold are contained in Riemann's **curvature tensor**. Importantly, a Riemannian metric is only infinitesimally Euclidean, but in general not locally or globally. The curvature quantifies that deviation from being Euclidean.

For a more detailed discussion and for the proofs of various results, we refer to [58, 62].

As indicated, tensor calculus is about coordinate representations of geometric objects and the transformations of those representations under coordinate changes. Tensor calculus thus has to reconcile the invariant properties of geometric objects with their non-invariant local descriptions.

We consider a d-dimensional differentiable manifold M. Tensor calculus involves intricate, but well thought out notational schemes for efficiently representing and making transparent the transformation behavior of infinitesimal geometric quantities under coordinate changes. A useful convention is the Einstein summation convention

$$a^i b_i := \sum_{i=1}^{d} a^i b_i. \tag{5.3.9}$$

That is, a summation sign is omitted when the same index occurs twice in a product, once as an upper and once as a lower index. This rule is not affected by the possible presence of other indices; for example,

$$A^i_j b^j = \sum_{j=1}^{d} A^i_j b^j. \tag{5.3.10}$$

The conventions about when to place an index in an upper or lower position will be given subsequently. We also recall the Kronecker symbol

$$\delta^i_k := \begin{cases} 1 & \text{when } i = k \\ 0 & \text{when } i \neq k. \end{cases} \tag{5.3.11}$$

The manifold M is locally modeled after $\mathbb{R}^d$, and so, locally, it can be represented by coordinates $x = (x^1, \ldots, x^d)$ taken from some open subset of $\mathbb{R}^d$. These coordinates, however, are not canonical, and we may as well choose other ones, $y = (y^1, \ldots, y^d)$, with $x = f(y)$ for some homeomorphism f. Since our manifold M is differentiable, we can cover it by local coordinates for which all coordinate transitions are diffeomorphisms. Differential geometry then investigates how various expressions representing objects on M like tangent vectors transform under coordinate changes. Here and in the sequel, all objects defined on a differentiable manifold will be assumed to be differentiable themselves. This is checked in local coordinates, but since coordinate transitions are diffeomorphic, the differentiability property does not depend on the choice of coordinates.

The simplest object in the present context is a (differentiable) function $\phi : M \to \mathbb{R}$. Of course, the value of ϕ at some point $p \in M$ does not depend on the choice of local coordinates. Thus, changing coordinates by $x = f(y)$, we have $\phi(x) = \phi(f(y))$. That is, instead of applying the function ϕ in the x-coordinates, we need to utilize the function $\phi \circ f$ in the y-coordinates. In more abstract terms, changing coordinates from x to y pulls the function ϕ defined in the x-coordinates back to $f^\star \phi$ defined for the y-coordinates, with $f^\star \phi(y) = \phi(f(y))$. We shall now discuss what happens to the derivatives of ϕ under such coordinate changes.

Here, an operator that takes the derivative of a function in a given direction is called a tangent vector. As a formula, when $\psi : M \to \mathbb{R}$ is a function, then in the coordinates given by y, a tangent vector at the point p with coordinate representation y_0 is of the form

$$W = w^i \frac{\partial}{\partial y^i}, \tag{5.3.12}$$

and it operates on ψ as

$$W(\psi)(y_0) = w^i \frac{\partial \psi}{\partial y^i}\Big|_{y=y_0}. \tag{5.3.13}$$

The index i in $\frac{\partial}{\partial y^i}$ is considered as a lower index, and the summation convention (5.3.9) is applied.

When now $\psi = f^\star \phi$, we want to transform the vector W to the x-coordinates, i.e., push it forward, so that

$$(f_\star W)(\phi) = W(f^\star \phi) \tag{5.3.14}$$

and the value of the derivative is the same in the x- and the y-coordinates. By the chain rule, we have

$$W(f^\star \phi) = w^k \frac{\partial}{\partial y^k} \phi(f(y)) = w^k \frac{\partial x^i}{\partial y^k} \frac{\partial \phi}{\partial x^i}, \tag{5.3.15}$$

and therefore, since this has to hold for all functions ϕ,

$$f_\star W = w^k \frac{\partial x^i}{\partial y^k} \frac{\partial}{\partial x^i}. \tag{5.3.16}$$

We have thus derived the transformation behavior of tangent vectors under coordinate changes.

The essential principle here is that in order that a quantity, here the derivative of a function, be invariant under coordinate changes, according to (5.3.14), the corresponding operator has to be transformed appropriately.

The tangent vectors at $p \in M$ form a vector space, called the tangent space $T_p M$ of M at p. A basis of $T_p M$ is given by the $\frac{\partial}{\partial y^i}$, considered as derivative operators at the point p represented by y_0 in the local coordinates, as in (5.3.13).

A vector field is then defined as $W(y) = w^i(y) \frac{\partial}{\partial y^i}$, that is, by having a tangent vector at each point of M, with differentiable coefficients $w^i(y)$, of course. The vector space of vector fields on M is written as $\Gamma(TM)$. (In fact, $\Gamma(TM)$ is a module over the ring $C^\infty(M)$.)

Later, we shall need the Lie bracket $[V, W] := VW - WV$ of two vector fields $V(y) = v^i(y) \frac{\partial}{\partial y^i}$, $W(y) = w^j(y) \frac{\partial}{\partial y^j}$. $[V, W]$ operates on a function ψ as

$$[V, W]\psi(y) = v^i(y) \frac{\partial}{\partial y^i} (w^j(y) \frac{\partial}{\partial y^j} \psi(y)) - w^j(y) \frac{\partial}{\partial y^j} (v^i(y) \frac{\partial}{\partial y^i} \psi(y))$$

$$= (v^i(y) \frac{\partial w^j(y)}{\partial y^i} - w^i(y) \frac{\partial v^j(y)}{\partial y^i}) \frac{\partial \psi(y)}{\partial y^j}. \tag{5.3.17}$$

In particular, for coordinate vector fields, we have

$$\left[\frac{\partial}{\partial y^i}, \frac{\partial}{\partial y^j} \right] = 0. \tag{5.3.18}$$

Returning to the x-coordinates and considering a single tangent vector, $V = v^i \frac{\partial}{\partial x^i}$ at some point x_0, we define a covector or cotangent vector $\omega = \omega_i dx^i$ at this point as an object dual to V, with the rule

$$dx^i \left(\frac{\partial}{\partial x^j} \right) = \delta^i_j \tag{5.3.19}$$

yielding

$$\omega_i dx^i \left(v^j \frac{\partial}{\partial x^j} \right) = \omega_i v^j \delta^i_j = \omega_i v^i. \tag{5.3.20}$$

This expression depends only on the coefficients v^i and ω_i at the point p. We write this as $\omega(V)$, the application of the covector ω to the vector V, or as $V(\omega)$, the application of V to ω.

Like the tangent vectors, the cotangent vectors at p constitute a vector space, which is then called the cotangent space $T_p^\star M$.

For a differentiable function ϕ, we have the differential of ϕ,

$$d\phi = \frac{\partial \phi}{\partial x^i} dx^i. \tag{5.3.21}$$

When V is a tangent vector, we then have

$$V(\phi) = d\phi(V). \tag{5.3.22}$$

Here, for simplicity of notation, we have left out the point p at which we perform these operations. That point does not matter, as long as all objects are considered at the same point. Subsequently, this will lead us to the concepts of vector fields and 1-forms. First, however, we need to look at the transformation behavior

$$dx^i = \frac{\partial x^i}{\partial y_j} dy^j \tag{5.3.23}$$

required for the invariance of $\omega(V)$. Thus, the coefficients of ω in the y-coordinates are given by the identity

$$\omega_i dx^i = \omega_i \frac{\partial x^i}{\partial y_j} dy^j. \tag{5.3.24}$$

Again, a covector $\omega_i dx^i$ is pulled back under a map f:

$$f^\star(\omega_i dx^i) = \omega_i \frac{\partial x^i}{\partial y^j} dy^j. \tag{5.3.25}$$

A 1-form then assigns a covector to every point in M, and thus, it is locally given as $\omega_i(x) dx^i$.

The transformation behavior of a tangent vector as in (5.3.16) is called contravariant, the opposite one of a covector as (5.3.24) covariant. We may then also consider higher order tensors. For instance, an object of the form $a^i_j \frac{\partial}{\partial x^i} \otimes dx^j$ transforms as $a^i_j \frac{\partial y^k}{\partial x^i} \frac{\partial x^j}{\partial y^\ell} \frac{\partial}{\partial y^k} \otimes dy^\ell$, while $a_{ij} dx^i \otimes dx^j$ becomes $a_{ij} \frac{\partial x^i}{\partial y^k} \frac{\partial x^j}{\partial y^\ell} dy^k \otimes dy^l$. In fact, the tensor product symbol $\otimes$ is often left out in the notation of tensor calculus; for instance, one simply writes $a_{ij} dx^i dx^j$. In general, an object whose coefficients have r upper and s lower indices and with the corresponding transformation behavior is called an r times contravariant and s times covariant tensor; for instance, $a_{ij} dx^i dx^j$ is of type $(0, 2)$.

A differentiable manifold specifies how one can pass from the local structure as given by the concept of a manifold to an infinitesimal structure. We shall now see how in turn we can utilize such an infinitesimal structure to arrive at local and even global constructions.

5.3.2 Riemannian Geometry

So far, we have considered the transformation behavior under coordinate changes of infinitesimal objects at one and the same point p. In other words, we have connected the structure of a differentiable manifold with linear algebra.

Geometry, however, asks for more, for being able to perform quantitative measurements. Within the framework of a differentiable manifold with its infinitesimal aspects as just developed, this requirement can be decomposed into two aspects. One concerns infinitesimal measurements, that is, assigning lengths to tangent vectors and quantifying angles between tangent vectors. The other concerns the comparison of measurements at different points. In fact, this turns out to be a special instance of a more general question, how to relate the infinitesimal geometries at different points. The first question can also be considered in more generality; how to impose geometric structures as known from the vector space $\mathbb{R}^d$ onto the tangent spaces of a differentiable manifold in a consistent manner.

A direct way to address the first requirement consists simply in imposing a Euclidean structure on each tangent space $T_p M$. This is achieved by a tensor $g_{ij} dx^i \otimes dx^j$ with symmetric and positive definite $(g_{ij})_{i,j=1,\ldots,d}$. We then have a Euclidean product of tangent vectors:

$$\langle V, W \rangle := g_{ij} v^i w^j \tag{5.3.26}$$

for $V = v^i \frac{\partial}{\partial x^i}$, $W = w^i \frac{\partial}{\partial x^i}$. Since v^i and w^i transform contravariantly, g_{ij} has to transform doubly covariantly (as indicated by the lower position of the indices), so that the product as a scalar quantity remains invariant under coordinate transformations.

There are also some tensor calculus conventions involving a metric tensor $G = (g_{ij})_{i,j}$. The inverse metric tensor is written as $G^{-1} = (g^{ij})_{i,j}$, that is, by raising the indices. In particular

$$g^{ij} g_{jk} = \delta^i_k := \begin{cases} 1 & \text{when } i = k \\ 0 & \text{when } i \neq k. \end{cases} \tag{5.3.27}$$

We then have the general convention for raising and lowering indices

$$v^i = g^{ij} v_j \text{ and } v_i = g_{ij} v^j. \tag{5.3.28}$$

In an analogous manner, one may also impose other structures on tangent spaces, like symplectic or complex structures in the case of even d. These structures then have to satisfy the appropriate transformation rules under coordinate changes.

Given a metric, we possess notions of length and angle. For a tangent vector V, its length is given as

$$\|V\| := \sqrt{\langle V, V \rangle}, \tag{5.3.29}$$

as in (5.1.2). Likewise, as in (5.1.3), the angle α between two tangent vectors V, W at a point p is defined by

$$\cos \alpha \, \|V\| \|W\| := \langle V, W \rangle. \tag{5.3.30}$$

In particular, we then have a notion of orthogonality. The tangent vectors V, W at some point p are orthogonal to each other if

$$\langle V, W \rangle = 0. \tag{5.3.31}$$

In a differentiable manifold M, when we have some linear subspace L of a tangent space $T_p M$, we can look at those tangent vectors that are not contained in L, but when we also have a Riemannian metric, we can then distinguish those vectors that are orthogonal to L. With this, we can then also define the gradient of a differentiable function ϕ as the direction of steepest ascent. This works as follows. We recall the differential (5.3.21) of ϕ, the 1-form $d\phi = \frac{\partial \phi}{\partial x^i} dx^i$. According to (5.3.22), for every tangent vector V, we have $V(\phi) = d\phi(V)$. The gradient of ϕ is then that vector $\nabla \phi$ that satisfies

$$\langle \nabla \phi, V \rangle = d\phi(V) \tag{5.3.32}$$

for every tangent vector V. Untangling (5.3.32) in local coordinates, we obtain

$$\nabla \phi = g^{ij} \frac{\partial \phi}{\partial x^i} \frac{\partial}{\partial x^j}. \tag{5.3.33}$$

This is most easily checked by inserting (5.3.33) into the left hand side of (5.3.32) and verifying that (5.3.32) is indeed satisfied. In particular, the gradient $\nabla \phi$ is orthogonal to the level sets of ϕ in the following sense. Let $\{\phi \equiv \kappa\}$, for some $\kappa \in \mathbb{R}$, be a level set. Let p be a point in this level set, i.e., $\phi(p) = \kappa$. A tangent vector $V \in T_p M$ is then tangent to this level set if $V(\phi) = 0$, that is, if the value of ϕ does not change to first order in the direction of V. But then, by (5.3.21), (5.3.32), we have that

$$\langle \nabla \phi, V \rangle = 0. \tag{5.3.34}$$

It is important to point out that without a Riemannian metric, we can define the differential $d\phi$, but not the gradient for a differentiable function ϕ on a differentiable manifold.

Given a metric, we can also define and compute the length of a differentiable curve $c : [0, 1] \rightarrow M$ as

$$L(c) := \int_0^1 \langle \dot{c}(t), \dot{c}(t) \rangle^{1/2} dt, \tag{5.3.35}$$

that is, by integrating the length of its tangent vector along the curve.

We can then utilize this to define the distance between points $p, q \in M$ as

$$d(p, q) := \inf\{L(c) : c : [0, 1] \rightarrow M \text{ with } c(0) = p, c(1) = q\}. \tag{5.3.36}$$

After checking some technical details, one finds that equipped with this distance function $d(., .)$, M becomes a metric space, that is, d is positive, symmetric, and satisfies the triangle inequality. One also checks—which is a good exercise—that the topology introduced by that metric $d(., .)$ coincides with the original topology of M as a manifold.

Of course, we can then also ask whether any two points $p, q \in M$ can be connected by a shortest curve, that is, whether there exists a curve c_0 with $c_0(0) = p, c_0(1) = q$ and

$$L(c_0) = d(p, q). \tag{5.3.37}$$

The Hopf-Rinow theorem assures us that such a curve does indeed exist if M is complete. It is a called a *geodesic curve*, and it satisfies the equations

$$\ddot{c}_0^k(t) + \Gamma_{ij}^k(c_0(t))\dot{c}_0^i(t)\dot{c}_0^j(t) = 0 \quad \text{for } k = 1, \ldots, d. \tag{5.3.38}$$

with

$$\Gamma_{ij}^k = \frac{1}{2} g^{kl}(g_{il,j} + g_{jl,i} - g_{ij,l}), \tag{5.3.39}$$

where

$$g_{ij,k} := \frac{\partial}{\partial x^k} g_{ij}. \tag{5.3.40}$$

In particular, the Hopf-Rinow theorem tells us that the infimum in (5.3.36) is, in fact, a minimum, that is

$$d(p, q) := \min\{L(c) : c : [0, 1] \to M \text{ with } c(0) = p, c(1) = q\}. \tag{5.3.41}$$

There is one small technical caveat here. A solution of (5.3.38) need not be the shortest connection of its endpoints. It is only locally shortest, that is, for any sufficiently small $\epsilon > 0$ (which may depend on c) and $t_1, t_2 \in [0, 1]$ with $|t_1 - t_2| < \epsilon$, the restricted curve $c_{[t_1, t_2]}$ is the (unique) shortest connection between $c(t_1)$ and $c(t_2)$. In general, however, geodesic connections between two points in a manifold are not unique. In fact, on a compact Riemannian manifold, there always exist infinitely many geodesic connections between any two points p and q. Most of these geodesic connections, however, are not shortest connections between p and q. Shortest connections, in fact, are generically unique. Of course, all these claims require proof, but we refer to the literature, e.g. [58] and the references therein.

In the preceding, we have also seen the basic paradigm of the *calculus of variations* (see [64]). The problem of minimizing the integral (5.3.35), a global quantity, is transformed into the differential equation (5.3.38), an infinitesimal principle. The underlying rationale is that if a curve c yields a minimum for the global quantity (5.3.35), then it also has to be locally minimizing, that is, every portion of the curve $c_{[t_1, t_2]}$ for $0 \le t_1 < t_2 \le 1$ then also has to be the shortest connection between its endpoints $c(t_1)$ and $c(t_2)$. And if the curve c is always locally minimizing, then it also has to be infinitesimally minimizing, which leads to the differential equation (5.3.38).

There is still another direction in which the preceding can lead our thinking. We have used the Riemannian metric and the length functional (5.3.35) induced by that Riemannian metric to obtain the metric $d(.,.)$ in (5.3.36). And on a complete Riemannian metric, the distance $d(p, q)$ between two points equals the length of the shortest (geodesic) connection c between them. We can, however, also try to reverse this and start with some metric $d(.,.)$ on some set X. The metric then induces a topology on X, with a basis of open sets given by the open metric balls $U(x, r), x \in X, r \ge 0$.

as in Example 3 in Sect. 4.1. We can then define the length of a continuous curve $c : [0, 1] \to X$ as

$$L(c) := \sup\{\sum_{i=1}^{n} d(c(t_{i-1}), c(t_i)) : 0 = t_0 < t_1 < \cdots < t_n = 1, n \in \mathbb{N}\}.$$
(5.3.42)

The curve c is then called a *geodesic* if

$$L(c_{|[\tau_1, \tau_2]}) = d(c(\tau_1), c(\tau_2)) \text{ whenever } |\tau_1 - \tau_2| \le \epsilon, \qquad (5.3.43)$$

for some $\epsilon > 0$. Thus, a geodesic realizes the distance between its sufficiently close points.

The metric space (X, d) is called a *geodesic space* if for any two points $p, q \in X$, there exists a shortest geodesic joining them, i.e., a continuous[2] curve $c : [0, 1] \to X$ with $c(0) = p, c(1) = q$ and

$$L(c) = d(p, q). \qquad (5.3.44)$$

In particular, a metric space that is geodesic has to be path connected, that is, any two points can be joined by some curve.

So far, we have worked with arbitrary parametrizations of our curves on the unit interval $[0, 1]$. For many purposes, however, it is more convenient to work with a particular parametrization. We can parametrize a curve proportionally to arclength, that is, we consider the homeomorphism $\sigma : [0, 1] \to [0, 1]$ with the property that the curve $\gamma := c \circ \sigma$ satisfies

$$L(\gamma_{|[0,t]}) = tL(\gamma)(= tL(c)) \text{ for all } t \in [0, 1]. \qquad (5.3.45)$$

On a geodesic space, a midpoint $m(p, q)$ of p and q as defined in Definition 2.1.9, see (2.1.56), has to satisfy

$$m(p, q) = \gamma(\frac{1}{2}) \qquad (5.3.46)$$

for a shortest geodesic $\gamma : [0, 1] \to X$ connecting p and q and parametrized proportionally to arclength. We note that since the shortest geodesic between p and q need not be unique, they may have more than one such midpoint.

These concepts will be taken up below in Sect. 5.3.3.

It might also be worth pointing out that we have just seen an instance of a general mathematical strategy. When one, perhaps somewhat special structure, here a Riemannian metric, implies some crucial and fundamental property, here that any two points can be connected by a shortest curve, one can then take that property as a starting axiom in its own right and develop the theory on the basis of that axiom, disregarding the original structure that led us to that axiom. This often leads to a considerable clarification of the conditions and structures that one really needs for a certain important property and to the identification of the right level of abstraction.

---

[2] The epithet "continuous" is used here only for emphasis, as all curves are implicitly assumed to be continuous.

We now return to Riemannian geometry. An isometry i between two Riemannian manifolds M, g and N, γ can be defined either infinitesimally, in the sense that for all tangent vectors V, W at any point $p \in M$, we have

$$\langle V, W \rangle_g = \langle i_\star V, i_\star W \rangle_\gamma \qquad (5.3.47)$$

or equivalently by the requirement that for any curve c in M,

$$L_g(c) = L_\gamma(i \circ c), \qquad (5.3.48)$$

where, of course, the subscript g or γ indicates w.r.t. which metric the expressions are computed.

So far, we have computed derivatives of functions. We have also talked about vector fields $V(x) = v^i(x) \frac{\partial}{\partial x^i}$ as objects that depend differentiably on their arguments x. This naturally raises the question of how to compute their derivatives. However, this encounters the problem that in contrast to functions, the representation of such objects depends on the choice of local coordinates, and we have described in some detail that and how they transform under coordinate changes.

It turns out that on a differentiable manifold, there is in general no single canonical way of taking derivatives of vector fields or other tensors in an invariant manner. There are, in fact, many such possibilities, and they are called connections or covariant derivatives. Only when we have additional structures, like a Riemannian metric, can we single out a particular covariant derivative on the basis of its compatibility with the metric. For our purposes, however, we also need other covariant derivatives, and so, we now develop this notion. As usual, we begin with

Axiom 9 *A connection on a differentiable manifold is a scheme for the comparison of the infinitesimal geometries at different points. This scheme is itself based on an infinitesimal principle.*

Again, we need to elaborate the formal details of this axiom. Let M be a differentiable manifold. We recall that $\Gamma(TM)$ denotes the space of vector fields on M. An (affine) connection or covariant derivative on M is a linear map

$$\nabla : \Gamma(TM) \otimes_\mathbb{R} \Gamma(TM) \to \Gamma(TM)$$
$$(V, W) \mapsto \nabla_V W$$

satisfying:

(i) ∇ is tensorial in the first argument:

$$\nabla_{V_1 + V_2} W = \nabla_{V_1} W + \nabla_{V_2} W \quad \text{for all } V_1, V_2, W \in \Gamma(TM)$$
$$\nabla_{fV} W = f \nabla_V W \quad \text{for all } f \in C^\infty(M), V, W \in \Gamma(TM);$$

(ii) ∇ is $\mathbb{R}$-linear in the second argument:

$$\nabla_V(W_1 + W_2) = \nabla_V W_1 + \nabla_V W_2 \quad \text{for all } V, W_1, W_2 \in \Gamma(TM)$$

and it satisfies the product rule

$$\nabla_V(fW) = V(f)W + f\nabla_V W \quad \text{for all } f \in C^\infty(M), V, W \in \Gamma(TM). \qquad (5.3.49)$$

$\nabla_V W$ is called the covariant derivative of W in the direction V. By (i), for any $x_0 \in M$, $(\nabla_V W)(x_0)$ only depends on the value of V at x_0. By way of contrast, it also depends on the values of W in some neighborhood of x_0, as it naturally should as a notion of a derivative of W. The example on which this is modeled is the Euclidean connection given by the standard derivatives, that is, for $V = V^i \frac{\partial}{\partial x^i}$, $W = W^j \frac{\partial}{\partial x^j}$,

$$\nabla_V^{eucl} W = V^i \frac{\partial W^j}{\partial x^i} \frac{\partial}{\partial x^j}.$$

However, this is not invariant under nonlinear coordinate changes, and since a general manifold cannot be covered by coordinates with only linear coordinate transformations, we need the above more general and abstract concept of a covariant derivative.

Let U be a coordinate chart in M, with local coordinates x and coordinate vector fields $\frac{\partial}{\partial x^1}, \dots, \frac{\partial}{\partial x^d}$ ($d = \dim M$). We then define the Christoffel symbols of the connection ∇ via

$$\nabla_{\frac{\partial}{\partial x^i}} \frac{\partial}{\partial x^j} =: \Gamma_{ij}^k \frac{\partial}{\partial x^k}. \tag{5.3.50}$$

(It will become apparent below why we are utilizing here the same symbols as in (5.3.38), (5.3.39).)

Thus,

$$\nabla_V W = V^i \frac{\partial W^j}{\partial x^i} \frac{\partial}{\partial x^j} + V^i W^j \Gamma_{ij}^k \frac{\partial}{\partial x^k}. \tag{5.3.51}$$

In order to understand the nature of the objects involved, we can also leave out the vector field V and consider the covariant derivative ∇W as a 1-form. In local coordinates

$$\nabla W = W_{;i}^j \frac{\partial}{\partial x^j} dx^i, \tag{5.3.52}$$

with

$$W_{;i}^j := \frac{\partial W^j}{\partial x^i} + W^k \Gamma_{ik}^j. \tag{5.3.53}$$

If we change our coordinates x to coordinates y, then the new Christoffel symbols,

$$\nabla_{\frac{\partial}{\partial y^l}} \frac{\partial}{\partial y^m} =: \tilde{\Gamma}_{lm}^n \frac{\partial}{\partial y^n}, \tag{5.3.54}$$

are related to the old ones via

$$\tilde{\Gamma}_{lm}^n(y(x)) = \left\{ \Gamma_{ij}^k(x) \frac{\partial x^i}{\partial y^l} \frac{\partial x^j}{\partial y^m} + \frac{\partial^2 x^k}{\partial y^l \partial y^m} \right\} \frac{\partial y^n}{\partial x^k}. \tag{5.3.55}$$

In particular, due to the term $\frac{\partial^2 x^k}{\partial y^l \partial y^m}$, the Christoffel symbols do not transform as a tensor. However, if we have two connections $^1\nabla$, $^2\nabla$, with corresponding Christoffel symbols $^1\Gamma_{ij}^k$, $^2\Gamma_{ij}^k$, then the difference $^1\Gamma_{ij}^k - {}^2\Gamma_{ij}^k$ does transform as a tensor. Expressed more abstractly, this means that the space of connections on M is an affine space.

For a connection ∇, we define its torsion tensor via

$$T(V, W) := \nabla_V W - \nabla_W V - [V, W] \quad \text{for } V, W \in \Gamma(TM). \tag{5.3.56}$$

Inserting our coordinate vector fields $\frac{\partial}{\partial x^i}$ as before, we obtain

$$T_{ij} := T\left(\frac{\partial}{\partial x^i}, \frac{\partial}{\partial x^j}\right) = \nabla_{\frac{\partial}{\partial x^i}} \frac{\partial}{\partial x^j} - \nabla_{\frac{\partial}{\partial x^j}} \frac{\partial}{\partial x^i}$$

since coordinate vector fields commute, i.e., $[\frac{\partial}{\partial x^i}, \frac{\partial}{\partial x^j}] = 0$

$$= \left(\Gamma^k_{ij} - \Gamma^k_{ji}\right) \frac{\partial}{\partial x^k}. \tag{5.3.57}$$

We call the connection ∇ torsion-free or symmetric if $T \equiv 0$. By the preceding computation, this is equivalent to the symmetry

$$\Gamma^k_{ij} = \Gamma^k_{ji} \quad \text{for all } i, j, k. \tag{5.3.58}$$

Let $c(t)$ be a smooth curve in M, and let $V(t) := \dot{c}(t) \ (= \dot{c}^i(t)\frac{\partial}{\partial x^i}(c(t)))$ in local coordinates) be the tangent vector field of c. In fact, we should instead write $V(c(t))$ in place of $V(t)$, but we consider t as the coordinate along the curve $c(t)$. Thus, in those coordinates $\frac{\partial}{\partial t} = \frac{\partial c^i}{\partial t}\frac{\partial}{\partial x^i}$, and in the sequel, we shall frequently and implicitly make this identification, that is, switch between the points $c(t)$ on the curve and the corresponding parameter values t. Let $W(t)$ be another vector field along c, i.e., $W(t) \in T_{c(t)}M$ for all t. We may then write $W(t) = w^i(t)\frac{\partial}{\partial x^i}(c(t))$ and form

$$\nabla_{\dot{c}(t)} W(t) = \dot{w}^i(t)\frac{\partial}{\partial x^i} + \dot{c}^i(t)w^j(t)\nabla_{\frac{\partial}{\partial x^i}}\frac{\partial}{\partial x^j}$$

$$= \dot{w}^i(t)\frac{\partial}{\partial x^i} + \dot{c}^i(t)w^j(t)\Gamma^k_{ij}(c(t))\frac{\partial}{\partial x^k} \tag{5.3.59}$$

(the preceding computation is meaningful as we see that it depends only on the values of W along the curve $c(t)$, but not on other values in a neighborhood of a point on that curve).

This represents a (nondegenerate) linear system of d first-order differential operators for the d coefficients $w^i(t)$ of $W(t)$. Therefore, for given initial values $W(0)$, there exists a unique solution $W(t)$ of

$$\nabla_{\dot{c}(t)} W(t) = 0. \tag{5.3.60}$$

We observe the following facts:

1. Since the system (5.3.60) is linear w.r.t. W, the solution depends linearly on the initial values. That is, when $W_1(t)$ and $W_2(t)$ are the solutions with initial values $W_1(0)$ and $W_2(0)$, resp., then the solution with initial values $\alpha_1 W_1(0) + \alpha_2 W_2(0)$ is obtained as $\alpha_1 W_1(t) + \alpha_2 W_2(t)$. In particular,

$$W(0) \mapsto W(1) \tag{5.3.61}$$

   is a *linear map* from $T_{c(0)}M$ to $T_{c(1)}M$.

2. The system (5.3.60) is autonomous. Therefore, if we reparametrize the curve c, that is, consider $\tilde{c}(t) := c(\gamma(t))$ for some diffeomorphism $\gamma : [0, 1] \to [0, 1]$, then the solution of $\nabla_{\dot{\tilde{c}}(t)} W(t) = 0$ is given by $W(\gamma(t))$.

3. The system (5.3.60) depends *nonlinearly* on the curve c, because in (5.3.59), the Christoffel symbols in general are nonconstant functions of c. Therefore, in general, $W(t)$ will depend not only on the initial

values $W(0)$, but also on the curve c. Moreover, because of the nonlinear structure, this dependence is not easily computed, and, in fact, in general, it cannot be computed explicitly at all.

The solution $W(t)$ of (5.3.60) is called the *parallel transport* of $W(0)$ along the curve $c(t)$. We also say that $W(t)$ is covariantly constant along the curve c. This now furnishes a scheme for comparing the infinitesimal geometries at different points $p, q \in M$. We take a smooth curve c : $[0, 1] \to M$ with $c(0) = p, c(1) = q$. We can then identify the tangent vector $W = W(0) \in T_p M$ with $W(1) \in T_q M$ where $W(t)$ denotes the parallel transport along the curve c. Thus, from an infinitesimal principle, the integration of the system of differential equations (5.3.60), we obtain a global comparison between tangent spaces at different points.

The result of this comparison, that is, $W(1) \in T_q M$ for given $W(0) \in T_p M$ will in general depend on the curve c, as noted in 3.

In particular, we may ask what happens when we take $q = p$ and parallel transport $W(0)$ along a nontrivial curve c with $c(0) = c(1) = p$, or for short, a nontrivial loop based at p. Thus, instead of different points, we take the same point, but in between we follow some nontrivial loop, so that the identity of the point should make no difference for the comparison principle. We then get a vector $W(1) \in T_p M$ that is in general different from $W(0)$. We can, of course, do this for every $W \in T_p M$, and, according to 1., we obtain a *linear* map

$$L_c : T_p M \to T_p M$$
$$W(0) \mapsto W(1). \tag{5.3.62}$$

Moreover, when we then parallel transport the vector $V(0) := W(1)$ as $V(t)$ along another curve c' with $c'(0) = c'(1) = p$, we obtain another vector $V(1) = L_{c'}(V(0))$, and we have the group law

$$(L_{c'} \circ L_c)(W) = L_{c'}(L_c(W)) = L_{c' \cdot c}(W) \text{ and } L_{c^{-1}} = (L_c)^{-1}, \tag{5.3.63}$$

where $c' \cdot c(t) := c(2t)$ for $0 \le t \le 1/2$ and $= c'(2t - 1)$ for $1/2 \le t \le 1$, i.e., $c' \cdot c$ is the composition of the curves c and c', and $c^{-1}(t) := c(1 - t)$, that is, c^{-1} is the curve c traversed in the opposite direction. Here, we are applying 2. above. Thus, we have a homomorphism from the groups of loops based at p with the composition law to the group of linear automorphisms of the vector space $T_p M$. The image of this homomorphism is called the holonomy group H_p of the connection ∇ at p. The holonomy groups at different points are conjugate to each other. In fact, for $p, q \in M$, a closed loop γ at q and a curve c with $c(0) = p, c(1) = q$, we have

$$L_\gamma = L_c \circ L_{c^{-1} \cdot \gamma \cdot c} \circ L_{c^{-1}} \tag{5.3.64}$$

where we observe that analogous to (5.3.62), we presently have a linear map $L_c : T_p M \to T_q M$.

In general, however, the holonomy groups are simply isomorphic to the general linear group $Gl(d, \mathbb{R})$. Therefore, they are not useful for quantifying the path dependence of parallel transport. Therefore, we now undertake a different consideration. We define the curvature tensor R by

$$R(V, W)Z := \nabla_V \nabla_W Z - \nabla_W \nabla_V Z - \nabla_{[V,W]} Z, \tag{5.3.65}$$

or in local coordinates

$$R^k_{lij} \frac{\partial}{\partial x^k} := R(\frac{\partial}{\partial x^i}, \frac{\partial}{\partial x^j})\frac{\partial}{\partial x^l} \quad (i, j, l = 1, \ldots, d). \qquad (5.3.66)$$

The curvature tensor can be expressed in terms of the Christoffel symbols and their derivatives via

$$R^k_{lij} = \frac{\partial}{\partial x^i}\Gamma^k_{jl} - \frac{\partial}{\partial x^j}\Gamma^k_{il} + \Gamma^k_{im}\Gamma^m_{jl} - \Gamma^k_{jm}\Gamma^m_{il}. \qquad (5.3.67)$$

We also note that, as the name indicates, the curvature tensor R is, like the torsion tensor T, but in contrast to the connection ∇ represented by the Christoffel symbols, a tensor. This means that when one of its arguments is multiplied by a smooth function, we may simply pull out that function without having to take a derivative of it. Equivalently, it transforms as a tensor under coordinate changes; here, the upper index k stands for an argument that transforms as a vector, that is contravariantly, whereas the lower indices l, i, j express a covariant transformation behavior.

The curvature tensor quantifies to what degree covariant derivatives do not commute. Intuitively, it quantifies the difference between infinitesimally transporting the vector Z first in the direction of V and then in the direction of W compared with transporting it first in the direction of W and then in the direction of V. Or equivalently, it compares Z with the result of transporting it first in the direction V, then in the direction W, then in the direction $-V$, and finally in the direction $-W$, that is, by transporting it around an infinitesimal rectangle. When the end result is Z again, the curvature $R(V, W)Z$ is 0. When the result is different from Z, the corresponding curvature is $\neq 0$. In that sense, the curvature quantifies the infinitesimal path dependence of parallel transport.

As we have observed, the curvature R constitutes a tensor. In particular, whether it vanishes or not does not depend on the choice of coordinates. Therefore, the curvature tensor can yield geometric invariants, that is, quantities that do not depend on the choice of coordinates, but only on the underlying geometry, here that of the connection ∇.

Now, let M carry a Riemannian metric $g = \langle \cdot, \cdot \rangle$. We may then ask about the compatibility of a connection with this Riemannian structure, in the sense that parallel transport preserves the lengths of and the angles between tangent vectors. We therefore say that ∇ is a Riemannian connection if it satisfies the metric product rule

$$Z\langle V, W\rangle = \langle \nabla_Z V, W\rangle + \langle V, \nabla_Z W\rangle. \qquad (5.3.68)$$

One can show that for any Riemannian metric g, there exists a unique torsion-free Riemannian connection, the so-called Levi-Cività connection ∇^g. It is given by

$$\langle \nabla^g_V W, Z\rangle = \frac{1}{2}\{V\langle W, Z\rangle - Z\langle V, W\rangle + W\langle Z, V\rangle$$
$$- \langle V, [W, Z]\rangle + \langle Z, [V, W]\rangle + \langle W, [Z, V]\rangle\}. \qquad (5.3.69)$$

The Levi-Cività connection ∇^g respects the metric as required above; that is, if $V(t)$, $W(t)$ are parallel vector fields along a curve $c(t)$, then

$$\langle V(t), W(t)\rangle \equiv const. \qquad (5.3.70)$$

Thus, products between tangent vectors remain invariant under parallel transport.

The curvature tensor of the Levi-Cività connection of a Riemannian manifold then yields invariants of the Riemannian metric. This was one of the fundamental insights of Bernhard Riemann.

The Christoffel symbols of ∇^g can be expressed through the metric; in local coordinates, with $g_{ij} = \langle \frac{\partial}{\partial x^i} \frac{\partial}{\partial x^j} \rangle$, we use the abbreviation

$$g_{ij,k} := \frac{\partial}{\partial x^k} g_{ij} \tag{5.3.71}$$

and have

$$\Gamma_{ij}^k = \frac{1}{2} g^{kl} (g_{il,j} + g_{jl,i} - g_{ij,l}), \tag{5.3.72}$$

or, equivalently,

$$g_{ij,k} = g_{jl} \Gamma_{ik}^l + g_{il} \Gamma_{jk}^l = \Gamma_{ikj} + \Gamma_{jki}. \tag{5.3.73}$$

We note that (5.3.72) is the same as (5.3.39). This is another way of seeing the compatibility between the connection ∇^g and the metric g, as (5.3.72) comes from (5.3.50) and therefore is defined through the connection whereas (5.3.39) directly comes from the metric.

From (5.3.67) and (5.3.72), we see that the curvature tensor of the Levi-Cività connection of a Riemannian metric can be expressed in terms of (first and) second derivatives of that Riemannian metric g_{ij}. It turns out that there are no invariants that are given solely in terms of first derivatives of the metric. In fact, as Riemann had already observed, for any point p on a Riemannian manifold, we may introduce local coordinates such that

$$g_{ij}(p) = \delta_{ij}, \text{ and } g_{ij,k}(p) = 0 \text{ for all } i, j, k. \tag{5.3.74}$$

Thus, in particular, all first derivatives of the metric can be made to vanish at the arbitrary point p. Thus, also all Christoffel symbols then vanish at p. In general, however, they can only be made to vanish at a single point, but not for instance locally in some neighborhood of that point. Nevertheless, (5.3.74) has an important consequence that can often simplify computations considerably. Namely, when we do any tensor computation involving first derivatives of the metric, we may then assume that (5.3.74) is satisfied at the point p where we happen to carry out the computation. The result of the computations can then be simply transformed back into arbitrary coordinates by the rules for the transformations of co- or contravariant tensors. That is, we can carry out the computation in the simplified coordinates of (5.3.74) and obtain the result in other coordinates from the transformation behavior of the tensor in question.

Since the first derivatives of the metric can be made to vanish by a suitable coordinate transformation, invariants cannot possibly be computed from such first derivatives. In contrast, the second derivatives cannot all be made to disappear, as they are constrained by the curvature tensor which has a coordinate-independent meaning. Historically, it was the other way around. Gauss for surfaces in three-dimensional Euclidean space and Riemann in general discovered that certain combinations of second derivatives of the metric tensor, as computed in local coordinates, yield expressions that remain invariant under coordinate changes. Such expressions were then called curvature from the way they were originally derived in terms of certain curvatures of curves on a surface.

A curve $c(t)$ in M is called autoparallel or geodesic if

$$\nabla_{\dot{c}} \dot{c} = 0. \tag{5.3.75}$$

In local coordinates, (5.3.75) becomes

$$\ddot{c}^k(t) + \Gamma_{ij}^k(c(t))\dot{c}^i(t)\dot{c}^j(t) = 0 \quad \text{for } k = 1, \ldots, d. \tag{5.3.76}$$

Formally, this is the same as (5.3.38). We note, however, that the Christoffel symbols in (5.3.38) were defined through the metric whereas here, they come from the connection. Thus, (5.3.38) and (5.3.76) coincide only for the Levi-Cività connection, that is, when the two definitions of the Christoffel symbols yield the same result. (5.3.38) stems from the length minimizing property of the curve, that is, from a metric concept. In contrast, (5.3.76) expresses an analogue of straightness in Euclidean space. In Euclidean space, a curve is shortest if and only if it is straight. In the general geometric context, these two properties only coincide for the Levi-Cività connection, that is, when the metric and the connection are compatible.

Equation (5.3.76) constitutes a system of second-order ODEs, and given $p \in M$, $V \in T_pM$, and since M is complete, one can show that there exists a geodesic

$$c_V : [0, \infty) \to M$$

with $c_V(0) = p, \dot{c}_V(0) = V$. Without the assumption of completeness, such a geodesic need only exist on some interval $[0, \delta)$, with $\delta > 0$ depending on p and V.

We point out once more that (5.3.75) is a *nonlinear* system of differential equations for the curve c, in contrast to (5.3.60) which is a *linear* system for the vector field W along the curve c. Thus, when applying (5.3.60) to the tangent field of the curve c itself, it becomes nonlinear.

5.3.3 Curvature as a Measure of Nonlinearity

Definition 5.3.6 A connection ∇ on the differentiable manifold M is called *flat* if each point in M possesses a neighborhood U with local coordinates for which all the coordinate vector fields $\frac{\partial}{\partial x^i}$ are parallel, that is,

$$\nabla \frac{\partial}{\partial x^i} = 0. \tag{5.3.77}$$

Theorem 5.3.4 *A connection ∇ on M is flat if and only if its curvature and torsion vanish identically.*

In most cases, only torsionfree connections are considered anyway. Therefore, curvature is the crucial quantity whose vanishing implies flatness. We recall that curvature was introduced as a measure of the path dependence of parallel transport. Thus, when curvature vanishes, that is, when parallel transport is path independent, we can introduce covariantly constant coordinate vector fields. And this condition, (5.3.77), is characteristic of Euclidean space. One also calls Euclidean space flat space, and this explains the terminology employed in the definition. "Flat" here means "non-curved".

Proof When the connection is flat, all $\nabla_{\frac{\partial}{\partial x^i}} \frac{\partial}{\partial x^j} = 0$, and so, all Christoffel symbols $\Gamma_{ij}^k = 0$, and therefore, also T and R vanish, as they can be expressed in terms of the Γ_{ij}^k, see (5.3.57), (5.3.67).

For the converse direction, we shall find local coordinates y for which

$$\nabla dy = 0. \tag{5.3.78}$$

Here, we are using the connection, again denoted by ∇ on the cotangent bundle T^*M induced by the connection ∇ on TM. This connection is defined by the property that for a vector field V and a one-form ω, we have

$$d(V, \omega) = (\nabla V, \omega) + (V, \nabla \omega), \tag{5.3.79}$$

that is, by a product rule. In contrast to the product rule for a metric connection that uses the metric $\langle ., .\rangle$, here we use the pairing between vectors and one-forms which is given by duality. In local coordinates, this connection is characterized by the Christoffel symbols:

$$\nabla_{\frac{\partial}{\partial x^i}} dx^j = -\Gamma_{ik}^j dx^k. \tag{5.3.80}$$

For such coordinates y, the coordinate vector fields $\frac{\partial}{\partial y^i}$ then are covariantly constant, i.e., satisfy (5.3.77), because $(dy^j, \frac{\partial}{\partial y^i}) = \delta_i^j$, hence $0 = d(dy^j, \frac{\partial}{\partial y^i}) = (\nabla dy^j, \frac{\partial}{\partial y^i}) + (dy^j, \nabla \frac{\partial}{\partial y^i})$ by (5.3.79) and $\nabla dy^j = 0$ by (5.3.78).

For given coordinates, we have $dy = \frac{\partial y}{\partial x^i} dx^i$. We shall proceed in two steps. We first construct a covariantly constant (vector valued) 1-form $\mu_i dx^i$, and then show that the μ_i can be represented as derivatives, $\mu_i = \frac{\partial y}{\partial x^i}$. For the first step, we shall use the vanishing of the curvature, and for the second, the vanishing of the torsion. In both steps, the decisive ingredient will be the Theorem of Frobenius. That theorem says that we can locally integrate an overdetermined system of ordinary differential equations where all first derivatives of a function are prescribed when that system is compatible with the commutation of second derivatives; see e.g. Appendix A in [35].

The equation for the first step,

$$\nabla \mu_i dx^i = 0 \tag{5.3.81}$$

by (5.3.80) is equivalent to the system

$$\frac{\partial}{\partial x^j} \mu_i + \Gamma_{ji}^k \mu_k = 0 \quad \text{for all } i, j. \tag{5.3.82}$$

In vector notation, this becomes

$$\frac{\partial}{\partial x^j} \mu + \Gamma_j \mu = 0. \tag{5.3.83}$$

We now use the Theorem of Frobenius, as mentioned. For a smooth solution, the second derivatives should commute, that is, we should have $\frac{\partial}{\partial x^i}\frac{\partial}{\partial x^j}\mu = \frac{\partial}{\partial x^j}\frac{\partial}{\partial x^i}\mu$ for all i,j. By (5.3.83), this implies

$$[\Gamma_i, \Gamma_j] + \frac{\partial}{\partial x^i}\Gamma_j - \frac{\partial}{\partial x^j}\Gamma_i = 0 \tag{5.3.84}$$

holds for all i,j. The Theorem of Frobenius then says that this necessary condition is also sufficient, that is, we can solve (5.3.83) locally if and only if (5.3.84) holds. With indices, this is

$$\frac{\partial \Gamma_{j\ell}^k}{\partial x^i} - \frac{\partial \Gamma_{i\ell}^k}{\partial x^j} + \Gamma_{im}^k \Gamma_{j\ell}^m - \Gamma_{jm}^k \Gamma_{i\ell}^m = 0 \quad \text{for all } i,j. \tag{5.3.85}$$

By (5.3.67), this means that the curvature tensor vanishes. We can thus solve (5.3.82) for the μ_i. In order that these μ_i are derivatives $\frac{\partial y}{\partial x^i}$, the necessary and sufficient condition (again, by the theorem of Frobenius) is

$$\frac{\partial}{\partial x^i}\mu_j = \frac{\partial}{\partial x^j}\mu_i \quad \text{for all } i,j, \tag{5.3.86}$$

which by (5.3.82) in turn is equivalent to the condition $\Gamma_{ij}^k = \Gamma_{ji}^k$ for all i,j,k, that is, the vanishing of the torsion T, see the derivation of (5.3.58). This completes the proof. $\square$

Equation (5.3.77) is an infinitesimal condition. We now want to measure the deviation from flatness in a local sense. We assume that the connection ∇ is the Levi-Città connection of a Riemannian metric $\langle .,.\rangle$; in particular, it is torsionfree. The crucial aspect here is the behavior of geodesics. We recall the geodesic equation (5.3.75), and we assume that we have a family of geodesics $c(.,s)$ depending on a parameter s. We then have

$$\nabla_{\frac{\partial c}{\partial t}} \frac{\partial c(t,s)}{\partial t} = 0. \tag{5.3.87}$$

Since we assume that all the curves $c(.,s)$ are geodesic, hence all satisfy (5.3.87), we can then also take the covariant derivative w.r.t. s to get

$$
\begin{aligned}
0 &= \nabla_{\frac{\partial c}{\partial s}}\nabla_{\frac{\partial c}{\partial t}}\frac{\partial c(t,s)}{\partial t} \\
&= \nabla_{\frac{\partial c}{\partial t}}\nabla_{\frac{\partial c}{\partial s}}\frac{\partial c(t,s)}{\partial t} + R(\frac{\partial c}{\partial s}, \frac{\partial c}{\partial t})\frac{\partial c}{\partial t} \quad \text{by (5.3.65)} \\
&= \nabla_{\frac{\partial c}{\partial t}}\nabla_{\frac{\partial c}{\partial t}}\frac{\partial c(t,s)}{\partial s} + R(\frac{\partial c}{\partial s}, \frac{\partial c}{\partial t})\frac{\partial c}{\partial t} \quad \text{since } \nabla \text{ is torsionfree.} \quad (5.3.88)
\end{aligned}
$$

Thus, for any s, the vector field $X(t) := \frac{\partial c(t,s)}{\partial s}$ satisfies the so-called Jacobi equation

$$\nabla_{\frac{\partial c}{\partial t}}\nabla_{\frac{\partial c}{\partial t}} X(t) + R(X, \frac{\partial c}{\partial t})\frac{\partial c}{\partial t} = 0. \tag{5.3.89}$$

A solution X of (5.3.89) is called a *Jacobi field* along the geodesic $c(.,s)$. We note that (5.3.89) is a linear equation for the vector field X; in fact, it is the linearization of the geodesic equation, as we have obtained it by differentiating the equation for geodesics w.r.t. the parameter s. For a Jacobi field X, we obtain

$$\frac{d^2}{dt^2}\frac{1}{2}\langle X, X\rangle = \langle \nabla_{\frac{\partial c}{\partial t}} X, \nabla_{\frac{\partial c}{\partial t}} X\rangle - \langle R(X, \frac{\partial c}{\partial t})\frac{\partial c}{\partial t}, X\rangle$$

$$\geq -\langle R(X, \frac{\partial c}{\partial t})\frac{\partial c}{\partial t}, X\rangle. \tag{5.3.90}$$

Therefore, an upper bound for the curvature term $\langle R(X, \frac{\partial c}{\partial t})\frac{\partial c}{\partial t}, X\rangle$ yields a lower bound for the second derivative of the squared norm of X. We wish to explore the geometric consequences of this fact. In order to take into account the particular structure of this curvature term, we first introduce the important.

Definition 5.3.7 The *sectional curvature* of the plane spanned by the (linearly independent) tangent vectors $X, Y \in T_p M$ of the Riemannian manifold M is defined as

$$K(X \wedge Y) := \langle R(X, Y)Y, X\rangle \frac{1}{\|X \wedge Y\|^2} \tag{5.3.91}$$

$(\|X \wedge Y\|^2 = \langle X, X\rangle\langle Y, Y\rangle - \langle X, Y\rangle^2)$.

Note that we have chosen the normalization factor in (5.3.91) so that the sectional curvature does not depend on the lengths of the vectors X and Y nor on the angle between them, but only on the two-dimensional tangent plane that they span.

In coordinates, with $X = \xi^i \frac{\partial}{\partial x^i}$, $Y = \eta^i \frac{\partial}{\partial x^i}$, and putting

$$R_{ijkl} := \langle R(\frac{\partial}{\partial x^k}, \frac{\partial}{\partial x^l})\frac{\partial}{\partial x^j}, \frac{\partial}{\partial x^i}\rangle, \tag{5.3.92}$$

this becomes

$$K(X \wedge Y) = \frac{R_{ijk\ell}\xi^i\eta^j\xi^k\eta^\ell}{g_{ik}g_{j\ell}(\xi^i\xi^k\eta^j\eta^\ell - \xi^i\xi^j\eta^k\eta^\ell)}$$

$$= \frac{R_{ijk\ell}\xi^i\eta^j\xi^k\eta^\ell}{(g_{ik}g_{j\ell} - g_{ij}g_{k\ell})\xi^i\eta^j\xi^k\eta^\ell}. \tag{5.3.93}$$

Definition 5.3.8 The Riemannian manifold M is said to have constant sectional curvature K if

$$K(X \wedge Y) \equiv K \tag{5.3.94}$$

for all linearly independent tangent vectors $X, Y \in T_x M$ for all $x \in M$.

Euclidean space then has constant sectional curvature 0. A sphere of radius 1 has constant curvature 1, while hyperbolic space has curvature -1. This can be verified by direct computation, but we shall obtain these values below in a more elegant way with the help of Jacobi fields. Spaces of constant curvature ρ then are obtained by scaling the sphere or hyperbolic space, depending on whether $\rho > 0$ or < 0. For instance, the sphere of radius $\frac{1}{\sqrt{\rho}}$ has curvature $\rho > 0$. In fact, these curvature properties also characterize those spaces, in the following sense. Any Riemannian manifold

$\boxed{\text{Sphere}}$

$\boxed{\text{Hyperbolic space}}$

$\boxed{\begin{array}{l}\text{Sphere,}\\ \text{hyperbolic space}\end{array}}$

Sphere,
hyperbolic space

with curvature $\rho \equiv$ const. is locally isometric to a sphere, Euclidean space, or hyperbolic space with that curvature ρ. Such a space is also called a space form.

The Rauch comparison theorems then control a solution of the Jacobi equation (5.3.89) on a Riemannian manifold of sectional curvature

$$\lambda \leq K \leq \mu \qquad (5.3.95)$$

locally by the solutions of the Jacobi equation on the spaces of constant curvature λ and μ, resp. These comparison results follow from comparison theorems for ordinary differential equations. Essentially, locally, a solution of (5.3.89) is then smaller than the solution for $K \equiv \lambda$ and larger than the solution for $K \equiv \mu$, for corresponding initial values. That is, a lower curvature bound controls a Jacobi field from above, and an upper curvature bound controls it from below.

The Jacobi equation (for vector fields orthogonal to the tangent vector field of the underlying geodesic) for constant sectional curvature ρ is

$$\ddot{f}(t) + \rho f(t) = 0 \qquad (5.3.96)$$

for $\rho \in \mathbb{R}$. The reason why the Jacobi equation (5.3.89) which is a vector equation reduces to the scalar equation (5.3.96) in the constant curvature is that for constant sectional curvature ρ, we have

$$R(X, Y)Y = \rho \|Y\|^2 X. \qquad (5.3.97)$$

Also, we are using orthonormal coordinate vector fields $e_i(t)$ that are parallel along c, i.e., $\nabla_{\dot{c}(t)} e_i(t) = 0$. Thus, for $X(t) = x^i(t)e_i(t) \in T_{c(t)}M$, we have $\nabla_{\dot{c}(t)} X(t) = \frac{d}{dt}x^i(t)e_i(t) =: \dot{X}(t)$ and likewise $\nabla_{\dot{c}(t)}\nabla_{\dot{c}(t)} X(t) = \ddot{X}(t)$.

The solutions of (5.3.96) are simply

$$c_\rho(t) := \begin{cases} \cos(\sqrt{\rho}\,t) & \text{if } \rho > 0, \\ 1 & \text{if } \rho = 0, \\ \cosh(\sqrt{-\rho}\,t) & \text{if } \rho < 0, \end{cases} \qquad (5.3.98)$$

and

$$s_\rho(t) := \begin{cases} \frac{1}{\sqrt{\rho}}\sin(\sqrt{\rho}\,t) & \text{if } \rho > 0, \\ t & \text{if } \rho = 0, \\ \frac{1}{\sqrt{-\rho}}\sinh(\sqrt{-\rho}\,t) & \text{if } \rho < 0. \end{cases} \qquad (5.3.99)$$

They have initial values $f(0) = 1$, $\dot{f}(0) = 0$, resp. $f(0) = 0$, $\dot{f}(0) = 1$.

Sphere,
hyperbolic space

The solutions of these equations describe the behavior of families of geodesics on (scaled) spheres or hyperbolic spaces, and we turn once more to those basic examples of Riemannian manifolds.

On $\mathbb{R}^{n+1}$, we consider the two quadratic forms

$$\langle x, x \rangle_+ := (x^1)^2 + \cdots + (x^n)^2 + (x^{n+1})^2 \text{ and} \qquad (5.3.100)$$

$$\langle x, x \rangle_- := (x^1)^2 + \cdots + (x^n)^2 - (x^{n+1})^2 \qquad (5.3.101)$$

and recall the sphere (5.3.7)

$$S^n := \{x \in \mathbb{R}^{n+1} : \langle x, x \rangle_+ = 1\} \qquad (5.3.102)$$

and the hyperbolic space (5.3.8)

$$H^n := \{x \in \mathbb{R}^{n+1} : \langle x, x \rangle_- = -1, x^{n+1} > 0\}. \qquad (5.3.103)$$

For $x \in S^n$, $V \in T_x S^n$ satisfies

$$\langle x, V \rangle_+ = 0, \qquad (5.3.104)$$

and for $x \in H^n$, $V \in T_x H^n$, we analogously have

$$\langle x, V \rangle_- = 0. \qquad (5.3.105)$$

Therefore, by Sylvester's theorem,[3] the restriction of $\langle ., . \rangle_-$ to $T_x H^n$ yields a positive definite quadratic form. $\langle ., . \rangle_+$ and $\langle ., . \rangle_-$ then induce Riemannian metrics on S^n and H^n, resp. Let $O(n+1)$ and $O(n, 1)$ be the subgroups of the general linear group $Gl(n+1)$ that leave $\langle ., . \rangle_+$ and $\langle ., . \rangle_-$ invariant, resp. They then also leave S^n and H^n invariant, if we restrict in the latter case to the subgroup of $O(n, 1)$ that maps the positive x^{n+1}-axis to itself. Since they leave the quadratic forms and their level spaces S^n and H^n, resp., invariant, they operate by isometries on those spaces. Since geodesics are mapped to geodesics by isometries, they should remain invariant when their endpoints are fixed under such an isometry. In fact, this is in general only correct locally, that is, when these endpoints are sufficiently close together and the geodesic is their shortest connection. In that case, such shortest geodesics are unique, and they do indeed have to remain invariant. From this, we can conclude that the geodesics are precisely the intersections of S^n and H^n with two-dimensional linear subspaces of $\mathbb{R}^{n+1}$, because those subspaces remain invariant under reflections, which are elements of the respective isometry groups. Thus, for $x \in S^n$, $V \in T_x S^n$, $\langle V, V \rangle_+ = 1$, the geodesic $c : \mathbb{R} \to S^n$ with $c(0) = x$, $\dot{c}(0) = V$ is given by

$$c(t) = (\cos t)x + (\sin t)V. \qquad (5.3.106)$$

Analogously, the geodesic $c : \mathbb{R} \to H^n$ with $c(0) = x \in H^n$, $\dot{c}(0) = V \in T_x H^n$, $\langle V, V \rangle_- = 1$ is given by

$$c(t) = (\cosh t)x + (\sinh t)V. \qquad (5.3.107)$$

In fact, since we have $\langle x, x \rangle_- = -1$, $\langle x, V \rangle_- = 0$, $\langle V, V \rangle_- = 1$, we get

$$\langle \dot{c}(t), \dot{c}(t) \rangle_- = -\sinh^2 t + \cosh^2 t = 1, \qquad (5.3.108)$$

and analogously on S^n.

If we then have $W \in T_x H^n$ with $\langle W, V \rangle_- = 0$, $\langle W, W \rangle_- = 1$, we obtain a family of geodesics

$$c(t, s) := \cosh t x + \sinh t (\cos s V + \sin s W). \qquad (5.3.109)$$

---

[3] This theorem says that every homogeneous quadratic polynomial is reducible by real orthogonal substitutions to the form of a sum of positive and negative squares. See [75], p.577.

This family yields the Jacobi field

$$X(t) := \sinh t\, W \qquad (5.3.110)$$

at $s = 0$. It satisfies

$$\ddot{X}(t) - X(t) = 0. \qquad (5.3.111)$$

From the Jacobi equation (5.3.89), we then conclude that the hyperbolic space H^n possesses sectional curvature $\equiv -1$. Analogously, the sphere S^n has sectional curvature $\equiv 1$.

| Hyperbolic space |

| Sphere |

Likewise, the scaled sphere (5.3.7)

$$S^n(\rho) := \left\{ x \in \mathbb{R}^{n+1} : \langle x, x \rangle_+ = \frac{1}{\sqrt{\rho}} \right\} \qquad (5.3.112)$$

then has sectional curvature $\equiv \rho$, and analogously, the scaled hyperbolic space $H^n(\rho)$ has sectional curvature $\equiv -\rho$.

| Sphere, |
| hyperbolic space |

The solutions of (5.3.98) and (5.3.99) thus give the characteristic behavior of families of geodesics on (scaled) spheres or hyperbolic spaces. On a sphere, we have the families of geodesics going through two antipodal points, say north and south pole, and the distance between them behaves like the sin function s_ρ for $\rho > 0$. On hyperbolic space, in contrast, geodesics diverge exponentially, as the sinh function s_ρ for $\rho < 0$. Euclidean space, where geodesics diverge linearly, is intermediate between the cases of positive and negative curvature.

In fact, these properties are characteristic of Riemannian manifolds with curvature controls, and they lead to an axiomatic approach to curvature. This was first conceived in the Vienna school; in particular, the contribution of Wald [112] exhibits a key idea. We can put this into a general perspective. A configuration of three points in a metric space (X, d) has to obey the triangle inequality, but otherwise there mutual distances can be arbitrary. We call a configuration of three points (x_1, x_2, x_3) in X a *triangle*. For such a triangle in (X, d), there exist points $\overline{x}_1, \overline{x}_2, \overline{x}_3 \in \mathbb{R}^2$ with the same distances

$$d(x_i, x_j) = |\overline{x}_i - \overline{x}_j|, \qquad \text{for every } i, j = 1, 2, 3. \qquad (5.3.113)$$

We say that the points $\overline{x}_1, \overline{x}_2, \overline{x}_3 \in \mathbb{R}^2$ represent an isometric embedding of the triangle (x_1, x_2, x_3) into the Euclidean plane. The term "isometry" here refers to the fact that distances are preserved. Such a triple of points $(\overline{x}_1, \overline{x}_2, \overline{x}_3)$ is called a *comparison triangle* for the triangle (x_1, x_2, x_3), and it is unique up to isometries (a result of classical Euclidean geometry, as the reader may recall). Likewise, we can also find comparison triangles in hyperbolic instead of Euclidean space, and also in a sphere, if the distances $d(x_i, x_j)$ are not too large. This means that such a triangle does not exhibit essential geometric information about the underlying space (X, d). This changes if we consider configurations of four points $x_1, \ldots, x_4$ with their mutual distances. In particular, as Wald [112] observed, a configuration of four points on a smooth surface S can be isometrically embedded into a unique constant curvature surface, unless particular conditions hold, like when some of the points coincide or when they all lie on the same geodesic curve. That is, we find a unique K and a surface S_K of constant curvature

K with its metric denoted by d_K on which there are points $\overline{x}_i, i = 1, 2, 3, 4$ with

$$d(x_i, x_j) = d_K(\overline{x}_i - \overline{x}_j), \qquad \text{for every } i, j = 1, 2, 3, 4. \qquad (5.3.114)$$

Wald then says that the surface S has curvature $\leq k$ ($\geq k$) if for all such configurations the comparison surface S_K has $K \leq k$ ($K \geq k$). This works well for surfaces, which was Wald's aim. For higher dimensional spaces, however, arbitrary four-point configurations seem to be too general. Just consider a tetrahedron in Euclidean 3-space, that is, four points with all distances between different points equal to each other. Such a configuration can be isometrically embedded into some sphere of at ppropriate curvature depending on the distance in the tetrahedron, but not into the Euclidean plane. Therefore, we should only work with specific four-point configurations. In fact, as we shall see in a moment, things work well if we start with a triangle and choose the fourth point on the shortest geodesic between two of the vertices of that triangle. The critical distance which yields information about the curvature of the underlying space is then the distance from that fourth point to the remaining third vertex of the triangle.

We thus turn to the general theories of curvature inequalities in metric spaces that were developed by Busemann [19] and Alexandrov [1]. We now describe the essential principles of this approach (see e.g. [12, 18, 57]). For simplicity, we only treat a special case, that of generalized nonpositive curvature. We use the concept of a geodesic space as defined in Sect. 5.3.2.

Definition 5.3.9 The geodesic space (X, d) is said to have curvature bounded from above by 0 in the sense of Busemann if for every $p \in X$ there exists some $\rho_p > 0$ such that for all $x, y, z \in B(p, \rho_p)$

$$\frac{d(m(x, y), m(x, z))}{d(y, z)} \leq \frac{1}{2} \qquad (5.3.115)$$

where the midpoints $m(x_1, x_2)$ have been defined in 2.1.56 and characterized in (5.3.46).

In a similar manner, one can also define upper curvature bounds other than 0, and also lower curvature bounds for metric spaces, although the technical details are somewhat more complicated.

In a certain sense, Busemann's definition provides a negative view of nonpositive curvature, in the sense that geodesics are diverging faster than in the Euclidean case. The distance between y and z is at least twice as large as the distance between $m(x, y)$ and $m(x, z)$, the two points that are half as far away from x than y and z, on the same geodesics. A more positive view, that is more restrictive, but which implies stronger properties, was provided by Alexandrov. He started from the observation that in a space of negative curvature, the geodesic from y to z comes at least as close to x as the corresponding geodesic would in Euclidean space. More precisely,

Definition 5.3.10 The geodesic space (X, d) is said to have curvature bounded from above by 0 in the sense of Alexandrov if for every $p \in X$

there exists some $\rho_p > 0$ such that for all $x_1, x_2, x_3 \in B(p, \rho_p)$ and any shortest geodesic $c : [0, 1] \to X$ with $c(0) = x_1, c(1) = x_2$, parametrized proportionally to arclength, for all $0 \le t \le 1$

$$d^2(x_3, c(t)) \le (1 - t)d^2(x_1, x_3) + td^2(x_2, x_3) - t(1 - t)d^2(x_1, x_2).$$
$$(5.3.116)$$

In fact, it suffices to require (5.3.116) for $t = \frac{1}{2}$, that is, for the midpoint $m(x_1, x_2) = c(\frac{1}{2})$ of x_1 and x_2. (5.3.116) then becomes

$$d^2(x_3, m(x_1, x_2)) \le \frac{1}{2}d^2(x_1, x_3) + \frac{1}{2}d^2(x_2, x_3) - \frac{1}{4}d^2(x_1, x_2).$$
$$(5.3.117)$$

We again consider a triangle, that is, a configuration of three points (x_1, x_2, x_3) in X a *triangle* and a comparison triangle $(\overline{x}_1, \overline{x}_2, \overline{x}_3)$ in $\mathbb{R}^2$ satisfying (5.3.113).

Let $\overline{c} : [0, 1] \to \mathbb{R}^2$ be the straight line segment with $\overline{c}(0) = \overline{x}_1, \overline{c}(1) = \overline{x}_2$, that is the analogue of the geodesic curve c from x_1 to x_2. Then an equivalent formulation of (5.3.116) is

$$d(x_3, c(t)) \le \|\overline{x}_3) - \overline{c}(t)\| \text{ for } 0 \le t \le 1, \qquad (5.3.118)$$

and in particular for $t = \frac{1}{2}$ as in (5.3.117).

The Fig. 5.1 depicts the relationship between the distances in a triangle in a space of nonpositive curvature in the sense of Alexandrov and a Euclidean comparison triangle.

We now want to reformulate the condition in Definition 5.3.10, or more precisely its version 5.3.117, in such a manner that it becomes meaningful even for metric spaces (X, d) that are not geodesic, like discrete metric spaces. Let $x_1, x_2 \in X$. We put

$$\rho_{(x_1, x_2)}(x) := \max_{i=1,2} d(x, x_i) \text{ and } r(x_1, x_2) := \inf_{x \in X} \rho_{(x_1, x_2)}(x). \quad (5.3.119)$$

For every x, we have that

$$x \in B(x_1, \rho_{(x_1, x_2)}(x)) \cap B(x_2, \rho_{(x_1, x_2)}(x)) \qquad (5.3.120)$$

and thus, the intersection of these two balls centered at x_1 and x_2 with the same radius $r = \rho_{(x_1, x_2)}(x)$ is nonempty. When x_1 and x_2 possess a

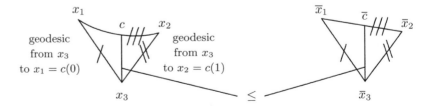

Fig. 5.1 Comparison between a *triangle* in a space of nonpositive curvature in the sense of Alexandrov and the *triangle* with the same lengths of corresponding sides (indicated by *slashes*) in the Euclidean plane. Redrawn from [57]

midpoint $m(x_1, x_2)$, then, as we have observed in Lemma 2.1.3

$$r(x_1, x_2) = \frac{1}{2} d(x_1, x_2), \qquad (5.3.121)$$

and

$$m(x_1, x_2) \in B(x_1, r(x_1, x_2)) \cap B(x_2, r(x_1, x_2)) \qquad (5.3.122)$$

and so, again, that intersection is nonempty. Moreover, the infimum in (5.3.119) is realized by $x = m(x_1, x_2)$. Thus, the smallest radius needed to have a nonempty intersection of two closed balls contains some very basic information about a metric space. We now proceed to look at radii that guarantee a nonempty intersection of three balls, and we shall find that this contains curvature information.

We seek the smallest radius r_3 such that the ball $B(x_3, r_3)$ intersects the two balls $B(x_1, r_{12})$ and $B(x_2, r_{12})$. Since those two balls intersect only in $m(x_1, x_2)$, that other ball then also has to contain $m(x_1, x_2)$. Therefore, that radius has to be equal to the distance between x_3 and that midpoint. Thus, we conclude that $r_3 = d(x_3, m(x_1, x_2))$ is the smallest radius r' of a closed ball $B(x_3, r')$ centered at x_3 such that

$$B(x_1, r_{12}) \cap B(x_2, r_{12} \cap B(x_3, r') \neq \emptyset. \qquad (5.3.123)$$

This observation can now be related to our curvature bound (5.3.117), because $m(x_1, x_2) = c(\frac{1}{2})$ in the notation of Definition 5.3.10. The inequality (5.3.117) thus becomes

$$r_3 \leq \bar{r}_3 \qquad (5.3.124)$$

for the Euclidean analogue $\bar{r}_3$ of r_3.

Along these lines, one can also formulate another version of nonpositive curvature that treats all three points x_1, x_2, x_3 of the triangle equally and that consequently is easier to formulate. This was discovered in [7]. Again, we consider a comparison triangle $\bar{x}_1, \bar{x}_2, \bar{x}_3 \in \mathbb{R}^2$ with the same distances

$$d(x_i, x_j) = |\bar{x}_i - \bar{x}_j|, \qquad \text{for every } i, j = 1, 2, 3.$$

We then define the functions

$$\rho_{(x_1, x_2, x_3)}(x) := \max_{i=1,2,3} d(x, x_i), \qquad x \in X,$$

and,

$$\rho_{(\bar{x}_1, \bar{x}_2, \bar{x}_3)}(\bar{x}) := \max_{i=1,2,3} \|\bar{x} - \bar{x}_i\|, \qquad \bar{x} \in \mathbb{R}^2.$$

The numbers

$$r(x_1, x_2, x_3) := \inf_{x \in X} \rho_{(x_1, x_2, x_3)}(x) \quad \text{and} \quad r(\bar{x}_1, \bar{x}_2, \bar{x}_3) := \min_{\bar{x} \in \mathbb{R}} \rho_{(\bar{x}_1, \bar{x}_2, \bar{x}_3)}(\bar{x})$$

$$(5.3.125)$$

are called the *circumradii* of the respective triangles.

Definition 5.3.11 (*Nonpositive curvature*) Let (X, d) be a metric space. We say that $\mathrm{Curv} X \leq 0$ if, for each triangle (x_1, x_2, x_3) in X, we have

$$r(x_1, x_2, x_3) \leq r(\bar{x}_1, \bar{x}_2, \bar{x}_3), \qquad (5.3.126)$$

where $\bar{x}_i$ with $i = 1, 2, 3$ are the vertices of an associated comparison triangle.

Again, by obvious modifications, one can formulate other curvature bounds. As shown in [7], on a Riemannian manifold, the condition $\mathrm{Curv}X \leq 0$ is again equivalent to nonpositive sectional curvature. Also, nonpositive Alexandrov curvature implies nonpositive curvature in the sense of Definition 5.3.11 whereas nonpositive Busemann curvature doesn't in general.

From the preceding, it is clear that the geometric content of Definition 5.3.11 is the following. When the infimum in (5.3.125) is achieved, $r(x_1, x_2, x_3)$ is the smallest number r with the property that

$$\bigcap_{i=1,2,3} B(x_i, r) \neq \emptyset, \tag{5.3.127}$$

and this radius must not be larger than the corresponding radius for a Euclidean comparison triangle. (When the infimum in (5.3.125) is not achieved, we only have $\bigcap_{i=1,2,3} B(x_i, r(x_1, x_2, x_3) + \varepsilon) \neq \emptyset$ for any $\varepsilon > 0$, but the geometric intuition is essentially the same.)

Thus, while for two balls in a geodesic space $B(x_1, r) \cap B(x_2, r) \neq \emptyset$ whenever $r \geq \frac{1}{2}d(x_1, x_2)$, and therefore, the smallest such number does not have much geometric content, the intersection patterns of three balls reflect an important geometric property. The condition (5.3.127) does not involve any point x explicitly. It thus embodies the principle that three balls in X should have a nonempty intersection whenever the corresponding balls in the Euclidean plane with the same distances between their centers intersect nontrivially. We can apply this principle to any metric space and search for the minimal radius for which the balls centered at three given points have a nonempty intersection. Curvature bounds such as (5.3.126) quantify the dependence of such a minimal radius on the distances between the points involved, as compared to the Euclidean situation.

Since for discrete metric spaces, small balls might just contain their center and nothing else, it might be appropriate to utilize the following modification.

Definition 5.3.12 Let $\varepsilon > 0$. A metric space (X, d) has ε-*relaxed nonpositive curvature* if

$$\bigcap_{i=1,2,3} B(x_i, r) \neq \emptyset \text{ whenever } r \geq r(\overline{x}_1, \overline{x}_2, \overline{x}_3) + \varepsilon \tag{5.3.128}$$

for every triangle $(x_1, x_2, x_3) \in X$.

The essential point here is that ε is independent of the distances between the points x_1, x_2, x_3 constituting the triangle. Therefore, when these distances are smaller than ε, the condition (5.3.128) is automatically satisfied. In contrast, when these distances are large compared with ε, the condition is essentially the same as (5.3.127).

The formulation of the nonpositive condition in terms of intersection patterns of balls, as in (5.3.127) and (5.3.128), should also remind us of the construction of the Čech complex from intersection patterns of balls in Sect. 3.1 which could then be analyzed in Sect. 4.6 with the tools of algebraic topology. Our nonpositive curvature condition can then be considered as a geometric refinement of this topological scheme. Of course, we could also consider intersection patterns of more than three balls. While for geodesic spaces this does not seem to contain much further geometric content, for discrete spaces a version of (5.3.128) for more than three balls could yield useful geometric information.

Whether two balls intersect does not contain specific geometric information about the underlying space, because the criterion is simply that the sum of their radii be at least as large as the distance between their centers. When, however, the space also carries a measure, we can look at the measure of the intersection of two balls as a function of their radii and the distance between their centers. This contains information about the Ricci curvature when the space is a Riemannian manifold. Here, the Ricci curvature in the direction of a tangent vector X is defined as the average of the sectional curvatures of all the tangent planes containing X. Therefore, one can use this in general metric spaces that are equipped with a measure to define generalized Ricci curvature bounds. While here, we do not go into the details, we only remark that it is natural that for Ricci curvature, we not only need a metric, but also a measure, since the Ricci curvature in the Riemannian case had been defined as an average over sectional curvatures, and taking averages implicitly invokes a measure.

For a survey of sectional and Ricci curvature inequalities in metric spaces and their geometric implications, we refer to [9] and the references provided there.

5.4   Schemes

After having treated manifolds and metric aspects, we now turn to one of the other possibilities for conceptualizing space, as discussed in Sect. 5.1, that of a scheme. In this section, I shall heavily draw upon the references [33, 46, 102] which are recommended for a more detailed, extensive, and precise treatment. My aim here is only to describe the essential ideas.

The starting idea is the following one. We consider a topological space $(X, \mathcal{O}(X))$, and we assume that the topology is rich enough, that is, contains enough open sets so that each point yields a closed set (if not, we should replace "point" by "closed set" in the sequel). We have the notion of continuous functions $f : X \to \mathbb{R}$,[4] and we may impose the condition that the continuous functions separate points, that is, for any $a \neq b \in X$, there exists a continuous function f with $f(a) \neq f(b)$. By subtracting a constant—which is always continuous—, we may then assume that $f(a) = 0$.

Now, the continuous functions on X form a ring $C(X)$ as one can add and multiply their values in $\mathbb{R}$. This ring is commutative and contains an identity (always denoted as 1), the function $f \equiv 1$. In general, however, $C(X)$ is not a field because any f with $f(a) = 0$ for some $a \in X$ has no multiplicative inverse. In fact, given $a \in X$, the functions $f \in C(X)$ with $f(a) = 0$ constitute an ideal I_a in the ring $C(X)$ (see Sect. 5.4.1 below

---

[4]For the present discussion, other rings would do as well, but we stick to $\mathbb{R}$ here for concreteness. $\mathbb{C}$ will become important below.

for the definition and properties of ideals).[5] And if we impose the above separation property, then $I_a \neq I_b$ for $a \neq b \in X$. That is, when $\mathcal{I}$ is the set of ideals in $C(X)$, we obtain an injective map

$$X \to \mathcal{I}$$
$$a \mapsto I_a. \tag{5.4.129}$$

This means that we can recover the points in X from algebraic objects, ideals in a ring. According to the standard mathematical strategy that we have already encountered several times, we can then turn things around and start with those algebraic objects as our primary objects. That is, we consider a ring (commutative, with 1) together with its set of ideals, and then try to construct some topological space whose points are given by ideals. Obviously, there are some immediate problems, and therefore, ultimately, if this is going to work at all, it needs to be refined. In particular, not every ideal in $C(X)$ corresponds to a point of X. For instance, we can take any closed subset A of X. A then defines the ideal I_A of those continuous functions that are 0 at all the points of A. Thus, we can hope at best to recover the closed subsets of X. However, when A contains two different points $a \neq b$, then $I_A \subsetneq I_a, I_b$. More generally, whenever $B \subset A$, for closed subsets A, B, then $I_A \subset I_B$. Thus, we have a contravariant functor from the closed subsets of X to the ideals in $C(X)$.

Therefore, we can try to identify those ideals that correspond to points in X, as opposed to larger subsets, as those that are maximal, that is, not contained in some larger ideal (different from the entire ring). This, however, brings us into trouble with functorial properties. When $h : R_1 \to R_2$ is a ring homomorphism, the preimage of any ideal in R_2 is an ideal in R_1—so far, so good—but the preimage of a maximal ideal need not be maximal itself. Of course, this corresponds to the simple topological fact that a continuous map $F : X_1 \to X_2$ between topological spaces need not be injective, but can map some (closed, for our purposes) set $A \subset X_1$ containing more than one point to a single point a in X_2. When $f \in C(X_2)$ satisfies $f(a) = 0$, then $f \circ F \in C(X_1)$ satisfies $f(x) = 0$ for all $x \in A$, that is, it vanishes at A. Therefore, the pullback of a function that vanishes at $a \in X_2$ under the map F then becomes a function that vanishes on the entire set A. One possible resolution of this difficulty might consist in considering all closed subsets of the topological space X, as opposed to single points only. This would fit well with the functorial properties of continuous maps. A continuous map pulls closed sets back to closed ones, and the induced map on the ideal sets is then a transformation of the corresponding ideals.

There may be too many such closed subsets, however, to make this really feasible. Another approach could consist in restricting the class of functions, that is, not consider all continuous functions, but rather assume that the space X carries some additional structure and to restrict to the class of functions that preserve that structure. In fact, we shall see that this will

---

[5]In complex and algebraic geometry, there is a way to circumvent this problem, namely, to look at meromorphic functions, that is, also for functions assuming the value ∞ in a controlled manner.

help us to identify the strength of this approach. More precisely, we shall see that this approach is constitutive for modern algebraic geometry.

Before we proceed, it will be useful to recall some algebra:

5.4.1 Rings

We consider a commutative ring R with an identity, denoted as 1, and all rings occurring in the sequel will likewise automatically be assumed to be commutative and have a 1.

$a \in R$ is called a (multiplicative) unit if there exists some $b \in R$ with $ab = 1$. A ring is a field if all its elements $\neq 0$ are units.

We say that $a \neq 0$ divides c if there exists some $b \neq 0$ with $ab = c$. $a \neq 0$ in R is called a proper zero divisor if there exists some $b \neq 0$ in R with $ab = 0$, that is, if a divides 0. (In general, when talking about division concepts, 0 is always automatically excluded, without further explicit mentioning in the sequel.) A ring without proper zero divisors (and which is nontrivial in the sense that $1 \neq 0$) is called an integral domain. The integers $\mathbb{Z}$ are, of course, an example. From an integral domain, one can construct a quotient field. One takes pairs (a, b) of elements with $b \neq 0$ and considers pairs $(a_1, b_1), (a_2, b_2)$ as equivalent if $a_1 b_2 = a_2 b_1$. Every field is an integral domain, but not conversely, as the example $\mathbb{Z}$ again shows. On the other hand, the rings $\mathbb{Z}_q$ for q not a prime number are not integral domains. In fact, let $q = mn$ with $m, n > 1$ be a nontrivial product. Then $mn = 0$ mod q, and therefore, m and n divide 0.

$\boxed{\mathbb{Z}}$

$\boxed{\mathbb{Z}_q}$

An integral domain R is called a unique factorization domain if every $a \in R$ that is not a unit is a finite product, unique up to order and units, of irreducible elements. Here, $b \in R$ is called irreducible if it is not a unit and if the only divisors of b are units u or of the form bu for some unit u (it is clear that such elements u and bu trivially are divisors of b). The uniqueness of the factorization is equivalent to the condition that whenever an irreducible p divides a product ab, then it divides one of the factors. (The proof, which is elementary, is left to the reader or can be looked up, e.g. in [121] or [73].)

A particular and important class of unique factorization domains are the Euclidean domains. These are the integral domains where we can associate an integer $\nu(a)$ to every element a, with the properties that $\nu(b) \leq \nu(a)$ whenever b divides a, and for any a, b with $b \neq 0$, there exist elements q, r with $a = bq + r$ and $\nu(r) \leq \nu(b)$. The ring of integers is a Euclidean domain with $\nu(n) := |n|$. In a Euclidean domain, any nonzero elements a, b have a greatest common divisor d, and d is a linear combination of a and b, $d = \alpha a + \beta b$, for some $\alpha, \beta \in R$, see again [73, 121] for simplicity. In particular, a Euclidean domain is a unique factorization domain. For, when an irreducible p divides ab, but not a, then the greatest common divisor between p and a is 1, hence $1 = \alpha a + \beta p$. Therefore, $b = \alpha ab + \beta pb$, and since p is assumed to divide ab, it then also divides b.

We recall that an ideal I in R is a nonempty set of elements of R that forms a subgroup of R as an abelian group and satisfies $aI \subset I$ for all $a \in R$ (Definition 2.1.21). In particular, I is then a subring of R (but not every subring is an ideal, as is shown by the integer subring $\mathbb{Z}$ of the rationals $\mathbb{Q}$). When an ideal contains 1, it has to coincide with the ring R itself. Thus, the ideals of interest do not contain 1. Thus, they cannot contain a unit either. In particular, a field does not possess any nontrivial ideals, that is, ideals different from the field itself and the zero ideal (0), the ideal containing only 0. The intuitive idea that will be behind the subsequent constructions is that the presence of nontrivial ideals encodes the deviation of a ring from being a field. And, as we shall explore in a moment, nontrivial ideals can be generated by nontrivial noninvertible elements of a ring.

The intersection of ideals is again an ideal.

From an ideal I of R, we obtain the quotient ring R/I. The kernel $\mathrm{Ker}(h)$ of a ring homomorphism $h : R_1 \to R_2$ is an ideal in R_1. h then induces a ring isomorphism $R_1/\mathrm{Ker}(h) \to R_2$. More generally, when I_2 is an ideal in R_2, then its preimage $I_1 := h^{-1}(I_2)$ is an ideal in R_1. $\mathrm{Ker}(h)$ is then the preimage of the zero ideal $(0) \subset R_2$, and therefore the smallest one among all the preimages of ideals.

For any $a \in R$, aR, i.e., the set of elements of the form $ab, b \in R$, is an ideal, called the principal ideal generated by a. As already mentioned, for this ideal to be nontrivial, a should not be a unit, nor should it be 0. More generally, a family $\{a_\lambda\}_{\lambda \in \Lambda}$ generates the ideal I if every $b \in I$ can be written as a finite sum $b = \sum_{i=1}^{n} b_i a_{\lambda_i}$ with $b_i \in R$. We then also write $I = (a_\lambda)_{\lambda \in \Lambda}$. By definition, the empty set generates the zero ideal. An ideal is said to be finitely generated if it possesses a finite set of generators.

Definition 5.4.1 The ring R (commutative with 1, as always) is said to be *Noetherian* if every ideal is finitely generated.

The Noetherian property is very important, because it is needed for most results in the theory of schemes. The ring $C(M)$ of continuous functions on a manifold, unfortunately, is not Noetherian. For $p \in M$, the continuous functions f with $f(p)$ constitute an ideal I_p, as remarked above, but this ideal is not finitely generated. In fact, I_p is not even finite dimensional, and so, we cannot find any finite set of functions $f_1, \ldots, f_n$ so that any $f \in I_p$ is a linear combination of those functions. What we can only hope for is to expand such an f into an infinite series, like a Fourier series, but this does not qualify for the Noetherian property.

We now come to a fundamental example of Noetherian rings. Given a ring R, we consider the polynomial ring $R[X]$, the ring of polynomials $a_0 + a_1 X + \ldots a_n X$, with $a_0, \ldots, a_n \in R, n \in \mathbb{N} \cup \{0\}$, in the indeterminate X. When $a_n \neq 0$, we say that the polynomial has degree n. Analogously, we can consider the ring of polynomials with k indeterminates, $R[X_1, \ldots, X_k]$. The ring of polynomials $K[X]$ with coefficients in a field K and a single indeterminate is a Euclidean domain, with $\nu(f) :=$ degree of f for $f \neq 0$ and $\nu(0) = -1$.

We have the Hilbert basis theorem:

Theorem 5.4.1 *If the ring R is Noetherian, then so are the polynomial rings $R[X_1, \ldots, X_k]$.*

From an ideal I, we can construct another ideal, its radical $\mathrm{Rad}(I)$, the set of all $a \in R$ with $a^n \in I$ for some $n \in \mathbb{N}$. The nilradical $\mathrm{Rad}(0)$ consists of all nilpotent elements of R (an element a is called nilpotent if $a^n = 0$ for some $n \in \mathbb{N}$). The ring R is called reduced if $\mathrm{Rad}(0) = (0)$. For any ring R, we have the reduced ring $R_{\mathrm{red}} := R/\mathrm{Rad}(0)$.

An ideal $I \neq R$ of R is called maximal if there is no other ideal $I' \neq I, R$ with $I \subset I'$. The ideal I is maximal iff R/I is a field. This is easily follows from the fact that every a that is not a unit (nor 0) generates a nontrivial ideal. In other words, the zero ideal (0) is the only maximal ideal of a field, and this characterizes a field.

Maximal ideals, however, are not good for functorial purposes, because the preimage of a maximal ideal in R_2 under a homomorphism $h : R_1 \to R_2$ need not be maximal in R_1. For instance, let h embed a ring R without proper zero divisors that is not a field into a field K. Then the zero ideal (0) is maximal in K, but its preimage, the zero ideal (0) of R is not maximal in R.

This functorial problem goes away when we consider prime ideals instead of maximal ones. Here, an ideal $I \neq R$ in R is called prime if whenever $ab \in I$ for elements $a, b \in R$, then $a \in I$ or $b \in I$. In the integers $\mathbb{Z}$, the prime ideals are precisely the ideals $p\mathbb{Z}$ generated by a prime number p, and the zero ideal (0). When q is prime, then $\mathbb{Z}_q$ is a field (as discussed in Sect. 2.1.6), and hence the only prime ideal is (0), but when q is not prime, and p is a prime number dividing q, then the multiples of p (modulo q) also constitute a prime ideal. For instance, $\mathbb{Z}_4$ has the prime ideals $\{0\}$ and $\{0, 2\}$.

I is a prime ideal iff R/I is an integral domain. For, if $h : R \to R/I$ is the natural homomorphism, then $ab \in I$ iff $h(a)h(b) = 0$, and if R/I is an integral domain, this can happen only if $h(a) = 0$ or $h(b) = 0$.

In particular, any maximal ideal is prime because any field is an integral domain. Also, (0) is a prime ideal in the ring R iff R is an integral domain.

As indicated, if $h : R_1 \to R_2$ is a homomorphism, and if $I_2 \subset R_2$ is a prime ideal, then $I_1 := h^{-1}(I_2)$ is prime. For, if $ab \in I_1$, then $h(a)h(b) \in I_2$ and since the latter is prime, $h(a) \in I_2$ or $h(b) \in I_2$, and hence $a \in I_1$ or $b \in I_1$ so that I_1 is indeed prime.

Let K be a field and consider the polynomial ring $R := K[X_1, \ldots, X_n]$. Let $a_1, \ldots, a_n \in K$ and consider all the polynomials $f \in R$ with $f(a_1, \ldots, a_n) = 0$. They form a prime ideal I in R since K is a field. Since we can write every element $g \in R$ as $g(X) = a + \sum_i g_i(X)(X - a_i)$ for some elements $g_i \in R$, we see that $g(a_1, \ldots, a_n) = 0$ iff $a = 0$. Therefore, I is generated by the polynomials $X - a_i$. On the other hand, any ideal that contains both I and some element g with $g(a_1, \ldots, a_n) \neq 0$ then has to contain some $a \neq 0$, hence 1 since K is a field, and hence the ideal has to be the entire ring R. Therefore, the prime ideal I is maximal. In particular, R/I is a field, in fact, the ground field K.

In a Euclidean domain, for instance the integers $\mathbb{Z}$ or the polynomial ring $K[X]$ in one indeterminate over a field, every prime ideal is maximal. For, if I is a prime ideal in a Euclidean domain, it is of the form pR for some element p (a prime), and if $a \notin I$, then the greatest common divisor between a and p is 1, hence $\alpha a + \beta p = 1$ for some $\alpha, \beta \in R$, and therefore the image of a in the quotient R/I has an inverse. Thus, R/I is a field. And thus, the ideal I is maximal.

However, in the polynomial ring $K[X_1, X_2]$ with two indeterminates, there exist prime ideals that are not maximal. In fact, the ideal I generated by the irreducible polynomial X_1 is prime (since a field trivially is a unique factorization domain). It is contained in the ideal generated by the two polynomials X_1 and X_2. The latter ideal, however, does not contain the constants and therefore still is not the entire ring $K[X_1, X_2]$. Thus, I is not maximal.

Considering again the ring of continuous functions $C(X)$ on a topological space X, then for any $p \in X$, the ideal I_p of functions vanishing at p is prime because whenever a product fg vanishes at p, then at least one of the factors has to vanish there. For an arbitrary closed set $A \subset X$, the ideal I_A of functions vanishing on A need not be prime, because A could be a nontrivial union of two closed sets A_1, A_2, both of them smaller than A itself. We can then take a function f_1 vanishing on A_1, but not on all of A, and analogously f_2 for A_2. Then $f_1 f_2 \in I_A$, but neither of the factors is.

A ring R is called local if it possesses an ideal $J \neq R$ that contains all other ideals. Given a ring R and a prime ideal I, we can construct a local ring R_I as follows. We consider pairs $(a, b), a \in R, b \in R \setminus I$, with the identification

$$(a_1, b_1) \sim (a_2, b_2) \text{ if there exists some } c \in R \setminus I \text{ with } c(a_1 b_2 - a_2 b_1) = 0. \tag{5.4.130}$$

(Since R is not assumed to be an integral domain, we need to account for the presence of proper zero divisors.) On such pairs, we have the familiar operations from the calculus of fractions

$$(a_1, b_1) + (a_2, b_2) = (a_1 b_2 + a_2 b_1, b_1 b_2) \tag{5.4.131}$$

$$(a_1, b_1)(a_2, b_2) = (a_1 a_2, b_1 b_2). \tag{5.4.132}$$

The equivalence classes of pairs with these operations constitute a ring, the local ring R_I of the prime ideal I. That the ring is indeed local is seen as follows. We have the homomorphism $i : R \to R_I, i(a) = (a, 1). i(a)$ is invertible in R_I precisely when $a \notin I$. Every element in R_I is of the form $(i(a), i(b))$, with $b \notin I$. The elements $(i(a), i(b))$, with $b \notin I$, but with $a \in I$, constitute an ideal J, and since every element of R_I that is not contained in J has an inverse, this ideal J has to contain all other ideals.

When $R = C(X)$, $p \in X$ and I_p is the prime ideal of functions vanishing at p as before, then the local ring R_{I_p} consists of the pairs (f, g) with $g(p) \neq 0$. Thus, even though we may not be able to divide by such a g globally, as it may vanish somewhere, we can divide by it locally, in the vicinity of the point p. In that sense, the local ring R_I is closer to being a field than the original ring R as all elements not in the ideal I are now invertible.

We can also put this construction in a kind of dual manner. A nonempty subset $S \subset R$ that does not contain 0 and is closed under multiplication is called multiplicative. Examples are $S = \{g^n, n \in \mathbb{N}\}$ for a single g that is not nilpotent, or $S = R \setminus I$ for some prime ideal I. For a multiplicative set S, we can construct its ring of fractions, also denoted as R_S (this, unfortunately, is a confusing notation in view of the earlier notation R_I; the excuse is that an ideal is not multiplicative as it contains 0), consisting of all pairs $(a, s), a \in R, s \in S$, with

$$(a_1, s_1) \sim (a_2, s_2) \text{ if there exists } s \in S \text{ with } s(a_1 s_2 - a_2 s_1) = 0, \tag{5.4.133}$$

again with the usual operations on fractions. For $S = \{g^n, n \in \mathbb{N}\}$ for a single non-nilpotent g, we also simply write R_g. This is just the ring obtained by adjoining an inverse to g. The prime ideals of R_g are precisely those prime ideals of R that do not contain g, under the homomorphism that sends $a \in R$ to $(a, 1) = (ga, g) \in R_g$.

From this construction, we also see that the elements of R that belong to all its prime ideals are precisely the nilpotent elements. As it is clear that a nilpotent element belongs to all prime ideals, we only need to verify that for an element $g \in R$ that is not nilpotent, we can find some prime ideal to which it does not belong. For this purpose, we look at the ring R_g just constructed. As just noted, the homomorphism $R \to R_g$ induces an embedding $\mathrm{Spec}\, R_g \to \mathrm{Spec}\, R$ whose image consists of those prime ideals in R that do not contain g.

In other words, the intersection of all prime ideals of R equals its nilradical $\mathrm{Rad}(0)$. Similarly, the intersection of all prime ideals containing $g \in R$ is given by the elements f with $f^n = ga$ for some $n \in \mathbb{N}, a \in R$. This is seen by applying the preceding reasoning to the ring $R/(g)$.

5.4.2 Spectra of Rings

Definition 5.4.2 The set of prime ideals of a ring R is called the spectrum of R, $\mathrm{Spec}\, R$.

We equip $\mathrm{Spec}\, R$ with a topology, by taking as the closed sets the sets of the form $V(E)$ where E is any subset of R and $V(E)$ consists of all the prime ideals containing E. Of course, when I is the ideal generated by E, then $V(E) = V(I)$. The $V(E)$ satisfy the axioms for closed sets because of the obvious properties

$$V\left(\bigcup_{\lambda \in \Lambda} E_\lambda\right) = \bigcap_{\lambda \in \Lambda} V(E_\lambda) \tag{5.4.134}$$

$$V(E_{12}) = V(E_1) \cup V(E_2), \tag{5.4.135}$$

with E_{12} being the intersection of the ideals generated by E_1 and E_2 and where Λ is an arbitrary index set. Thus, arbitrary intersections and finite unions of closed sets are closed, as they should. This topology is called the spectral topology, and henceforth we consider $\mathrm{Spec}\, R$ as a topological space.

For instance, for the ring $\mathbb{Z}$ of integers, when we take $m \in \mathbb{Z}$, then the prime ideals containing m are precisely those generated by the prime divisors of m. More generally, for a subset $\{m_1, m_2, \ldots\}$ of $\mathbb{Z}$, the prime ideals containing that subset are those generated by the common prime divisors of all the m_i. Thus, the closed sets in $\mathrm{Spec}\,\mathbb{Z}$ contain the prime ideals generated by the common prime divisors of some set of integers. In $\mathbb{Z}_q$, if m

is relatively prime to q, that is, their largest common divisor is 1, then m is not contained in any prime ideal, that is, $V(\{m\}) = \emptyset$. In contrast, if m and q have the common prime divisors $p_1, \ldots, p_k$, then m is contained in the prime ideals generated by those. Thus, the closed sets of $\mathrm{Spec}\,\mathbb{Z}_q$ contain the common prime divisors of q and some collection of positive integers $< q$.

$\boxed{\mathbb{Z}_q}$

We can already make the following observation. When we take $E = I$ for some prime ideal $I \subset R$, then, since $V(I)$ then consists of all the prime ideals containing I, we have $V(I) = I$ precisely if I is a maximal ideal. Thus, a prime ideal I becomes a closed set in $\mathrm{Spec}\,R$ if and only if I is maximal. Thus, not all points of $\mathrm{Spec}\,R$ are closed w.r.t. the spectral topology. In particular, the zero ideal (0) is not closed (unless R is a field).

Since, as observed above, preimages of prime ideals under ring homomorphisms are prime, every homomorphism $h : R_1 \to R_2$ induces a mapping

$$h^\star : \mathrm{Spec}\,R_2 \to \mathrm{Spec}\,R_1. \tag{5.4.136}$$

Since we always have

$$(h^\star)^{-1} V(E) = V(h(E)), \tag{5.4.137}$$

the preimages of closed sets are closed, and therefore $h^\star$ is continuous.

$\boxed{\mathbb{Z}}$

Let us look at the example of the integers $\mathbb{Z}$. We recall that $\mathrm{Spec}\,\mathbb{Z}$ consists of the ideals $p\mathbb{Z}$, p a prime number, and (0). Since any integer $\neq 0$ is contained in finitely many prime ideals, the closed sets in $\mathrm{Spec}\,\mathbb{Z}$ are the finite ones (and, of course, $\mathrm{Spec}\,\mathbb{Z}$ itself, as generated by 0). Therefore, the open sets are the complements of the finite sets. Therefore, the intersection of any two nonempty open sets is nonempty itself. In particular, $\mathrm{Spec}\,\mathbb{Z}$ cannot satisfy any Hausdorff property, and in fact, this is characteristic of spectra of rings. Actually, in some spectra, there even exist non-closed points. The closure of a point in $\mathrm{Spec}\,R$, that is, of a prime ideal I in R, is $V(I) = \bigcap_{E \supset I} V(E)$. Thus, when the ideal is not maximal, then $V(I)$ is larger than I, and hence I is not closed. Thus, when R contains prime ideals that are not maximal, then $\mathrm{Spec}\,R$ contains such non-closed points.

On the other hand, we have

Lemma 5.4.1 $\mathrm{Spec}\,R$ *is compact.*

Proof Let g be a non-nilpotent element of R (the nilpotent elements are precisely those that are contained in all prime ideals, as observed above), and consider the open set

$$D(g) := \mathrm{Spec}\,R - V(g), \tag{5.4.138}$$

that is, the prime ideals not containing g. In fact,

$$D(g) = \mathrm{Spec}\,R_g, \tag{5.4.139}$$

because the ideals of R_g correspond to those ideals in R that do not contain g, as remarked above. Moreover, $\mathrm{Spec}\,R_f \cap \mathrm{Spec}\,R_g = \mathrm{Spec}\,R_{fg}$ (since a prime ideal contains a product iff it contains one of the factors), so that the collection $D(g)$ is closed under finite intersection. These open sets form a basis of the topology. In fact, for an open set $U = \operatorname{Spec} R - V(E) = \operatorname{Spec} R - \bigcap_{g \in E} V(g) = \bigcup_{g \in E} \operatorname{Spec} R_g$. It therefore suffices to show that any covering by such sets $D(g)$ contains a finite subcover. Thus, let

$$\operatorname{Spec} R = \bigcup_{\lambda} D(g_\lambda). \qquad (5.4.140)$$

This means that

$$\bigcap_{\lambda} V(g_\lambda) = V(J) = \emptyset \qquad (5.4.141)$$

where J is the ideal generated by the g_λ. Thus, there is no prime ideal containing J, hence $J = R$. But in this case, we can find $g_{\lambda_1}, \ldots, g_{\lambda_k}$ and elements $h_1, \ldots, h_k$ with

$$g_{\lambda_1} h_1 + \cdots + g_{\lambda_k} h_k = 1 \qquad (5.4.142)$$

that is, the ideal generated by $g_{\lambda_1}, \ldots, g_{\lambda_k}$ is R, and hence

$$\operatorname{Spec} R = \bigcup_{i=1,\ldots,k} D(g_{\lambda_i}), \qquad (5.4.143)$$

and we have found a finite subcover. $\qquad\qquad\qquad\qquad\qquad\qquad\square$

5.4.3 The Structure Sheaf

We want to construct a sheaf $\mathcal{O}$ on the topological space $X = \operatorname{Spec} R$ for some ring R. This construction will rest on the basic principle that has been employed for the definition of the topology on $\operatorname{Spec} R$ when R was some ring of functions on a topological space Y. The basic sets are the "points", that is, the prime ideals as obtained from functions vanishing at that "point". Such a "point" is a closed set iff it is given by a maximal ideal. We are using the quotation marks here because by this construction we shall in general get more "points" than the ordinary points of the underlying topological space Y. First of all, Y itself is a "point", as it corresponds to the prime ideal generated by 0, that is, the function vanishing identically on all of Y. Moreover (suppressing some important technical details), when Y is a complex manifold or an algebraic variety, and if we consider rings of holomorphic functions, then also irreducible subvarieties other than points are given by prime ideals of holomorphic functions. We shall leave out the quotation marks from now and shall speak about points in this more general sense. Thus, while the basic closed sets are given by those points that correspond to maximal ideals, the basic open sets are then given by complements of such points, that is, by minimal sets, where certain functions f are nonzero. When a function f is nonzero, we can divide by it and consider quotients $\frac{g}{f}$. Therefore, on such a basic open set, we have more functions than on the entire space Y. This becomes particularly important when we look at compact complex manifolds or algebraic varieties. Those do not carry any global nonconstant holomorphic functions. However, open subsets do. Therefore, we can only hope to recover such a space from its holomorphic functions when we also consider such functions that are not defined everywhere, but only on the complement of some points. From a different perspective, one considers meromorphic functions, that is, functions that are holomorphic on the complement of a collection of points, but have poles at those points. Therefore, from this perspective, it is again natural to work with a topology whose open sets are those sets where suitable collections meromorphic functions have no poles.

We shall now formalize the preceding idea in the framework developed so far. We first consider the special case where R has no proper divisors of 0, that is, it is an integral domain. We let K be its field of fractions. For any open $U \subset X$, we let $\mathcal{O}(U)$ be the set of the $u \in K$ which for every $I \in U$, that is, for every prime ideal I in U, can be represented as $u = (a, b)$ with $a, b \in R$, with $b \notin I$. We also write this suggestively as $b(I) \neq 0$, to indicate that we consider a, b as functions on U, with b nonzero at the point I. It is straightforward to check that $\mathcal{O}$ yields a sheaf on X (and we shall verify this below in detail for the general case), called its structure sheaf.

We have $\mathcal{O}(\mathrm{Spec}\,R) = R$. This is seen as follows. If $u \in \mathcal{O}(\mathrm{Spec}\,R)$, then for every $I \in \mathrm{Spec}\,R$, there exist $a_I, b_I \in R$ with

$$u = (a_I, b_I), b_I(I) \neq 0. \tag{5.4.144}$$

The ideal J generated by all the b_I, $I \in \mathrm{Spec}\,R$ is therefore not contained in any prime ideal of R, hence $J = R$. Therefore, by a reasoning already applied above in the proof of Lemma 5.4.1, we can find $I_1, \ldots, I_k$ and $h_1, \ldots, h_k \in R$ with

$$b_{I_1} h_1 + \cdots + b_{I_k} h_k = 1. \tag{5.4.145}$$

Since, by (5.4.144), u can be represented as $u = (a_{I_j}, b_{I_j})$ for every j, by multiplying this representation by $b_{I_j} h_j$ and summing w.r.t. j, we obtain

$$u = a_{I_1} h_1 + \cdots + a_{I_k} h_k \in R, \tag{5.4.146}$$

and hence $\mathcal{O}(\mathrm{Spec}\,R) = R$, as claimed.

We now turn to the general case, and we put

$$\mathcal{O}(D(g)) = R_g, \tag{5.4.147}$$

recalling (5.4.138, 5.4.139). The idea behind this is again natural and simple. When R is the ring of continuous functions on some topological space X, and if $f \in R$, we consider the open set $D(f) = \{x \in X : f(x) \neq 0\}$, and on this open set, we have the well defined inverse function $\frac{1}{f}$, and so, on $D(f)$, we can adjoin this inverse to R to obtain the local ring R_f as the ring of continuous functions on $D(f)$.

If $D(f) \subset D(g)$, then all prime ideals that do not contain f do not contain g either, and then some power of f is a multiple of g, as observed at the end of Sect. 5.4.1. We can therefore define the restriction map $p_{D(g)D(f)}$ as the localization map $R_g \to R_{gf} = R_f$.

Lemma 5.4.2 *Let, as in Lemma 5.4.1, $D(g)$ be covered by open sets $D(g_\lambda)$. Then, if $h, k \in R_g$ are equal in each R_{g_λ}, they are equal.*

Conversely, if for each λ, there exists some $h_\lambda \in R_{g_\lambda}$ such that for any λ, μ, the images of h_λ and h_μ in $R_{g_\lambda g_\mu}$ are equal, then there exists some $h \in R_g$ whose image in R_{g_λ} is h_λ for all λ.

Proof We first consider the case where $g = 1$. Then $R_g = R$, $D(g) =$ Spec $R =: X$.

If $h, k \in R_g$ are equal in R_{g_λ}, then $(g_\lambda)^N (h - k) = 0$ for some power N. Since the cover can be assumed to be finite by the compactness of Lemma 5.4.1, $h - k$ is annihilated by a power of the product of the g_{λ_i}, $i = 1, \ldots, k$. Since the ideal generated by these elements is the entire ring, we have $h = k$ in R.

For the second part, if $h_\lambda = h_\mu$ in $R_{g_\lambda g_\mu}$, then $(g_\lambda g_\mu)^N h_\lambda = (g_\lambda g_\mu)^N h_\mu$ for all large N. Using again the compactness of Lemma 5.4.1, we can take the same N for all λ, μ. As before, the g_λ, and therefore also the $(g_\lambda)^N$ generate the entire ring R, that is,

$$1 = \sum_\lambda f_\lambda (g_\lambda)^N \text{ with } f_\lambda \in R. \tag{5.4.148}$$

We put

$$h = \sum_\lambda f_\lambda (g_\lambda)^N h_\lambda. \tag{5.4.149}$$

Then for each μ,

$$(g_m u)^N h = \sum_\lambda (g_m u)^N f_\lambda (g_\lambda)^N h_\lambda = \sum_\lambda (g_m u)^N f_\lambda (g_\lambda)^N h_\mu$$
$$= (g_m u)^N (\sum_\lambda f_\lambda (g_\lambda)^N) h_\mu = (g_m u)^N h_\mu.$$

This means that h is equal to h_μ on each $D(g_\lambda)$.

The case of a general g then follows by applying the previous case to $X' := D(g)$, $R' := R_g$, $g'_\lambda := g g_\lambda$. $\square$

Lemma 5.4.2 verifies the sheaf property of (5.4.147) w.r.t. to an open covering of Spec R. It is not hard to verify (but we omit this here for brevity) that this implies the sheaf property in general.

Definition 5.4.3 The sheaf given by (5.4.147) is called the structure sheaf of Spec R.

5.4.4 Schemes

Definition 5.4.4 An affine scheme $(X, \mathcal{O})$ is the spectrum of a commutative ring R with identity, $X = \text{Spec} R$, equipped with its topology and its structure sheaf $\mathcal{O}$ as constructed above.

The ring R is then given by $R = \mathcal{O}(X)$. An affine scheme can be characterized by the following properties, whose underlying ideas we also recall here once more:

1. For the points $x \in X$, the stalks $\mathcal{O}_x$ are local rings, that is, they possess a nontrivial ideal that contains all other ideals. Here, the stalk is obtained by dividing by all elements of R that do not vanish at x (in the sense

that they are not contained in the prime ideal corresponding to x), and the maximal ideal J_x in question is then given by those elements that do vanish at x. This ideal does indeed contain all other ones, because all elements not in J_x are invertible in the stalk $\mathcal{O}_x$ by its construction.

2. Open sets can be obtained from enlarging the ring R by dividing by suitable nontrivial elements. More precisely, let $f \in R$ be non-nilpotent. We then have the open set U_f with

$$\mathcal{O}(U_f) = R[f^{-1}], \qquad (5.4.150)$$

that is, U_f is the set of points $x \in X$ for which f maps to a unit of the stalk $\mathcal{O}_x$. Again, the idea is to obtain the open set U_f from f as the set of points where f can be inverted. In this way, we obtain the topology of X from the algebraic structure of the ring R.

3. From the stalk $\mathcal{O}_x$ and its maximal ideal J_x, we obtain an ideal in $\mathcal{O}(X)$ as its preimage. Again, the idea is to associate to the point x the ideal of those functions that are not invertible at x. By associating to $x \in X$ this ideal, we obtain a map

$$X \to \operatorname{Spec}\mathcal{O}(X). \qquad (5.4.151)$$

For an affine scheme, this is a homeomorphism of topological spaces.

Definition 5.4.5 A scheme X is a topological space, denoted by $|X|$ when there is danger of confusion, equipped with a sheaf $\mathcal{O}$ of rings, such that $|X|$ is covered by open sets U_α of the form

$$U_\alpha \cong |\operatorname{Spec}R_\alpha| \text{ for rings } R_\alpha \text{ with } \mathcal{O}(U_\alpha) \cong \mathcal{O}_{\operatorname{Spec}R_\alpha}. \qquad (5.4.152)$$

Here, $\cong$ means homeomorphic.

We also say that $(|X|, \mathcal{O})$ is locally affine when this condition holds.

We now consider the basic example. Let K be an algebraically closed field. The affine n-dimensional space over K is defined as

$$\mathcal{A}^n_K := \operatorname{Spec}K[x_1, \ldots, x_n], \qquad (5.4.153)$$

Affine
n-dimensional space
---

the spectrum of the ring of polynomials in n variables over K. Since K is algebraically closed, by Hilbert's Nullstellensatz, the quotient of the polynomial ring $K[x_1, \ldots, x_n]$ by any maximal ideal is K itself. The maximal ideals are therefore of the form

$$J = (x_1 - \xi_1, \ldots, x_n - \xi_n), \text{ for } \xi_1, \ldots, \xi_n \in K. \qquad (5.4.154)$$

Thus, the maximal ideals, that is, the closed points of the scheme $\mathcal{A}^n_K$ are identified with n-tupels $(\xi_1, \ldots, \xi_n)$ of elements of K. Irreducible polynomials $f(x_1, \ldots, x_n) \in K[x_1, \ldots, x_n]$ generate prime ideals and therefore yield other, non-closed points. The closure of such a point contains all $(\xi_1, \ldots, \xi_n)$ with $f(\xi_1, \ldots, \xi_n) = 0$. Thus, such a point corresponds to an irreducible subvariety Σ_f of K^n, and in addition to the closed points contained in it, its closure also contains the points corresponding to all the irreducible subvarieties of Σ_f, obtained by the simultaneous vanishing of other irreducible polynomials in addition to f. In particular, the point corresponding to the zero ideal (0) contains in its closure all the points corresponding to irreducible subvarieties of K^n, including all the closed points $(\xi_1, \ldots, \xi_n) \in K^n$.

Spaces of Relations

6

6.1 Relations and Simplicial Complexes

In this chapter, we shall combine some of the previous threads. In Sect. 3.2, we contrasted graphs as constituted of elements standing in relations to one another versus graphs as relations identifying elements. In Chap. 5, we considered spaces as consisting of points, or more generally, pointlike objects that can again be viewed as the elements of such spaces. On the other hand, at the end of Sect. 5.1, we also suggested a possible conceptualization of space as consisting of being assembled from simple local pieces. Here, we shall now take up that suggestion, but translate it into a framework of relations as being constitutive for a space. At the same time, this chapter will provide an introduction to basic concepts of algebraic topology. In contrast to Chap. 4, where we introduced cohomology theory as a tool for defining and analyzing sheaves, here we shall start with homology theory.

We need to recall the concept of a simplicial complex as introduced in Sect. 3.1. We consider a *finite*[1] set V whose elements are denoted by v or $v_0, v_1, v_2, \ldots$. We assume that finite tuples of elements can stand in a relation $r(v_0, v_1, \ldots, v_q)$. We assume that all the elements $v_0, \ldots, v_q$ in such a relation are different from each other. The tuple $(v_0, v_1, \ldots, v_q)$ is considered as ordered. The ordering will be relevant only up to the parity of a permutation. Thus, we have $(v_0, v_1, \ldots, v_q) = (v_{i_0}, \ldots, v_{i_q})$ for any *even* permutation $(i_0, \ldots, i_q)$ of $(0, \ldots, q)$, while we put $(v_0, v_1, \ldots, v_q) = -(v_{i_0}, \ldots, v_{i_q})$ for an *odd* permutation. In other words, we consider an odd permutation as a change of orientation of such a tuple. This will be an important aspect in the sequel. Put simply, the rule is that any change of orientation introduces a minus sign. This will then allow for suitable cancellations, and this is at the basis of (co)homology theory. In particular, this sign convention implies that

$$(v_0, v_1, \ldots, v_q) = 0 \text{ whenever } v_i = v_j \text{ for some } i \neq j, \qquad (6.1.1)$$

---

[1]The assumption of finiteness is made here to avoid technical complications and the need for additional technical assumptions that are not pertinent to the essential ideas of this chapter.

that is, whenever two of the members of such a tuple are the same. This is a useful convention.

r takes its values in some set or space R which we leave unspecified. We write $r(v_0, \ldots, v_q) = o$ in the absence of a relation, for some formal element $o \in R$.

We use the convention $r(\emptyset) = o$.

We shall now assume that $r(v_0, v_1, \ldots, v_q) \neq o$ iff $r(-(v_0, v_1, \ldots, v_q)) \neq o$, that is, the presence of a relation does not depend on the orientation of the tuple. We recall from Sect. 3.1 that such a relation structure defines a simplicial complex Σ when the following properties hold.

(i) $$r(v) \neq o, \qquad\qquad\qquad (6.1.2)$$

that is, each element stands in a nontrivial relation with itself.

(ii) If $r(v_0, \ldots, v_q) \neq o$, then also $r(v_{i_1}, \ldots, v_{i_p}) \neq o$ for any (different)

$$i_1, \ldots, i_p \in \{0, \ldots, q\}, \qquad\qquad (6.1.3)$$

that is, whenever some elements stand in a nontrivial relation, then this also holds for any subset of them.

The simplicial complex Σ is visualized by associating a q-dimensional simplex σ_q to every collection of elements with $r(v_0, \ldots, v_q) \neq o$. We put $|\sigma_q| := q$ and call it the *dimension*.

When we take the ordering up to parity of a permutation into account, we obtain an orientation for every such simplex. The convention (6.1.1) also eliminates degenerate simplices, that is, those where not all vertices are different.

In the sequel, however, we shall identify the simplicial complex *as a geometric object* with the collection of its unoriented simplices, that is, with all tuples $(v_0, \ldots, v_q)$ with $r(v_0, v_1, \ldots, v_q) \neq o$, considering all permutations of the vertices of a simplex as yielding the same unoriented simplex. This can be seen as a *geometric* representation. For instance, when we speak about a map $\Sigma \to S$, for some set S, we mean assigning an element of S to every (unoriented) simplex of Σ. Nevertheless, the orientations of the simplices will play decisive roles in the *algebraic* computations.

Definition 6.1.1 When $r(v_0, \ldots, v_q) \neq o$ and $\{v_{i_0}, \ldots, v_{i_p}\} \subset \{v_0, \ldots, v_q)\}$, we say that $(v_{i_0}, \ldots, v_{i_p})$ is a subsimplex of $(v_0, \ldots, v_q)$. If $p = q - 1$, we also speak of a face of $(v_0, \ldots, v_q)$.

The dimension of a simplicial complex is defined as the maximal dimension among its simplices.

In order to have a category **Simp** of simplicial complexes, we need to define the morphisms.

Definition 6.1.2 A *simplicial map* $s : \Sigma_1 \to \Sigma_2$ between simplicial complexes maps each vertex v of Σ_1 to a vertex $s(v)$ of Σ_2 and assigns to each simplex $(v_0, \ldots, v_q)$ of Σ_1 the simplex of Σ_2 spanned by the images $s(v_i), i = 0, \ldots, q$.

Note that the image $s(\sigma_q)$ of a q-simplex is of dimension lower than q when s is not injective on the set of vertices of σ_q.

6.2 Homology of Simplicial Complexes

We shall now give an introduction to the homology of simplicial complexes; references are [47, 87, 104, 108].

Let G be an abelian group. A q-chain is defined as a formal linear combination

$$c_q = \sum_{i=1}^{m} g_i \sigma_q^i \tag{6.2.4}$$

for elements g_i of G and q-simplices σ_q^i of Σ. The q-chains then form a group $C_q = C_q(\Sigma, G)$, with the group operation of G:

$$\sum_{i=1}^{m} g_i \sigma_q^i + \sum_{i=1}^{m} g_i' \sigma_q^i = \sum_{i=1}^{m} (g_i + g_i') \sigma_q^i. \tag{6.2.5}$$

Here, the $+$ sign in the last expression denotes the group operation of G whereas the addition expressed by $\sum$ is a formal one. I hope that the reader will not confuse these two different operations.

For any q for which Σ does not contain any q-simplices, we put, of course, $C_q(\Sigma, G) = 0$, the trivial group. We also note that the inverse of $\sum_{i=1}^{m} g_i \sigma_q^i$ in the group C_q is $\sum_{i=1}^{m} -g_i \sigma_q^i$, that is, taking the inverses of the g_i in G, but equivalently, we could write $\sum_{i=1}^{m} g_i(-\sigma_q^i)$, that is, giving all simplices involved the opposite orientation.

Definition 6.2.1 For a simplicial complex Σ, we let α_q be the number of unoriented q-simplices. The Euler characteristic of Σ is then

$$\chi(\Sigma) := \sum_{q} (-1)^q \alpha_q. \tag{6.2.6}$$

In fact, α_q is the rank of the group $C_q(\Sigma, \mathbb{Z})$, the group of q-chains with coefficients in the integers $\mathbb{Z}$. [2]

Definition 6.2.2 The boundary of an oriented q-simplex $\sigma_q = (v_0, v_1, \ldots, v_q)$ is the $(q - 1)$-chain

$$\partial \sigma_q := \sum_{i=0}^{q} (-1)^i (v_0, \ldots, \hat{v}_i, \ldots, v_q) \text{ for } q > 0, \tag{6.2.7}$$

---

[2]This simply means that $C_q(\Sigma, \mathbb{Z})$ is isomorphic to $\mathbb{Z}^{\alpha_q}$, the product of α_q copies of $\mathbb{Z}$; in fact, since $\mathbb{Z}$ is abelian, we should rather speak of a sum instead of a product and write $\mathbb{Z}^{\alpha_q} = \mathbb{Z} \oplus \cdots \oplus \mathbb{Z}$ with α_q summands.

and, of course, $\partial\sigma_0 = 0$ for a 0-chain. Here, $\hat{v}_i$ means that the vertex v_i is omitted.

The boundary of the q-chain $c_q = \sum_{i=1}^{m} g_i\sigma_q^i$ then is, by linearity,

$$\partial c_q := \sum_{i=1}^{m} g_i\partial\sigma_q^i. \tag{6.2.8}$$

When we want to emphasize that ∂ operates on q-chains, we shall write ∂_q.

2-simplex

In Fig. 6.1, $\sigma_2 = (\sigma_0^1, \sigma_0^2, \sigma_0^3)$ and $\sigma_1^1 = (\sigma_0^1, \sigma_0^2)$, $\sigma_1^2 = (\sigma_0^2, \sigma_0^3)$, $\sigma_1^3 = (\sigma_0^3, \sigma_0^1) = -(\sigma_0^1, \sigma_0^3)$, and hence

$$\partial\sigma_2 = \sigma_1^1 + \sigma_1^2 + \sigma_1^3$$
$$\partial\sigma_1^1 = \sigma_0^2 - \sigma_0^1$$
$$\ker\partial_2 = 0$$
$$\ker\partial_1 = [\sigma_1^1 + \sigma_1^2 + \sigma_1^3]$$
$$\ker\partial_0 = [\sigma_0^1, \sigma_0^2, \sigma_0^3]$$
$$\mathrm{im}\,\partial_2 = \ker\partial_1$$
$$\mathrm{im}\,\partial_1 = [\sigma_0^1 - \sigma_0^2, \sigma_0^2 - \sigma_0^3]$$
$$b_2 = 0 = b_1, \ b_0 = 1, \quad \text{and hence } \chi = 1.$$

(For the definition of the Betti numbers b_q, see Definition 6.2.5 below, and for their relation with the Euler characteristic χ, see Theorem 6.2.3.)

Definition 6.2.3 The q-chain c_q is called closed or, equivalently, a cycle, if

$$\partial_q c_q = 0. \tag{6.2.9}$$

The q-chain c_q is called a boundary if there exists some $(q+1)$-chain γ_{q+1} with

$$\partial_{q+1}\gamma_{q+1} = c_q. \tag{6.2.10}$$

Thus, the closed chains are given by the kernel of ∂_q, $\ker\partial_q$, and the boundaries by the image of ∂_{q+1}, $\mathrm{im}\,\partial_{q+1}$. In particular, both of them yield subgroups of C_q.

If the simplex σ_2 had not been present in Fig. 6.2, we would have had no contribution in dimension 2 and instead

$$b_1 = 1, \ b_0 = 1, \quad \text{and hence } \chi = 0 \text{ (see Theorem 6.2.3).}$$

Fig. 6.1 For an explanation, see the main text

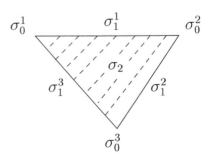

Thus, in the absence of σ_2, $[\sigma_1^1 + \sigma_1^2 + \sigma_1^3]$ is no longer a boundary, and hence we get a contribution to homology in dimension 1. In particular, this changes the Euler characteristic.

Theorem 6.2.1

$$\partial_{q-1}\partial_q = 0 \text{ for all } q. \tag{6.2.11}$$

We shall usually abbreviate this fundamental relation as

$$\partial^2 = 0. \tag{6.2.12}$$

Proof Because of (6.2.8), it suffices to show that

$$\partial\partial\sigma_q = 0 \tag{6.2.13}$$

for any oriented q-simplex. Since $C_s = 0$ for $s < 0$, we only need to consider the case $q \geq 2$. For $\sigma_q = (v_0, \ldots, v_q)$, we have

$$\partial\partial\sigma_q = \partial \sum_{i=0}^{q}(-1)^i (v_0, \ldots, \hat{v}_i, \ldots, v_q)$$

$$= \sum_{i=0}^{q}(-1)^i \partial(v_0, \ldots, \hat{v}_i, \ldots, v_q)$$

$$= \sum_{i=0}^{q}(-1)^i \left(\sum_{j=1}^{i-1}(-1)^j (v_0, \ldots, \hat{v}_j, \ldots, \hat{v}_i, \ldots, v_q) \right.$$

$$\left. + \sum_{j=i+1}^{q} (-1)^{j-1}(v_0, \ldots, \hat{v}_i, \ldots, \hat{v}_j, \ldots, v_q) \right)$$

$$= \sum_{j<i}(-1)^{i+j} (v_0, \ldots, \hat{v}_j, \ldots, \hat{v}_i, \ldots, v_q)$$

$$+ \sum_{j>i}(-1)^{i+j-1}(v_0, \ldots, \hat{v}_i, \ldots, \hat{v}_j, \ldots, v_q),$$

and exchanging i and j in the last sum gives the result. □

Corollary 6.2.1 im ∂_{q+1} *is a subgroup of* ker ∂_q, *and since G is abelian, it is a normal subgroup.*

This enables us to form the quotient group and to state

Definition 6.2.4 The quotient group

$$H_q(\Sigma, G) := \ker \partial_q / \operatorname{im} \partial_{q+1} \tag{6.2.14}$$

is called the qth homology group (with coefficients in G) of the simplicial complex Σ.

Remark Instead of taking the boundary $\partial \sigma_q$ of the q-simplex $\sigma_q = (v_0, \ldots, v_q)$ that maps it to a configuration of $(q-1)$-simplices, we could also have considered its augmentation, defined as follows

$$\alpha_q(v_0, \ldots, v_q) = \sum_v (v, v_0, \ldots, v_q) \tag{6.2.15}$$

where the sum is taken over all v for which $(v, v_0, \ldots, v_q)$ is a $(q+1)$-simplex in Σ. We then have

$$\alpha_{q+1} \alpha_q = 0, \tag{6.2.16}$$

or more shortly $\alpha^2 = 0$, because

$$\alpha_{q+1} \alpha_q (v_0, \ldots, v_q) = \sum_w \sum_v (w, v, v_0, \ldots, v_q)$$

where the first sum is over all w for which $(w, v, v_0, \ldots, v_q)$ is a $(q+2)$-simplex

$$= 0,$$

because whenever $(w, v, v_0, \ldots, v_q)$ is a simplex in Σ, then so is $(v, w, v_0, \ldots, v_q) = -(w, v, v_0, \ldots, v_q)$, that is, the terms in the double sum cancel in pairs.

Thus, for the augmentation α, it is even simpler to prove that its square vanishes than for the boundary δ. And since the vanishing of the square is the basic ingredient of the following considerations, we could work as well with α as with ∂ to get an equivalent theory. However, the geometric interpretation of what it means that $\partial \sigma = 0$ is perhaps somewhat easier than $\alpha \sigma = 0$, and so, we shall work out the theory for the operator ∂. Nevertheless, the reader is invited to pursue the considerations for the augmentation operator α as an exercise. But we shall now return to ∂.

For a cycle c_q, we denote by $[c_q] = [c_q]_\Sigma$ its equivalence class as an element of $H_q(\Sigma, G)$.

Theorem 6.2.2 $H_q(., G)$ *is a covariant functor from the category* **Simp** *of simplicial complexes to the category* **Ab** *of abelian groups.*

Proof We need to show that a simplicial map $s : \Sigma_1 \to \Sigma_2$ induces a group homomorphism $H_q(s) : H_q(\Sigma_1, G) \to H_q(\Sigma_2, G)$ with the appropriate composition properties. That we do indeed get such a homomorphism is the content of the following lemma. The rest of the proof is trivial. $\square$

Lemma 6.2.1 *The induced map* $C_q(s) : C_q(\Sigma_1, G) \to C_q(\Sigma_2, G)$ *on chains commutes with the boundary operators, that is, the following diagram commutes*

$$
\begin{array}{ccc}
C_q(\Sigma_1, G) & \xrightarrow{C_q(s)} & C_q(\Sigma_2, G) \\
\downarrow{\scriptstyle \partial} & & \downarrow{\scriptstyle \partial} \\
C_{q-1}(\Sigma_1, G) & \xrightarrow{C_{q-1}(s)} & C_{q-1}(\Sigma_2, G).
\end{array}
\tag{6.2.17}
$$

Hence we can pass to quotients to obtain maps $H_q(s)$.

Proof We have to show

$$\partial C_q(s)(\sigma_q) = C_{q-1}(s)(\partial \sigma_q) \tag{6.2.18}$$

for every q-simplex $\sigma_q = (v_0, \ldots, v_q)$. If all the images $s(v_i)$ are different, then this is obvious. When two images coincide, w.l.o.g. $s(v_0) = s(v_1)$, we have $C_q(s)(\sigma_q) = 0$, hence also $\partial C_q(s)(\sigma_q) = 0$. On the other hand, for $\partial \sigma_q = (v_1, v_2, \ldots, v_q) - (v_0, v_2, \ldots, v_q) + \sum_{i=2}^{q} (v_0, v_1, \ldots, \hat{v}_i, \ldots, v_q)$, the first two terms in the sum have the same image under $C_{q-1}(s)$ whereas the other terms get mapped to 0. Hence also $C_{q-1}(s)(\partial \sigma_q) = 0$ in this case.

Thus, closed chains are mapped to closed chains, and boundaries to boundaries, and we obtain the induced maps on the homology groups. $\qquad\square$

We consider the case $G = \mathbb{Z}$. Since $\mathbb{Z}$ is a finitely generated abelian group and Σ contains only finitely many simplices, then so is $C_q(\Sigma, \mathbb{Z})$ and hence also $H_q(\Sigma, \mathbb{Z})$. According to the classification of finitely generated abelian groups, then

$$H_q(\Sigma, \mathbb{Z}) \cong \mathbb{Z}^{b_q} \oplus \mathbb{Z}_{t_q^1} \cdots \oplus \mathbb{Z}_{t_q^{r_q}}, \tag{6.2.19}$$

that is, it is the direct sum of b_q copies of $\mathbb{Z}$ and r_q finite cyclic groups; in fact, we can arrange the *torsion coefficients* t_q^i in such a manner that t_q^i divides t_q^{i+1} for $i = 1, \ldots, r_q - 1$.

Definition 6.2.5 $b_q = b_q(\Sigma)$ is called the qth Betti number of Σ.

In the sequel, we shall discuss the important

Theorem 6.2.3

$$\chi(\Sigma) = \sum_q (-1)^q b_q(\Sigma). \tag{6.2.20}$$

Proof This result will follow from Corollary 6.2.5 or Corollary 6.3.1; see also Corollary 6.3.3. $\qquad\square$

In the discussion of Fig. 6.1, we have computed the Betti numbers of the two-dimensional simplex σ_2. When we take out the interior, we get its boundary $\partial \sigma_2$

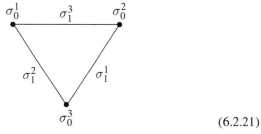

$$\tag{6.2.21}$$

For this simplicial complex, now im $\partial_2 = 0$, and hence we get the Betti numbers

$$b_1 = 1, \quad b_0 = 1, \text{ and hence } \chi = b_1 - b_0 = 0 \tag{6.2.22}$$

for the Euler characteristic of (6.2.20).

In fact, this extends to other dimensions as follows

Lemma 6.2.2 *For the q-dimensional simplex σ_q, we have*

$$b_p = 0 \text{ for } p > 0 \text{ and } b_0 = 1, \text{ hence } \chi = 1. \qquad (6.2.23)$$

For its boundary $\partial \sigma_q$, that is, for the $(q-1)$-dimensional simplicial complex that consists of all simplices of σ_q of dimension $< q$, we have

$$b_{q-1} = 1 = b_0 \text{ and } b_p = 0 \text{ for all other } p \qquad (6.2.24)$$

and hence

$$\chi(\partial \sigma_q) = \begin{cases} 0 & \text{for even } q \\ 2 & \text{for odd } q. \end{cases} \qquad (6.2.25)$$

The proof is, of course, not principally difficult, but a little lengthy, and omitted.

Since H_q is a quotient group of C_q, typically the b_q are much smaller than the α_q which are the ranks of the free abelian group $C_q(\Sigma, \mathbb{Z})$. Thus, in the computation of the Euler characteristic, some cancellation takes place. This cancellation is encoded in the homology of Σ, as will be elaborated in the sequel.

In order to get there, we need some more technical preparations. In particular, we shall need to generalize the previous constructions in the following manner.

Definition 6.2.6 Let Δ be a subcomplex of the simplicial complex Σ. Since $C_q(\Delta, G) \subset C_q(\Sigma, G)$, we can form the quotient group $C_q(\Sigma, \Delta; G) := C_q(\Sigma, G)/C_q(\Delta, G)$, and since the boundary operator ∂_q maps $C_q(\Delta)$ to $C_{q-1}(\Delta)$ (if a subcomplex is contained in Δ, then so is its boundary), we obtain an induced operator

$$\partial_q^{rel} : C_q(\Sigma, \Delta; G) \to C_{q-1}(\Sigma, \Delta; G), \qquad (6.2.26)$$

called the relative boundary operator. The quotient group

$$H_q(\Sigma, \Delta; G) := \ker \partial_q^{rel}/\operatorname{im} \partial_{q+1}^{rel} \qquad (6.2.27)$$

is called the qth relative homology group of Σ relative Δ (with coefficients in the abelian group G), and its dimension (as in (6.2.19)) is denoted by b_q^{rel}.

The elements of $\ker \partial_q^{rel}$ are also called (relative) cycles and those of $\operatorname{im} \partial_{q+1}^{rel}$ (relative) boundaries. For a cycle c_q, we denote by $[c_q]_{(\Sigma, \Delta)}$ its equivalence class as an element of $H_q(\Sigma, \Delta; G)$.

You should think here of the subcomplex Δ as being absent or taken out. Everything that gets mapped to Δ disappears.

The pairs of simplicial complexes that we shall consider below in Sect. 6.3 will be of the form (P_0, P_1) as in the definition of an exchange pattern. In the figures in this and the next section, we denote our simplicial pair by (Σ_0, Σ_1) in place of (Σ, Δ). In Fig. 6.2 (see Fig. 6.1 for the notation),

Fig. 6.2 For an explanation, see the main text

$$\Sigma_0 = \Sigma$$
$$\Sigma_1 = \Sigma \setminus \sigma_2$$
$$\ker \partial_2^{rel} = [\sigma_2]$$
$$\partial_1^{rel} = \partial_0^{rel} = 0$$
$$b_2^{rel} = 1, \quad b_1^{rel} = 0 = b_0^{rel}$$
$$\ker \partial_1^{\Sigma_1} = [\sigma_1^1 + \sigma_1^2 + \sigma_1^3]$$
$$b_1(\Sigma_1) = 1, \quad b_0(\Sigma_1) = 1.$$

Lemma 6.2.3 *The maps*

$$j_* : H_q(\Sigma) \to H_q(\Sigma, \Delta)$$
$$[c_q]_\Sigma \mapsto [c_q]_{(\Sigma, \Delta)} \tag{6.2.28}$$

and

$$\partial_* : H_q(\Sigma, \Delta) \to H_{q-1}(\Delta)$$
$$[c_q]_{(\Sigma, \Delta)} \mapsto [\partial c_q]_\Delta \tag{6.2.29}$$

are well defined.

Proof Since a boundary in Σ is also a relative boundary, j_* is well defined. If c_q is a relative q-cycle, then $\partial c_q \subset \Delta$ and since $\partial \partial c_q = 0$, it is a cycle in Δ (although it need not be a boundary in Δ as c_q is not assumed to be contained in Δ.) $\qquad\square$

Theorem 6.2.4 *The homology sequence of the pair* (Σ, Δ),

$$\cdots \xrightarrow{j_*} H_{q+1}(\Sigma, \Delta) \xrightarrow{\partial_*} H_q(\Delta) \xrightarrow{i_*} H_q(\Sigma)$$
$$\xrightarrow{j_*} H_q(\Sigma, \Delta) \xrightarrow{\partial_*} \cdots \tag{6.2.30}$$

(where i_* is the map $H_q(i)$ induced by the inclusion $i : \Delta \to \Sigma$ and j_*, ∂_* are defined in Lemma 6.2.3) is exact in the sense that the kernel of each map in the sequence coincides with the image of the preceding map.

Let us describe this homology sequence in words: A relative homology class, represented by a $(q + 1)$-chain with boundary in Δ is mapped to the cohomology class of that boundary (that boundary is closed because of $\partial^2 = 0$, but while by construction it is the boundary of a chain in Σ, it need not be the boundary of a chain in Δ, and may therefore represent a nontrivial element of the qth homology of Δ). Next, a homology class of Δ becomes a homology class in Σ under the inclusion $i : \Delta \to \Sigma$. Finally, a homology class in Σ trivially is also a homology class relative to Δ. From these observations, it is already clear that the composition of any two subsequent maps in the sequence is 0, that is, the image of any map is contained in the kernel of the next one. We now verify this formally and check the remaining details.

Proof We need to show the following three relations

$$\ker(i_* : H_q(\Delta) \to H_q(\Sigma)) = \mathrm{im}\,(\partial_* : H_{q+1}(\Sigma, \Delta) \to H_q(\Delta)),$$
(6.2.31)

$$\ker(j_* : H_q(\Sigma) \to H_q(\Sigma, \Delta)) = \mathrm{im}\,(i_* : H_q(\Delta) \to H_q(\Sigma)),$$
(6.2.32)

$$\ker(\partial_* : H_q(\Sigma, \Delta) \to H_{q-1}(\Delta)) = \mathrm{im}\,(j_* : H_q(\Sigma) \to H_q(\Sigma, \Delta)).$$
(6.2.33)

(6.2.31): We have $i_*\partial_*[c_{q+1}]_{\Sigma,\Delta} = i_*[\partial c_{q+1}]_\Delta = [\partial c_{q+1}]_\Sigma = 0$ as this is a boundary in Σ. Thus $\mathrm{im}\,\partial_* \subset \ker i_*$.

Conversely, let $[c_q]_\Delta \in \ker i_*$. Thus, c_q is a cycle in Δ which is a boundary in Σ, i.e., there exists some chain γ_{q+1} in Σ with $\partial\gamma_{q+1} = c_q$. Since c_q is in Δ, γ_{q+1} is a relative cycle, hence yields an element $[\gamma_{q+1}] \in H_{q+1}(\Sigma, \Delta)$. Thus, $[c_q]_\Delta = \partial_*[\gamma_{q+1}]$ is in $\mathrm{im}\,\partial_*$, hence $\ker i_* \subset \mathrm{im}\,\partial_*$.

(6.2.32): We have $j_*i_*[c_q]_\Delta = [c_q]_{(\Sigma,\Delta)} = 0$ if $c_q \subset \Delta$. Hence $\mathrm{im}\,i_* \subset \ker j_*$.

Conversely, let $[c_q]_\Sigma \in \ker j_*$. Then there exists a $(q+1)$-chain γ_{q+1} in Σ such that $c_q - \partial\gamma_{q+1} =: c_q^0 \subset \Delta$ (c_q is a boundary relative to Δ, as it represents $0 \in H_q(\Sigma, \Delta)$). Thus, $[c_q]_\Sigma = [c_q^0]_\Sigma = i_*[c_q^0]_\Delta \in \mathrm{im}\,i_*$. Hence $\ker j_* \subset \mathrm{im}\,i_*$.

(6.2.33): We have $\partial_*j_*[c_q]_\Sigma = [\partial c_q]_\Delta = 0$ because $\partial c_q = 0$. Hence $\mathrm{im}\,j_* \subset \ker \partial_*$.

Conversely, let $[c_q]_{(\Sigma,\Delta)} \in \ker \partial_*$. Then $[\partial c_q]_\Delta$ is a boundary in Δ, that is, there exists a $\gamma_q \subset \Delta$ with $\partial c_q = \partial\gamma_q$. Then $[c_q]_{(\Sigma,\Delta)} = [c_q - \gamma_q]_{(\Sigma,\Delta)}$ (because γ_q is trivial relative to Δ) $= j_*[c_q - \gamma_q]_\Sigma$. Hence $\ker \partial_* \subset \mathrm{im}\,j_*$. $\square$

The reader should convince herself that the preceding proof is not principally difficult. It simply consists in systematically disentangling the definitions of the various morphisms.

Theorem 6.2.5 *Let $s : (\Sigma_1, \Delta_1) \to (\Sigma_2, \Delta_2)$ be a simplicial map between pairs of simplicial complexes (i.e., s is a simplicial map from Σ_1 to Σ_2 with $s(\Delta_1) \subset \Delta_2$). As in Theorem 6.2.2, we get induced maps $H_q(\Sigma_1) \to H_q(\Sigma_2)$, $H_q(\Delta_1) \to H_q(\Delta_2)$, $H_q(\Sigma_1, \Delta_1) \to H_q(\Sigma_2, \Delta_2)$, all denoted by s_*. Then the following diagram commutes*

$$
\begin{array}{ccccccccc}
\cdots \xrightarrow{j_*} & H_{q+1}(\Sigma_1, \Delta_1) & \xrightarrow{\partial_*} & H_q(\Delta_1) & \xrightarrow{i_*} & H_q(\Sigma_1) & \xrightarrow{j_*} & H_q(\Sigma_1, \Delta_1) & \xrightarrow{\partial_*} \cdots \\
\downarrow & s_*\downarrow & & s_*\downarrow & & s_*\downarrow & & s_*\downarrow & \downarrow \\
\cdots \xrightarrow{j_*} & H_{q+1}(\Sigma_2, \Delta_2) & \xrightarrow{\partial_*} & H_q(\Delta_2) & \xrightarrow{i_*} & H_q(\Sigma_2) & \xrightarrow{j_*} & H_q(\Sigma_2, \Delta_2) & \xrightarrow{\partial_*} \cdots
\end{array}
$$
(6.2.34)

The *proof* of Theorem 6.2.5 proceeds as the proofs of Theorems 6.2.2, 6.2.4, and we therefore leave the details as an exercise.
We now come to the **excision theorem**.

Theorem 6.2.6 *Let Λ be a subset of, or more precisely, a collection of simplices contained in the subcomplex Δ of the simplicial complex Σ with the following properties:*

(i) *If the simplex σ_q is not contained in Λ, then none of its subsimplices is contained in Λ either.*
(ii) *A simplex σ_q contained in Λ must not be a subsimplex of a simplex in $\Sigma \setminus \Delta$.*

Then the inclusion $i : (\Sigma \setminus \Lambda, \Delta \setminus \Lambda) \to (\Sigma, \Delta)$ induces isomorphisms

$$i_* : H_q(\Sigma \setminus \Lambda, \Delta \setminus \Lambda) \to H_q(\Sigma, \Delta) \qquad (6.2.35)$$

for all q.

Proof By *(i)*, $\Sigma \setminus \Lambda$ and $\Delta \setminus \Lambda$ are subcomplexes of Σ. The elements of $C_q(\Sigma, \Delta)$ and $C_q(\Sigma \setminus \Lambda, \Delta \setminus \Lambda)$ can both be considered as linear combinations of oriented simplices of $\Sigma \setminus \Delta$; hence these chain groups are isomorphic. For any such $c \in C_q(\Sigma \setminus \Delta)$, by *(ii)* we have that $\partial c \in C_{q-1}(\Delta)$ if and only if $\partial c \in C_{q-1}(\Delta \setminus \Lambda)$. Therefore, also the homology groups H_q are isomorphic. $\qquad\square$

Let us consider a simple example

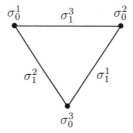

Here, Δ is the subcomplex consisting of the blue and the red simplices, that is, it contains $\sigma_0^1, \sigma_0^3, \sigma_0^2, \sigma_1^1, \sigma_1^3$, whereas Λ is the subset containing only the red simplices $\sigma_0^2, \sigma_1^3, \sigma_1^1$. Then $\Sigma \setminus \Lambda$ is the edge σ_1^2 together with its two vertices σ_0^1, σ_0^3, whereas $\Delta \setminus \Lambda$ has only the two vertices σ_0^1, σ_0^3. Then, according to our previous computations, $b_1(\Sigma \setminus \Lambda, \Delta \setminus \Lambda) = 1$ (because there is a single edge whose boundary is in $\Delta \setminus \Lambda$, hence is 0 in the relative homology) and $b_0(\Sigma \setminus \Lambda, \Delta \setminus \Lambda) = 0$ (because both vertices are in the part $\Delta \setminus \Lambda$ which is taken out). By the excision theorem, these then are also the Betti numbers of (Σ, Δ).

Corollary 6.2.2 *Let Δ_0, Δ_1 be subcomplexes of Σ with $\Sigma = \Delta_0 \cup \Delta_1$. Then we have isomorphisms*

$$i_* : H_q(\Delta_0, \Delta_0 \cap \Delta_1) \to H_q(\Sigma, \Delta_1) \qquad (6.2.36)$$

for all q.

Proof Let Λ be the subset of Δ_1 consisting of those simplices that are not subsimplices of simplices in Δ_0. Λ then satisfies the assumptions of Theorem 6.2.6, and we can apply that result with $\Delta_0 = \Sigma \setminus \Lambda$ and $\Delta_0 \cap \Delta_1 = \Delta_1 \setminus \Lambda$. $\square$

In order to put the preceding into perspective and also in order to achieve a generalization of Theorem 6.2.4 that will play an important role below, we shall now isolate the algebraic aspects of the constructions just presented. In this way, you can appreciate the fact that the geometric and the algebraic constructions are completely parallel.

Definition 6.2.7 A chain complex is a system $C = (C_q, \partial_q)_{q \in \mathbb{Z}}$ of abelian groups C_q and homomorphisms $\partial_q : C_q \to C_{q-1}$ with $\partial_q \circ \partial_{q+1} = 0$ for all q (we shall often write ∂ in place of ∂_q and call it the boundary operator). We then define cycles and boundaries and the homology groups $H_q(C)$ as before.

A chain map $f : C \to C'$ between chain complexes $C = (C_q, \partial_q)_{q \in \mathbb{Z}}$ and $C' = (C'_q, \partial'_q)_{q \in \mathbb{Z}}$ is a family of homomorphisms $f_q : C_q \to C'_q$ with $\partial'_q \circ f_q = f_{q-1} \circ \partial_q$ for all q. As a chain maps cycles to cycles and boundaries to boundaries, it induces homomorphisms $f_* = H_q(f) : H_q(C) \to H_q(C')$.

A sequence

$$0 \longrightarrow C' \xrightarrow{\alpha} C \xrightarrow{\beta} C'' \longrightarrow 0 \qquad (6.2.37)$$

(where 0 stands for the chain complex with the trivial group for every q) is called a short exact sequence if for all q the sequence

$$0 \longrightarrow C'_q \xrightarrow{\alpha_q} C_q \xrightarrow{\beta_q} C''_q \longrightarrow 0 \qquad (6.2.38)$$

is exact.

Here, since 0 stands for the trivial group, the exactness of (6.2.38) includes the statements that α_q is injective and that β_q is surjective. (6.2.38) is then called a short exact sequence of abelian groups.

Definition 6.2.8 A short sequence

$$0 \longrightarrow A' \xrightarrow{\alpha} A \xrightarrow{\beta} A'' \longrightarrow 0 \qquad (6.2.39)$$

of abelian groups is said to *split* if β has a right inverse, i.e., there exists a homomorphism $\beta'' : A'' \to A$ with $\beta \circ \beta'' = \mathrm{id}_{A''}$.

When (6.2.37) splits,

$$A \cong A' \oplus A''. \qquad (6.2.40)$$

In general, however, a short exact sequence need not split. Rather, the general case is accounted for by the following fundamental result.

Theorem 6.2.7 *Any short exact sequence*

$$0 \longrightarrow C' \xrightarrow{\alpha} C \xrightarrow{\beta} C'' \longrightarrow 0 \qquad (6.2.41)$$

induces a long exact sequence of homology groups

$$\ldots \longrightarrow H_{q+1}(C'') \xrightarrow{\partial_*} H_q(C') \xrightarrow{\alpha_*} H_q(C)$$

$$\xrightarrow{\beta_*} H_q(C'') \xrightarrow{\partial_*} \ldots \tag{6.2.42}$$

where the connecting homomorphism ∂_* is defined by

$$\partial_*[c''] = [\alpha^{-1}\partial\beta^{-1}c''] \in H_{q-1}(C') \text{ for } c'' \in H_q(C''). \tag{6.2.43}$$

Proof From the definition of chain complexes, we obtain the commutative diagram

$$
\begin{array}{ccccccccc}
0 & \longrightarrow & C'_{q+1} & \xrightarrow{\alpha_{q+1}} & C_{q+1} & \xrightarrow{\beta_{q+1}} & C''_{q+1} & \longrightarrow & 0 \\
\downarrow & & \partial'_{q+1}\downarrow & & \partial'_{q+1}\downarrow & & \partial'_{q+1}\downarrow & & \downarrow \\
0 & \longrightarrow & C'_q & \xrightarrow{\alpha_q} & C_q & \xrightarrow{\beta_q} & C''_q & \longrightarrow & 0 \\
\downarrow & & \partial'_q\downarrow & & \partial'_q\downarrow & & \partial'_q\downarrow & & \downarrow \\
0 & \longrightarrow & C'_{q-1} & \xrightarrow{\alpha_{q-1}} & C_{q-1} & \xrightarrow{\beta_{q-1}} & C''_{q-1} & \longrightarrow & 0
\end{array}
\tag{6.2.44}
$$

where each row is a short exact sequence of abelian groups. In the sequel, we shall often leave out the subscript q when it can be inferred from context. Let $c'' \in C''_q$ be a cycle, i.e., $\partial c'' = 0$. Since β_q is surjective by exactness, we find some $c \in C_q$ with $\beta(c) = c''$, and then

$$\beta(\partial c) = \partial''\beta(c) = \partial''c'' = 0. \tag{6.2.45}$$

By exactness, we find a unique $c' \in C'_{q-1}$ with $\alpha(c') = \partial c$, and then

$$\alpha(\partial'c') = \partial\alpha(c') = \partial^2 c = 0. \tag{6.2.46}$$

Since by exactness, α_{q-1} is injective, $\partial'c' = 0$. Therefore, c' is a $(q-1)$-cycle of C' and hence induces an element $[c'] \in H_q(C')$ which we denote by $\partial_*[c'']$. We need to show that ∂_* is well defined in the sense that this element of $H_q(C')$ depends only on the homology class of c''.

   To verify this, take $c_1 \in C_q$ with $[\beta(c_1)] = [c'']$, that is, there exists some $d'' \in C''_{q+1}$ with $\beta(c_1) = \beta(c) + \partial''d''$. By surjectivity of β again, we find $d \in C_{q+1}$ with $\beta(d) = d''$. Then $\beta(c_1) = \beta(c + \partial''d)$. By exactness again, we find $d' \in C'_q$ with $c_1 = c + \partial d + \alpha(d')$. Then $\partial c_1 = \partial c + \partial\alpha(d') = \alpha(c') + \alpha(\partial'd')$. Thus, $[\alpha^{-1}(\partial c')] = [c'] = [\alpha^{-1}(\partial c)]$ which shows, indeed, that the homology class of c' depends only on that of c''.   □

   We then have the following generalization of Theorem 6.2.5 that will play an important role below.

Corollary 6.2.3 *Let $\Sigma_2 \subset \Sigma_1 \subset \Sigma$ be complexes. We then have the long exact sequence of homology groups*

$$\ldots \xrightarrow{j_*} H_{q+1}(\Sigma, \Sigma_1) \xrightarrow{\partial_*} H_q(\Sigma_1, \Sigma_2) \xrightarrow{i_*} H_q(\Sigma, \Sigma_2)$$

$$\xrightarrow{j_*} H_q(\Sigma, \Sigma_1) \xrightarrow{\partial_*} \ldots \tag{6.2.47}$$

Proof We apply Theorem 6.2.7 to the short exact sequence

$$0 \longrightarrow C(\Sigma_1, \Sigma_2) \longrightarrow C(\Sigma, \Sigma_2) \longrightarrow C(\Sigma, \Sigma_1) \longrightarrow 0$$
$$(6.2.48)$$

which is obtained from the inclusions between our simplicial complexes.
$\square$

In fact, it is useful and insightful to develop a still more general picture. Let Σ_1, Σ_2 be subcomplexes of Σ. Then $\Sigma_1 \cap \Sigma_2$ and $\Sigma_1 \cup \Sigma_2$ are subcomplexes of Σ as well, and we have inclusions of chain complexes, $C(\Sigma_i) \subset C(\Sigma), i = 1, 2$, and $C(\Sigma_1 \cap \Sigma_2) = C(\Sigma_1) \cap C(\Sigma_2), C(\Sigma_1) + C(\Sigma_2) = C(\Sigma_1 \cup \Sigma_2)$, with the usual formal sum. With the inclusion maps $i_1 : \Sigma_1 \cap \Sigma_2 \subset \Sigma_1, i_2 : \Sigma_1 \cap \Sigma_2 \subset \Sigma_2, j_1 : \Sigma_1 \subset \Sigma_1 \cup \Sigma_2, j_2 : \Sigma_2 \subset \Sigma_1 \cup \Sigma_2$ and $i(c) := (C(i_1)c, -C(i_2)c), j(c_1, c_2) := C(j_1)c_1 + C(j_2)c_2$ (note the—sign in the definition of i), we obtain a short exact sequence of chain complexes

$$0 \longrightarrow C(\Sigma_1 \cap \Sigma_2) \xrightarrow{\ i\ } C(\Sigma_1) \oplus C(\Sigma_2) \xrightarrow{\ j\ } C(\Sigma_1 \cup \Sigma_2) \longrightarrow 0.$$
$$(6.2.49)$$

Definition 6.2.9 The long exact homology sequence obtained from (6.2.49) by Theorem 6.2.7,

$$\cdots \xrightarrow{\ j_*\ } \quad H_{q+1}(\Sigma_1 \cup \Sigma_2) \quad \xrightarrow{\ \partial_*\ } H_q(\Sigma_1 \cap \Sigma_2)$$

$$\xrightarrow{\ i_*\ } H_q(\Sigma_1) \oplus H_q(\Sigma_2) \qquad\qquad (6.2.50)$$

$$\xrightarrow{\ j_*\ } \quad H_q(\Sigma_1 \cup \Sigma_2) \quad \xrightarrow{\ \partial_*\ } \quad \cdots$$

is called the *Mayer-Vietoris sequence* of Σ_1 and Σ_2

When $\Sigma = \Sigma_1 \cup \Sigma_2$, the Mayer-Vietoris sequence tells us how to assemble the homology of Σ from the homologies of its pieces and their intersections.

More generally, when $(\Sigma_1, \Delta_1), (\Sigma_2, \Delta_2)$ are simplicial pairs contained in Σ, we not only have the short exact sequence (6.2.49), but also the short exact subsequence of the latter

$$0 \longrightarrow C(\Delta_1 \cap \Delta_2) \xrightarrow{\ i\ } C(\Delta_1) \oplus C(\Delta_2) \xrightarrow{\ j\ } C(\Delta_1 \cup \Delta_2) \longrightarrow 0$$
$$(6.2.51)$$

and the short exact quotient sequence (the reader is asked to verify this exactness)

$$0 \longrightarrow C(\Sigma_1 \cap \Sigma_2)/C(\Delta_1 \cap \Delta_2) \xrightarrow{\ i\ } C(\Sigma_1)/C(\Delta_1) \oplus C(\Sigma_2)/C(\Delta_2)$$

$$\xrightarrow{\ j\ } C(\Sigma_1 \cup \Sigma_2)/C(\Delta_1 \cup \Delta_2) \longrightarrow 0.$$
$$(6.2.52)$$

By Theorem 6.2.9 again, we obtain a long exact sequence of homology groups, called the *relative Mayer-Vietoris sequence*,

$$\cdots \xrightarrow{j_*} H_{q+1}(\Sigma_1 \cup \Sigma_2, \Delta_1 \cup \Delta_2) \xrightarrow{\partial_*} H_q(\Sigma_1 \cap \Sigma_2, \Delta_1 \cap \Delta_2)$$

$$\xrightarrow{i_*} H_q(\Sigma_1, \Delta_1) \oplus H_q(\Sigma_2, \Delta_2) \xrightarrow{j_*} H_q(\Sigma_1 \cup \Sigma_2, \Delta_1 \cup \Delta_2)$$

$$\xrightarrow{\partial_*} \qquad \cdots$$

$$(6.2.53)$$

For the pairs (Σ, Σ_2) and (Σ_1, Σ_1), this reduces to the sequence of Corollary 6.2.3 of the triple $(\Sigma, \Sigma_1, \Sigma_2)$, and when we furthermore have $\Sigma_2 = \emptyset$, we get the result of Theorem 6.2.4 back.

Definition 6.2.10 A *filtration* of the simplicial complex Σ is a family of subcomplexes $\Sigma = \Sigma_0 \supset \Sigma_1 \supset \cdots \supset \Sigma_n$.

Here is a simple example

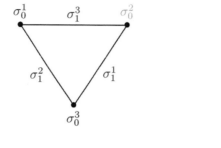

$$(6.2.54)$$

$\Sigma = \Sigma_0$ is everything, Σ_1 consists of all the colored simplices, that is, $\sigma_1^3, \sigma_1^1, \sigma_0^1, \sigma_0^3, \sigma_0^2$, Σ_2 consists of the red and the green simplices, that is, $\sigma_1^1, \sigma_0^3, \sigma_0^2$, and Σ_3 contains only the green vertex σ_0^2 and $\Sigma_4 = \emptyset$.

In the sequel, we wish to apply Corollary 6.2.3 iteratively to a filtration of a simplicial complex so as to build its homology up from the members of the filtration.

We shall need to address one more technical point. We start with the following observation. This observation will depend on the construction of tensor products of abelian groups which we now recall.

Definition 6.2.11 For abelian groups A and B, we let $F(A \times B)$ be the free abelian group generated by the set $A \times B$ (recall the construction of the chain groups). The tensor product $A \otimes B$ is then obtained by dividing out the equivalence relation $(a_1 + a_2, b) \sim (a_1, b) + (a_2, b)$, $(a, b_1 + b_2) \sim (a, b_1) + (a, b_2)$.

The tensor product can also be characterized by the property that for any abelian group C, the bilinear maps $A \times B \to C$ correspond to the homomorphisms $A \otimes B \to C$.

For an abelian group G and a field K, the tensor product $G \otimes K$ of abelian groups (taking the additive group of the field K) becomes a vector space over the field K, with the obvious operation $r(g \otimes s) = g \otimes (rs)$ for $g \in G, r, s \in K$. Furthermore, group homomorphisms $\rho : G_1 \to G_2$

induce vector space morphisms (i.e., linear maps) $\rho \otimes \text{id} : G_1 \otimes K \rightarrow G_2 \otimes K$.

Consequently, for a field K, from the homology groups $H_q(\Sigma, \mathbb{Z})$, we obtain vector spaces

$$H_q(\Sigma, \mathbb{Z}) \otimes K. \tag{6.2.55}$$

The fields we have in mind here are $K = \mathbb{R}$ and $K = \mathbb{Z}_2$. The point is that for instance $\mathbb{Z}_n \otimes \mathbb{R} = 0$, while $\mathbb{Z} \otimes \mathbb{R} = \mathbb{R}$, that is, by tensoring with $\mathbb{R}$, we get rid of the torsion in $H_q(\Sigma, \mathbb{Z})$.

In fact, from the universal coefficient theorem, in this manner, we can obtain homology groups with coefficients in K from those with coefficients in $\mathbb{Z}$. We shall now briefly explain this issue, without providing the technical details and proofs, however, as this shall not be needed in the sequel.

Definition 6.2.12 An exact sequence of abelian groups, with F being a *free* abelian group,

$$0 \longrightarrow R \longrightarrow F \longrightarrow A \longrightarrow 0 \tag{6.2.56}$$

is called a *free resolution* of A.

Example

$$0 \longrightarrow \mathbb{Z} \xrightarrow{n} \mathbb{Z} \longrightarrow \mathbb{Z}_n \longrightarrow 0 \tag{6.2.57}$$

is a free resolution of $\mathbb{Z}_n$.

There is an easy way to construct a free resolution of an abelian group A, called the standard resolution, as follows. We let $F(A)$ consist of the formal sums $\sum_{g \in A} \mu(g)g$, with $\mu(g) \in \mathbb{Z}$ and requiring that at most finitely many $\mu(g) \neq 0$. $F(A)$ is the free abelian group generated by the set A; we have already utilized this construction above for the chain groups. We then have a homomorphism $\rho : F(A) \rightarrow A$ that simply consists in replacing the formal $\sum$ in $F(A)$ by the group operation in A. With $R(A) := \ker \rho$, we obtain the standard resolution of A,

$$0 \longrightarrow R(A) \xrightarrow{i} F(A) \xrightarrow{\rho} A \longrightarrow 0. \tag{6.2.58}$$

When we tensor this exact sequence with another abelian group G, we preserve a part of this sequence; more precisely, we obtain the exact sequence

$$R(A) \otimes G \xrightarrow{i \otimes \text{id}_G} F(A) \otimes G \xrightarrow{\rho \otimes \text{id}_G} A \otimes G \longrightarrow 0. \tag{6.2.59}$$

We do not necessarily have the leftmost arrow of (6.2.58) in (6.2.59) as well because $i \otimes \text{id}_G$ need no longer be injective.

Definition 6.2.13 The kernel of $i \otimes \text{id}_G : R(A) \otimes G \rightarrow F(A) \otimes G$ is called the *torsion product* $\text{Tor}(A, G)$ of A and G.

In order to be able to compute some examples, we utilize the following lemma, without providing its (not too difficult) proof here.

Lemma 6.2.4 *For any free resolution*

$$0 \longrightarrow R \xrightarrow{i} F \xrightarrow{p} A \longrightarrow 0, \qquad (6.2.60)$$

we have

$$\ker(i \otimes \mathrm{id}_G) \cong \mathrm{Tor}(A, G). \qquad (6.2.61)$$

Thus, we can compute the torsion product from any free resolution of A, not necessarily the standard one.

Example

1. For a *free* abelian group,

$$\mathrm{Tor}(A, G) = 0, \qquad (6.2.62)$$

   because $0 \longrightarrow 0 \longrightarrow A \xrightarrow{\mathrm{id}_A} A \longrightarrow 0$ is a free resolution of the free group A.

2. $$\mathrm{Tor}(\mathbb{Z}_n, G) \cong \{g \in G : ng = 0\}, \qquad (6.2.63)$$

   because in the free resolution (6.2.57), the map i is given by multiplication by n. In particular, when G is torsion free,

$$\mathrm{Tor}(\mathbb{Z}_n, G) = 0. \qquad (6.2.64)$$

3. Also, by taking direct sums of free resolutions, we obtain

$$\mathrm{Tor}(A_1 \oplus A_2, G) \cong \mathrm{Tor}(A_1, G) \oplus \mathrm{Tor}(A_2, G). \qquad (6.2.65)$$

4. Combining the preceding items, we obtain

$$\mathrm{Tor}(A, \mathbb{R}) = 0 \qquad (6.2.66)$$

   for any finitely generated abelian group A.

The *universal coefficient theorem* (which we do not prove here) tells us

Theorem 6.2.8 *For a simplicial pair* (Σ, Δ), $q \in \mathbb{Z}$ *and an abelian group* G, *we have*

$$H_q(\Sigma, \Delta; G) \cong (H_q(\Sigma, \Delta; \mathbb{Z}) \otimes G) \oplus \mathrm{Tor}(H_{q-1}(\Sigma, \Delta; \mathbb{Z}), G). \quad (6.2.67)$$

For our purposes, the important consequence of Theorem 6.2.8 is that when the group G is a field, the homology groups $H_q(\Sigma, \Delta; G)$ are in fact vector spaces over G. The fields that we have in mind here are $\mathbb{R}$ and $\mathbb{Z}_2$. In the rest of this section, in fact, we shall take $G = \mathbb{R}$ (as opposed to our earlier choice $G = \mathbb{Z}$), so that we do not need to worry about torsion, see (6.2.66), and we shall hence work with the vector spaces $H_q(\Sigma, \Delta) = H_q(\Sigma, \Delta; \mathbb{R})$, without mentioning this explicitly.

We shall utilize the following simple result.

Lemma 6.2.5 *Let* $\cdots \rightarrow A_3 \xrightarrow{a_2} A_2 \xrightarrow{a_1} A_1 \longrightarrow 0$ *be an exact sequence of linear maps between vector spaces. Then for all* $k \in \mathbb{N}$

$$\dim A_1 - \dim A_2 + \dim A_3 - \cdots - (-1)^k \dim A_k + (-1)^k \dim \mathrm{im}\, a_k = 0. \qquad (6.2.68)$$

Proof If $\ell : V \to W$ is linear, then dim $V = \dim (\ker \ell) + \dim (\operatorname{im} \ell)$. The exactness implies that

$$\dim (\ker a_j) = \dim (\operatorname{im} a_{j+1}).$$

Hence by the exactness of the sequence

$$\dim A_j = \dim (\operatorname{im} a_{j-1}) + \dim (\operatorname{im} a_j)$$

and with

$$\dim A_1 = \dim (\operatorname{im} a_1)$$

the claim follows.   □

We now apply this lemma to the long exact sequence of Corollary 6.2.3 (with $G = \mathbb{R}$, to emphasize this once more). We put

$$b_q(\Sigma, \Delta) := \dim H_q(\Sigma, \Delta), \tag{6.2.69}$$

$$\nu_q(\Sigma, \Delta_1, \Delta_2) := \dim(\operatorname{im} \partial_{q+1}). \tag{6.2.70}$$

From the lemma, we get

$$\sum_{q=0}^{m} (-1)^q (b_q(\Sigma, \Sigma_1) - b_q(\Sigma, \Sigma_2) + b_q(\Sigma_1, \Sigma_2)) - (-1)^m \nu_m(\Sigma, \Sigma_1, \Sigma_2) = 0, \tag{6.2.71}$$

whence

$$(-1)^{m-1} \nu_{m-1}(\Sigma, \Sigma_1, \Sigma_2) \tag{6.2.72}$$
$$= (-1)^m \nu_m(\Sigma, \Sigma_1, \Sigma_2) - (-1)^m b_m(\Sigma, \Sigma_1) + (-1)^m b_m(\Sigma, \Sigma_2)$$
$$- (-1)^m b_m(\Sigma_1, \Sigma_2).$$

With

$$P_m(t, \Sigma, \Delta) := \sum_{q=0}^{m} b_q(\Sigma, \Delta) t^q, \tag{6.2.73}$$

$$P(t, \Sigma, \Delta) := \sum_{q \geq 0} b_q(\Sigma, \Delta) t^q, \tag{6.2.74}$$

$$Q_m(t, \Sigma, \Delta_1, \Delta_2) := \sum_{q=0}^{m} \nu_q(\Sigma, \Delta_1, \Delta_2) t^q, \tag{6.2.75}$$

$$Q(t, \Sigma, \Delta_1, \Delta_2) := \sum_{q \geq 0} \nu_q(\Sigma, \Delta_1, \Delta_2) t^q, \tag{6.2.76}$$

and multiplying (6.2.72) by $(-1)^q t^q$ and summing, we obtain

$$(-1)^m \nu_m(\Sigma, \Sigma_1, \Sigma_2) + (1 + t) Q_{m-1}(t, \Sigma, \Sigma_1, \Sigma_2)$$
$$= P_m(t, \Sigma, \Sigma_1) - P_m(t, \Sigma, \Sigma_2) + P_m(t, \Sigma_1, \Sigma_2) \tag{6.2.77}$$

and

$$(1 + t) Q(t, \Sigma, \Sigma_1, \Sigma_2) = P(t, \Sigma, \Sigma_1) - P(t, \Sigma, \Sigma_2) + P(t, \Sigma_1, \Sigma_2). \tag{6.2.78}$$

We obtain

Theorem 6.2.9 *Let* $\Sigma = \Sigma_0 \supset \Sigma_1 \supset \cdots \supset \Sigma_n$ *be a filtration of* Σ. *Then*

$$\sum_{j=1}^{n} P(t, \Sigma_{j-1}, \Sigma_j) = P(t, \Sigma_0, \Sigma_n) + (1+t)Q(t), \qquad (6.2.79)$$

with $Q(t) := \sum_{j=1}^{n-1} Q(t, \Sigma_{j-1}, \Sigma_j, \Sigma_n)$.

We observe that $Q(t)$ is a polynomial with nonnegative integer coefficients.

Proof We apply (6.2.78) for the triple $(\Sigma_{j-1}, \Sigma_j, \Sigma_n)$ and sum w.r.t. j. $\square$

In particular, choosing $t = -1$, we get

Corollary 6.2.4

$$\sum_{j=1}^{n} P(-1, \Sigma_{j-1}, \Sigma_j) = P(-1, \Sigma_0, \Sigma_n). \qquad (6.2.80)$$

We look at the example (6.2.54) which we recall here,

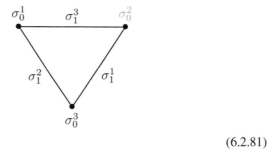

$$(6.2.81)$$

and for which we compute $b_1(\Sigma_0)(= b_1(\Sigma_0, \Sigma_4)) = 1, b_0(\Sigma_0) = 1$ and $b_1(\Sigma_0, \Sigma_1) = 1, b_0(\Sigma_0, \Sigma_1) = 0;$ $b_1(\Sigma_1, \Sigma_2) = 0, b_0(\Sigma_1, \Sigma_2) = 0;$ $b_1(\Sigma_2, \Sigma_3) = 0, b_0(\Sigma_2, \Sigma_3) = 0;$ $b_1(\Sigma_3)(= b_1(\Sigma_3, \Sigma_4)) = 0,$ $b_0(\Sigma_3) = 1$ which of course fits with the general formula.

Similarly, we obtain from (6.2.77)

$$(-1)^m \sum_{j=1}^{n} P_m(-1, \Sigma_{j-1}, \Sigma_j) \geq (-1)^m P(-1, \Sigma_0, \Sigma_n). \qquad (6.2.82)$$

Recalling the definitions (6.2.73), (6.2.74), we obtain from (6.2.82) and (6.2.80)

Corollary 6.2.5 *With*

$$c_q := \sum_{j} b_q(\Sigma_{j-1}, \Sigma_j), \qquad (6.2.83)$$

we have

$$c_m - c_{m-1} + \cdots + (-1)^m c_0 \geq b_m - b_{m-1} + \cdots + (-1)^m b_0 \text{ for all } m \geq 0, \qquad (6.2.84)$$

and

$$\sum_q (-1)^q c_q = \sum_q (-1)^q b_q. \tag{6.2.85}$$

In particular, when we construct a filtration by adding simplices one by one, we see that (6.2.85) includes the formula of Theorem 6.2.3.

Corollary 6.2.6

$$c_q \geq b_q \text{ for all } q. \tag{6.2.86}$$

Proof This follows from Corollary 6.2.5.   □

6.3   Combinatorial Exchange Patterns

In this section, we utilize the preceding results from simplicial homology theory to obtain a combinatorial analogue of Conley's homology theory for dynamical systems. For that purpose, we need to develop the concept of a combinatorial exchange pattern.

Definition 6.3.1 An *exchange pattern* of the relations defining the simplicial complex Σ means that for every q-simplex $\sigma_q = (v_0, \ldots, v_q)$ of Σ, we either do nothing or select vertices $w_1, \ldots, w_m$ for which $(w_\alpha, v_0, \ldots, v_q)$ is a $(q+1)$-simplex, that is, $r(w_\alpha, v_0, \ldots, v_q) \neq o$, and take the collection $P_0(\sigma_q)$ of all those $(q+1)$-simplices $(w_\alpha, v_0, \ldots, v_q)$ together with all their subsimplices as well as the collection $P_1(\sigma_q)$ of all q-simplices of the form $(w_\alpha, v_0, \ldots, \hat{v}_i, \ldots, v_q)$ for some $i = 0, \ldots, q$, again together with all their subsimplices. Equivalently, P_1 consists of all subsimplices of dimension $\leq q$ of P_0 with the exception of σ_q. (As it stands, $P_0(\sigma_q)$ and $P_1(\sigma_q)$ need not be simplicial complexes themselves, but later, we shall apply the procedure in such a manner that they will always be simplicial complexes.) We interpret such a simplex $(w_\alpha, v_0, \ldots, \hat{v}_i, \ldots, v_q)$ as one obtained by exchanging the element v_i in the relation defining σ_q against the element w_α leading to the relation $r(w_\alpha, v_0, \ldots, \hat{v}_i, \ldots, v_q) \neq o$. This exchange process is subject to the following constraint:

1. Whenever we have exchanged the relation defining some simplex, this simplex in turn cannot be used to exchange the relation of any of its faces.

We may iterate the process by selecting one of the q-simplices $(w_\alpha, v_0, \ldots, \hat{v}_i, \ldots, v_q)$ and repeating the process with that simplex. The resulting exchange set P_0 will then be the union of the exchange sets of those simplices, and the corresponding P_1 will consist of all proper subsimplices of P_0 without the exchanged ones, that is, without $(v_0, \ldots, v_q)$ and $(w_\alpha, v_0, \ldots, \hat{v}_i, \ldots, v_q)$.

We are allowed to exchange the relation of any simplex σ_q more than once, or conversely also not at all. In the latter case, we put $P_0(\sigma_q) = \sigma_q$, $P_1(\sigma_q) = \emptyset$. In order to ensure that the result of any exchange will still be a simplicial complex, later on we shall perform the exchanges w.r.t. decreasing dimension of the simplices involved, i.e., start with exchanges of the higher dimensional simplices and finally end up with exchanges of 0-dimensional simplices along 1-dimensional ones.

One important idea motivating this construction is the following. When $m = 1$, that is, when we utilize only one additional element to exchange the relation of σ_q, then we think of a cancellation of σ_q against the $(q+1)$-simplex (w_1, σ_q). In the sequel, we shall count q-simplices with the sign $(-1)^q$, and so, such a cancellation will not effect the result of such counting, but reduce the number of items to be counted by 2.

Here is a preliminary definition that may be helpful to develop some intuition, but which shall be refined subsequently.

Definition 6.3.2 An exchange orbit of index q is a finite sequence of simplices

$$\sigma_q^0, \tau_{q+1}^0, \sigma_q^1, \ldots, \sigma_q^m, \tau_{q+1}^m, \tag{6.3.87}$$

where, for $j = 0, \ldots, m-1$, τ_{q+1}^j is obtained by adding a vertex to σ_q^j, and $\sigma_q^{j+1} \neq \sigma_q^j$ is obtained by deleting a vertex from τ_{q+1}^j. In other words, σ_q^j and σ_q^{j+1} are different faces of τ_{q+1}^j. The exchange orbit is closed and called an exchange cycle if $\sigma_q^m = \sigma_q^0$. Two such closed orbits are equivalent when they contain the same simplices in the same cyclic order, that is, when they only differ by the choice of the starting simplex.

An exchange pattern can be readily visualized by drawing an arrow from a face into any simplex that is used for exchanging for any of its vertices. The condition in the definition then says that whenever an arrow is pointing out of some simplex, there cannot be any arrow pointing into it.

In Fig. 6.3 (see Fig. 6.1 for the notation)

$$\Sigma_0 = \Sigma$$
$$\Sigma_1' = \Sigma \setminus \{\sigma_2, \sigma_1^1\}$$
$$\ker \partial_2^{rel} = 0$$
$$\ker \partial_1^{rel} = [\sigma_1^1] = \operatorname{im} \partial_2^{rel}$$
$$b_2^{rel} = b_1^{rel} = b_0^{rel} = 0$$
$$\ker \partial_1^{\Sigma_1'} = 0, \quad \operatorname{im} \partial_1^{\Sigma_1'} = [\sigma_0^1 - \sigma_0^3, \sigma_0^3 - \sigma_0^2]$$
$$\ker \partial_0^{\Sigma_1'} = [\sigma_0^1, \sigma_0^2, \sigma_0^3]$$
$$b_1(\Sigma_1') = 0, \quad b_0(\Sigma_1') = 1.$$

Fig. 6.3 For an explanation, see the main text

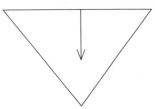

Fig. 6.4 For an explanation, see the main text

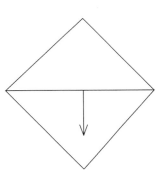

Fig. 6.5 For an explanation, see the main text; not satisfying the constraints developed for Definition 6.3.4 below

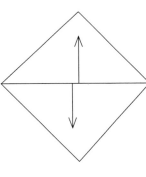

Fig. 6.6 For an explanation, see the main text; not satisfying the constraints developed for Definition 6.3.4 below

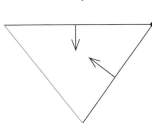

We now consider examples with two 2-simplices as in Fig. 6.4, distinguished by a corresponding superscript. For instance, in Fig. 6.5

$$\Sigma_0 = \Sigma$$
$$\Sigma_1' = \Sigma \setminus \{\sigma_2^1, \sigma_2^2, \sigma_1^1\}$$
$$\ker \partial_2^{rel} = [\sigma_2^1 + \sigma_2^2]$$
$$\ker \partial_1^{rel} = [\sigma_1^1] = \operatorname{im} \partial_2^{rel}$$
$$\partial_0^{rel} = 0,$$

and the topological invariants therefore are the same as those of Fig. 6.2.

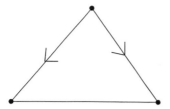

Fig. 6.7 For an explanation, see the main text; not satisfying the constraints developed for Definition 6.3.4 below

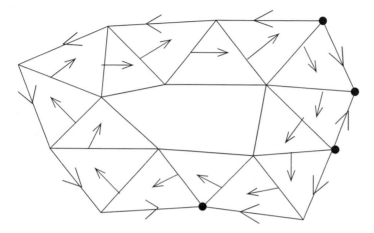

Fig. 6.8 A more complicated pattern. Note the cycle of *red edges* without arrows and the two vertices each with two incoming or two outgoing arrows

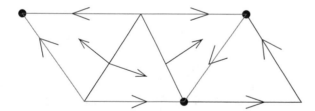

Fig. 6.9 Another more complicated pattern. The collection of *green edges* results from exchanging the two *interior edges* with arrows pointing into adjacent 2-simplices

In Fig. 6.6 (see Fig. 6.1 for the notation),

$$\Sigma_0 = \Sigma$$
$$\Sigma_1'' = \Sigma \setminus \{\sigma_2, \sigma_1^1, \sigma_1^2\}$$
$$\ker \partial_2^{rel} = 0$$
$$\ker \partial_1^{rel} = [\sigma_1^1, \sigma_1^2] \neq \operatorname{im} \partial_2^{rel}$$
$$\ker \partial_0^{rel} = 0 = \operatorname{im} \partial_1^{rel}$$
$$b_2^{rel} = 0, \quad b_1^{rel} = 1, b_0^{rel} = 0$$

$$\ker \partial_1^{\Sigma_1''} = 0, \quad \operatorname{im} \partial_1^{\Sigma_1''} = [\sigma_0^3 - \sigma_0^1]$$
$$\ker \partial_0^{\Sigma_1''} = [\sigma_0^1, \sigma_0^2, \sigma_0^3]$$
$$b_0(\Sigma_1'') = 2.$$

Similarly, for the one-dimensional simplicial complex in Fig. 6.7

$$b_1^{rel} = 1, \quad b_0^{rel} = 0$$

and with *rem* standing for remainder, $\quad b_1^{rem} = 0, \ b_0^{rem} = 1.$

By the Excision Theorem 6.2.6, for any simplicial complex Σ and any exchange pair (P_0, P_1), the homology of the pair $(\Sigma, (\Sigma \backslash P_0) \cup P_1)$ is the same as that of (P_0, P_1). We can do better, however, by breaking the exchange relation up into components. In order to identify the relevant components, we can proceed by decreasing dimension q. That is, we pick any top-dimensional simplex σ not used in an exchange if there is any such simplex. We then take $\Sigma_0 = \Sigma$, $\Sigma_1 = \Sigma \backslash \sigma$. We repeat this process until there is no further top-dimensional simplex not participating in any exchange, i.e., with no arrow pointing into it. We then take a top-dimensional simplex that has an arrow pointing into it, if any such simplex exists. Since σ participates in an exchange process, we take its orbit $P_0(\sigma)$, that is, all simplices (together with all their subsimplices) of the same dimension that can be reached from σ by following arrows backward or forward, together with all the faces from which those arrows originate. That is, $\sigma' \in P_0(\sigma)$ whenever there exists a sequence of simplices $\sigma^0 = \sigma, \sigma^1, \ldots, \sigma^n = \sigma'$ and faces τ^i of both σ^{i-1} and σ^i that use at least one of them for an exchange. Then any of the faces of $\bigcup_{i=0}^n \sigma^i$ different from the τ^i are not using any top-dimensional simplex for an exchange. We then let $P_1(\sigma)$ be the collection of those faces (together with all their subsimplices). In the present case, we take $\Sigma_0 = \Sigma$, $\Sigma_1 = \Sigma \backslash P_0(\sigma) \cup P_1(\sigma)$. We can repeat the process with other orbits, that is, having constructed Σ_j, we take a top-dimensional simplex σ in Σ_j (as Σ_j might already be of smaller dimension than Σ, a top-dimensional simplex of Σ_j might be of correspondingly smaller dimension) and take out its orbit $P_0(\sigma)$ without $P_1(\sigma)$ to obtain Σ_{j+1}. Again, in each dimension, we first cut out the simplices without arrows and then the nontrivial orbits. The order among the simplices without arrows or the order in which we eliminate the orbits in each dimension plays no role. It is only important to eliminate the simplices without arrows first, as otherwise the elimination of an orbit may lead to a space that is no longer a simplicial complex. This happens when for an orbit we utilize a face of a simplex without arrows. Also, it is important to carry the process out according to decreasing dimension. We proceed until we arrive at $\Sigma_n = \emptyset$.

Definition 6.3.3 The filtration resulting from the above process is called a filtration for the exchange pattern.

Note that the filtration for an exchange pattern need not be unique because we can eliminate the arrow-free simplices as well as the nontrivial orbits in each dimension in an arbitrary order. This will not affect the topological invariants c_q from the resulting excision process.

Corollary 6.3.1 *The inequalities of Corollary 6.2.5 hold.*

Proof This is clear because in each step the conditions of the Excision Theorem 6.2.6 are satisfied.   □

We have already observed above that Theorem 6.2.3 follows from Corollary 6.2.5, but now it is also a consequence of Corollary 6.3.1:

Corollary 6.3.2

$$\chi(\Sigma) = \sum_q (-1)^q b_q(\Sigma), \tag{6.3.88}$$

where the Euler number $\chi(\Sigma)$ has been defined in (6.2.6).

Proof We apply Corollary 6.3.1 to the exchange process where we simply take out the simplices of Σ one by one, according to decreasing dimension.   □

We shall now describe the theory of Forman [38, 39] in the framework that we have developed here.

Definition 6.3.4 A discrete vector field on the simplicial complex Σ is a collection of pairs of simplices $\tau^i_{q-1} \subset \sigma^i_q$, with the restriction that each simplex is contained in at most one pair.

By drawing an arrow from τ^i_{q-1} into σ^i_q for each pair, we create an exchange pattern in the sense of Definition 6.3.1. An orbit of index q of a vector field is then a sequence

$$\tau^1_{q-1}, \sigma^1_q, \tau^2_{q-1}, \sigma^2_q, \ldots, \tau^n_{q-1}, \sigma^n_q, \tau^{n+1}_{q-1} \tag{6.3.89}$$

such that $(\tau^i_{q-1}, \sigma^i_q)$ is always a pair for the vector field $\tau^i_{q-1} \neq \tau^{i+1}_{q-1}$, but both of them are faces of σ^i_q. The orbit is closed if $\tau^{n+1}_q = \tau^1_q$. Apart from this possibility, we assume that all the simplices involved are different from each other.

This then obviously represents a special case of our construction, and the preceding results apply.

Of particular interest in Forman's theory and for a variety of applications are the gradient vector fields of discrete Morse functions which we shall now describe. In fact, Forman proved that any vector field without closed orbits is such a gradient vector field.

Definition 6.3.5 A function $f : \Sigma \to \mathbb{R}$ (i.e., f assigns to every simplex of Σ a real number) is called a Morse function if for every simplex $\sigma_q \in \Sigma$

1. there is at most one simplex $\rho_{q+1} \supset \sigma_q$ with

$$f(\rho_{q+1}) \leq f(\sigma_q) \tag{6.3.90}$$

   and

2. at most one simplex $\tau_{q-1} \subset \sigma_q$ with

$$f(\tau_{q-1}) \geq f(\sigma_q). \tag{6.3.91}$$

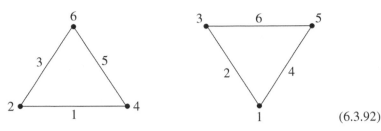

$$\tag{6.3.92}$$

The function on the graph on the left is not a Morse function, because the upper vertex (with the value 6) does not satisfy condition 1., whereas the function on the graph on the right is Morse.

When we draw an arrow from σ_q to ρ_{q+1} whenever $\sigma_q \subset \rho_{q+1}$ and $f(\rho_{q+1}) \leq f(\sigma_q)$, then we obtain a combinatorial exchange pattern. The above conditions then restrict the combinatorial exchange patterns that come from a Morse function. Condition 1 says that there cannot be more than one arrow emanating from any simplex. This is violated in Fig. 6.8 for the two red dots with two arrows into adjacent edges, and in Fig. 6.9 for the edge with two arrows into adjacent 2-simplices. Condition 2 does not hold in Fig. 6.6.

Lemma 6.3.1 *For a simplex σ_q, it is not possible that there both exist a simplex ρ_{q+1} with 1. and a simplex τ_{q-1} with 2.*

Proof When there exists a simplex ρ_{q+1} satisfying 1., then by applying 2. to ρ_{q+1}, for all other faces $\sigma'_q \neq \sigma_q$ of ρ_{q+1},

$$f(\rho_{q+1}) > f(\sigma'_q). \tag{6.3.93}$$

Each σ'_q in turn, by 2., has at most one face τ'_{q-1} with

$$f(\tau'_{q-1}) \geq f(\sigma'_q). \tag{6.3.94}$$

If now there existed some face τ_{q-1} of σ_q with (6.3.90), we would find some such σ'_q for which τ_{q-1} is also a face, but then we would have the chain of inequalities

$$f(\sigma'_q) < f(\rho_{q+1}) \leq f(\sigma_q) \leq f(\tau_{q-1}). \tag{6.3.95}$$

This, however, would violate 1. for τ_{q-1} because (6.3.95) means that both $f(\sigma_q)$ and $f(\sigma'_q)$ are smaller than $f(\tau_{q-1})$.   □

Definition 6.3.6 The simplex σ_q is critical for f if there neither exists a ρ_{q+1} with (6.3.90) nor a τ_{q-1} with (6.3.91).

In the right graph in (6.3.92), with its Morse function, the top edge (with the value 6) and the bottom vertex (with the value 0) are critical.

Lemma 6.3.1 says that for a non-critical simplex precisely one of the conditions 1. or 2. holds. Therefore, we may define

Definition 6.3.7 The gradient vector field associated to a Morse function f on the simplicial complex Σ is given by the pairs (σ_q, τ_{q-1}) with

$$f(\tau_{q-1}) \geq f(\sigma_q). \tag{6.3.96}$$

Since the function f is then decreasing along orbits, the gradient vector field of a Morse function does not possess any closed orbits.

Corollary 6.3.3 *The number μ_q of critical simplices of dimension q of a Morse function on Σ satisfies*

$$\mu_m - \mu_{m-1} + \cdots + (-1)^m \mu_0 \geq b_m - b_{m-1} + \cdots + (-1)^m b_0 \text{ for all } m \geq 0, \tag{6.3.97}$$

and

$$\sum_q (-1)^q \mu_q = \sum_q (-1)^q b_q. \tag{6.3.98}$$

Proof We apply Corollary 6.2.5 and observe the following points. The excision of a nontrivial orbit does not yield a contribution to the alternating sum of the relative Betti numbers c_q, because orbits of a Morse function cannot be closed, as just pointed out, and so, the simplices of dimension q and $q-1$ involved in such an orbit cancel in pairs, except for the last τ_{q-1}^{n+1} which is kept and not excised. The excision of a critical simplex of dimension q, in contrast, contributes 1 to the c_qs. $\qquad\square$

The reason why (6.3.98) holds is easy to understand. For a non-critical q-simplex σ_q, by Lemma 6.3.1, we either find a $(q+1)$-simplex with 1. or a $(q-1)$-simplex with 2., but not both. Therefore, we find precisely one simplex in an adjacent dimension that we can cancel with σ_q in the alternating sum defining the Euler characteristic $\chi(\Sigma)$ of Definition 6.2.1. Only the critical simplices remain and contribute to $\chi(\Sigma)$. That is, we can deduce directly from the definition of the Euler characteristic that $\chi(\Sigma) = \sum_q (-1)^q \mu_q$. Theorem 6.2.3 and Corollary 6.3.1 then tells us that we also have $\chi(\Sigma) = \sum_q (-1)^q b_q$.

Remark In the same manner that Forman conceived his theory just sketched as a discrete version of Morse theory, the framework developed here can be considered as a discrete version of Conley theory, see e.g. [60].

6.4   Homology of Topological Spaces

In this section, we shall sketch the homology theory of topological spaces, without providing the technical proofs, for which we refer to [104] or [108], for instance.

It may seem natural, and in fact, this was the first attempt towards a general homology theory, to try to approximate a topological space by a simplicial complex, by somehow subdividing it into sufficiently small pieces that are homeomorphic to simplices. This approach, however, encounters several problems. First of all, it is not clear whether a general topological space can be so approximated. Secondly, even for a manifold, where such a simplicial approximation is possible without problems, it is not clear to what extent the resulting homology theory might depend on the choice of simplicial approximation. One might then try to construct a common simplicial subdivision for any two such simplicial approximations, but this runs into combinatorial problems, and, in fact, is not possible in general. Thirdly, constructions within a homology theory for topological spaces should behave functorially under continuous mappings, but the image of a simplicial approximation of one space under a continuous map need not be one for the target space, for instance if the map happens to be not injective.

Therefore, this functorial behavior should already be built into the definitions.

This is the idea underlying **singular homology** which we shall now sketch.

We have to perform one conceptual shift. In the preceding sections, we had considered a simplex σ_q as a collection of $q + 1$ vertices, that is, as a combinatorial object. We now need to consider simplices as topological spaces.

$\boxed{q\text{-simplex}}$

Definition 6.4.1 The *(topological) simplex* Δ_q of dimension q is

$$\Delta_q := \left\{ x \in \mathbb{R}^{q+1} : x = \sum_{i=0}^{q} \lambda_i e^i, 0 \le \lambda_i \le 1, \sum \lambda_i = 1 \right\}, \quad (6.4.1)$$

where $e_0 = (1, 0, \ldots, 0), e_1 = (0, 1, 0, \ldots,), \ldots, e_q = (0, 0, \ldots, 0, 1)$ are the unit basis vectors of $\mathbb{R}^{q+1}$. Here, $(\lambda_0, \ldots, \lambda_q)$ are called the *barycentric coordinates* of the point $x = \sum_{i=0}^{q} \lambda_i e^i$.

The ith *face* Δ_{q-1}^i of Δ_q consists of all $x \in \Delta_q$ with $\lambda_i = 0$. The *(topological) boundary* $\dot{\Delta}_q$ is the union of all its faces. Thus, it consists of all $x \in \Delta_q$ for which at least one of the λ_j vanishes.

From now on, we shall consider simplices as topological as opposed to combinatorial objects, according to the preceding definition.

The basic example is this topological boundary $\dot{\Delta}_q$. The following lemma is geometrically obvious and readily checked.

Lemma 6.4.1 $\dot{\Delta}_q$ *is homeomorphic to the* $(q - 1)$-*dimensional sphere*

$$S^{q-1} = \{x = (x^1, \ldots, x^q) \in \mathbb{R}^q : \sum_{i=1}^{q} (x^i)^2 = 1\}. \quad (6.4.2)$$

$\boxed{\begin{array}{c}\text{sphere}\\ S^{q-1}\end{array}}$

This is illustrated in the following picture for $q = 2$, that is, for S^1.

Thus, the topological invariants of the sphere S^{q-1} can be computed from Lemma 6.2.2, according to the framework that will be developed in this section. In particular, we shall have

$$b_{q-1}(S^{q-1}) = b_0(S^{q-1}) = 1 \text{ and } b_p(S^{q-1}) = 0 \text{ for all other } p, \tag{6.4.3}$$

$$\chi(S^{q-1}) = \begin{cases} 2 & \text{for odd } q \\ 0 & \text{for even } q. \end{cases} \tag{6.4.4}$$

Similarly, the simplex Δ^q itself is homeomorphic to the q-dimensional ball

$$B^q = \left\{ x = (x^1, \ldots, x^q) \in \mathbb{R}^q : \sum_{i=1}^{q} (x^i)^2 \leq 1 \right\}, \tag{6.4.5}$$

and so, the Betti numbers of the ball are those of the simplex, that is,

$$b_0(B^q) = 1 \text{ and } b_p(B^q) = 0 \text{ for } p > 0, \text{ and hence } \chi(B^q) = 1. \tag{6.4.6}$$

We now start the general constructions of singular homology. The map

$$\delta^i_{q-1} : \Delta_{q-1} \to \Delta^i_{q-1} \subset \Delta_q \tag{6.4.7}$$

$$e_0 \mapsto e_0, \quad \ldots \quad, e_{i-1} \mapsto e_{i-1}, e_i \mapsto e_{i+1}, \quad \ldots \quad, e_{q-1} \mapsto e_q$$

maps the $(q-1)$-dimensional simplex bijectively and linearly on the ith face of the q-dimensional simplex. One checks that for $q \geq 2, 0 \leq k < j \leq q$

$$\delta^j_{q-1} \circ \delta^k_{q-2} = \delta^k_{q-1} \circ \delta^{j-1}_{q-2}. \tag{6.4.8}$$

Definition 6.4.2 Let X be a topological space. A singular q-simplex in X is a continuous map

$$\gamma_q : \Delta_q \to X. \tag{6.4.9}$$

Note that the map γ_q need not be injective. Thus, the image of Δ_q may look rather different from a simplex.

Analogously to the chain groups $C_q(X, G)$ (G being an abelian group) in simplicial homology, we now define the groups $S_q(X, G)$ of singular q-chains in X. Thus, a singular q-chain is a formal linear combination

$$s_q = \sum_{i=1}^{m} g_i \gamma^i_q \tag{6.4.10}$$

of finitely many singular q-simplices with coefficients $g_i \in G$.

A continuous map $f : X \to Y$ between topological spaces then induces a homomorphism $f_* : S_q(X, G) \to S_q(Y, G)$ between the singular chain groups for any q and G, because the continuous image of a singular q-chain is again a singular q-chain. Thus, we have functors $F_{q,G}$ from the category of topological spaces to the category of abelian groups, for every non-negative integer q and every abelian group G. We can also view the collection of these functors as a single functor F from the category of topological spaces to the category of singular chain complexes. We leave it to the reader to precisely define the latter category.

The boundary operator is given by

$$\partial = \partial_q : S_q(X) \to S_{q-1}(X) \tag{6.4.11}$$

$$\partial s_q = \sum_{i=0}^{q} (-1)^i s_q \circ \delta_{q-1}^i \tag{6.4.12}$$

for $q \geq 1$ ($\partial_q := 0$ for $q \leq 0$).

Again, the crucial result is

Lemma 6.4.2

$$\partial_q \circ \partial_{q+1} = 0. \tag{6.4.13}$$

Definition 6.4.3 The qth (singular) homology group of the topological space X with coefficients in G is

$$H_q(X, G) := \ker \partial_q / \operatorname{im} \partial_{q+1}. \tag{6.4.14}$$

Theorem 6.4.1 *A continuous map $f : X \to Y$ between topological spaces commutes with the boundary operator and therefore induces homomorphisms*

$$f_* := H_q(f, G) : H_q(X, G) \to H_q(Y, G). \tag{6.4.15}$$

In particular, when $h : X \to Y$ is a homeomorphism, then the induced maps h_* yield isomorphisms of the corresponding homology groups.

Corollary 6.4.1 *Homeomorphic topological spaces have the same Betti numbers. In particular, their Euler numbers coincide.*

Thus, we have a functor $G_{q,G}$ that maps the singular chain complex $S_q(X, G)$ to the abelian group $H_q(X, G)$ and that maps group homomorphisms given by singular chain maps to group homomorphisms. Again, we can view this as a general functor G. The compositions of the two functors $F_{q,G}$ and $G_{q,G}$ then give the homology functors $H_{q,G}$ that assigns to a topological space X its singular homology group $H_q(X, G)$, or more abstractly $H = G \circ F$ which assigns to X its collection of singular homology groups. H is the functor of real interest whereas F only plays an auxiliary role.

One then also defines homology groups of pairs (X, A) where $A \subset X$ are topological spaces. As for simplicial homology, one obtains exact homology sequences for pairs. Therefore, one has the same algebraic apparatus at one's disposal as for simplicial homology. The excision theorem now takes the following form

Theorem 6.4.2 *Let $U \subset A \subset X$ be topological spaces for which the closure $\bar{U}$ of U is contained in the interior A° of A. The inclusion map $i : (X \setminus U, A \setminus U) \to (X, A)$ then induces isomorphisms*

$$i_* : H_q(X \setminus U, A \setminus U) \to H_q(X, A). \tag{6.4.16}$$

The proof is technically more complicated than that of the simplicial excision theorem, because a singular chain in X need not be the sum of chains in A and $X \setminus U$. One therefore needs to construct suitable refinements by barycentric subdivision, but we spare the details.

Again, these constructions yield functors. For instance, we have the functor that assigns to every pair (X, A) of topological spaces its long exact sequence of singular homology groups.

Let us summarize some aspects that on one hand may appear rather obvious, but that on the other hand are conceptually important. Except when X is a finite topological space, the chain groups $S_q(X, G)$ are infinitely generated, or when G is a field, say $G = \mathbb{R}$, are infinite dimensional vector spaces. In order to extract useful information about the shape of X from them, one has to go to a reduced description that reflects only the essential aspects. These are the homology groups $H_q(X, G)$. In typical cases of interest, like compact manifolds, they are finitely generated, or when G is a field, finite dimensional vector spaces. When passing from $S(X, G)$ to $H(X, G)$, one considers all boundaries, that is, all chains s with $s = \partial t$ for some chain t, as trivial. Since $\partial \circ \partial = 0$, for such a chain necessarily $\partial s = 0$. There may, however, also exist chains σ that satisfy $\partial \sigma = 0$ without being boundaries themselves. These are the nontrivial chains that generate the homology groups, with the equivalence relation that σ_1 and σ_2 yield the same homology elements whenever $\sigma_1 - \sigma_2 = \partial \tau$ for some τ. For every non-negative integer q and every abelian group G, we then have the functor $H_q(., G)$ from the category of topological spaces to the category of abelian groups that assigns to every X its homology group $H_q(X, G)$. Since somehow an abelian group is a simpler object than a topological space, this functor extracts some of the topological information about X. However, some topological information is also lost by that functor. In particular, two topological spaces X and Y that have the same homology groups for every q and G can still be different as topological spaces, that is, need not be homeomorphic. One may then define more refined invariants of topological spaces that are allow for finer distinctions, but those are more complicated than the homology groups.

In fact, it is a general phenomenon that we construct a functor that attaches certain algebraic or other invariants to the objects in some category, then that functor typically is not injective, that is, the invariants cannot distinguish all the objects in the original category. The functor "forgets" some of the original structure. Nevertheless, good invariants like homology groups allow for many distinctions between different objects. For instance, compact two-dimensional manifolds are fully characterized by their homology groups. That is, two such manifolds with the same homology groups are homeomorphic. (The details of the classification of compact surfaces (without boundary) can be found in introductions to geometric topology, like [88], or (at least in the case of orientable surfaces) in textbooks on Riemann surfaces, for instance [61].) This, however, is no longer true in higher dimensions.

6.5   Homotopy Theory of Topological Spaces

In order to develop the appropriate general framework for the topological investigation of topological spaces, we shall now introduce homotopy theory.

We consider the category of pairs (X, A), $A \subset X$ closed, of topological spaces, with morphisms $f : (X, A) \to (Y, B)$ being continuous maps

$$f : X \to Y \text{ with } f(A) \subset B. \qquad (6.5.1)$$

As we are working with topological spaces, all maps in this section will be assumed to be continuous. We shall abbreviate $X = (X, \emptyset)$, which implicitly contains the fact that the category of topological spaces naturally is a subcategory of the category of pairs of topological spaces.

Two such morphisms f_0, f_1 are homotopic, that is, continuously deformable into each other, $f_0 \sim f_1 : (X, A) \to (Y, B)$, when, with $I = [0, 1]$, there exists an

$$F : (X, A) \times I \to (Y, B) \text{ with } F(x, t) = f_t(x) \text{ for } t = 0, 1. \quad (6.5.2)$$

According to our general assumption, F has to be continuous, and this is the crucial point.

Maps are homotopic when they can be continuously deformed into each other.

Lemma 6.5.1 *Homotopy $\sim$ is an equivalence relation on the set of continuous maps between pairs of topological spaces.*

Proof   1. Reflexivity: $f \sim f$ with $F(x, t) := f(x)$ for all x.
  2. Symmetry: If $f_0 \sim f_1$ with F as in (6.5.2), then $f_1 \sim f_0$ via $F'(x, t) := f(x, 1 - t)$.
  3. Transitivity: If $F_1 : f_0 \sim f_1$, $F_2 : f_1 \sim f_2$, put

$$F(x, t) := \begin{cases} F_1(x, 2t) & \text{for } 0 \le t \le 1/2 \\ F_2(x, 2t - 1) & \text{for } 1/2 \le t \le 1 \end{cases} \qquad (6.5.3)$$

  to get $f_0 \sim f_2$.   □

Thus, for $f : (X, A) \to (Y, B)$, we denote by $[f]$ its homotopy class, that is, its equivalence class w.r.t. $\sim$. Homotopic maps are thus considered to be equivalent within our category and should therefore not be distinguished.

We leave it to the reader to verify that composites of homotopic maps are homotopic. We then get a category of pairs whose objects are topological spaces and whose morphisms are homotopy classes of maps. This looks somewhat unsymmetric, however, as we have performed an identification among the morphisms, but not yet among the objects. Therefore, we shall now turn to the appropriate equivalence concept for pairs of topological spaces.

A homotopy equivalence between two pairs of spaces is defined by requiring that there exist

$$\phi : (X_1, A_1) \to (X_2, A_2), \ \psi : (X_2, A_2) \to (X_1, A_1) \tag{6.5.4}$$

$$\text{with } \phi \circ \psi \sim \text{id}_{X_2}, \psi \circ \phi \sim \text{id}_{X_1}.$$

We then also write $(X_1, A_1) \sim (X_2, A_2)$ and say that the two pairs are homotopy equivalent. We leave it as an exercise to verify that homotopy equivalence is an equivalence relation on the set of pairs of topological spaces.

Moreover, if $(X_1, A_1) \sim (X_2, A_2), (Y_1, B_1) \sim (Y_2, B_2), i = 1, 2$, then for $f : (X_1, A_1) \to (Y_1, B_1)$ and $g := \phi' \circ f \circ \psi$ where $\psi : (X_2, A_2) \to (X_1, A_1), \phi' : (Y_1, B_1) \to (Y_2, B_2)$ are the maps from the homotopy equivalences, we obtain $f \sim \psi' \circ g \circ \phi = \psi' \circ \phi' \circ f \circ \psi \circ \phi$. Therefore, $f_0 \sim f_1 : (X_1, A_1) \to (Y_1, B_1)$ iff $g_0 \sim g_1 : (X_2, A_2) \to (Y_2, B_2)$ (where g_i is related to f_i in the same manner as g was related to f). We then get the category of homotopy classes of pairs as objects and homotopy classes of maps as morphisms.

$A \subset X$ is called a strong deformation retract of X if there exists a continuous map $r : X \times [0, 1] \to X$ satisfying

$$r(x, t) = x \quad \forall x \in A \quad \text{(leave } A \text{ invariant)}$$
$$r(x, 0) = x \quad \forall x \in X \quad \text{(start with the identity)}$$
$$r(x, 1) \in A \quad \forall x \in X \quad \text{(end up in } A\text{)}.$$

Thus, we shrink the whole space X into its subset A while leaving all points in A itself unaffected.

Let us consider some examples, first for pairs of the form $(X, \emptyset), (Y, \emptyset)$. We let $B^2 = \{x = (x^1, x^2) \in \mathbb{R}^2 : (x^1)^2 + (x^2)^2 \leq 1\}$ be the closed unit disk and 0 the origin in $\mathbb{R}^2$. Then the identity map of B^2, $f_1 := id_{B^2}$, and the constant map f_0 that maps all of B^2 to 0 are homotopic: the homotopy is provided by $F(x, t) = tx$. This then also implies that the space B^2 is homotopically equivalent to the space consisting only of the origin, that is, of a single point. If we want to have nontrivial pairs here, we take $(B^2, \{0\})$ and $(\{0\}, \{0\})$. Another example of homotopically equivalent spaces is given by the cylinder $Z = \{(x_1, x_2, x_3) \in \mathbb{R}^3 : x_1^2 + x_2^2 = 1, -1 \leq x_3 \leq 1\}$ and the circle $S^1 = \{x = (x_1, x_2) \in \mathbb{R}^2 : x_1^2 + x_2^2 = 1\}$, with the required map from the cylinder to the circle simply collapsing the third coordinate x_3 to 0, and the map from the circle to the cylinder embedding the former into the latter as the circle $x_3 = 0$. Again, this also shows the homotopy equivalence of (Z, A) and (S^1, A), where A is any subset of S^1. The disk B^2 and the circle S^1, however, are not homotopically equivalent, as we see from Theorem 6.5.1 below, because their Euler numbers are different (1 for the disk (6.4.6), but 0 for the circle (6.4.4)). In any case, the intuition behind this result is that one would have to create a hole somewhere in B^2 to map it onto S^1 so that the composition of this map with the inclusion of S^1 into B^2 becomes homotopic to the identity of S^1. Cutting a hole, however, is not a continuous operation.

We can also interpret these examples in the sense that the origin (or, likewise any other point in B^2) is a strong deformation retract of the unit disk B^2, and so is the circle S^1 for the cylinder Z. This circle S^1, however, is not a strong deformation retract of the unit disk B^2.

Cylinder

Circle S^1

We also make the following observation. Homotopy equivalent spaces need not be homeomorphic as the example of the cylinder and the circle or the disk and the point shows. (There are some subtle aspects here, for instance that the dimension of a manifold is invariant under homeomorphisms, which we do not prove here; see for instance [36]. This result, first demonstrated by Brouwer, may be intuitively clear, but is, in fact, difficult to prove.) In particular, since Theorem 6.5.1 below tells us that the Betti numbers are homotopy invariants, there exist non-homeomorphic spaces with the same Betti numbers. In particular, the Betti numbers do not fully characterize the homeomorphism class of a topological space. In fact, as we shall point out after Theorem 6.5.1, the Betti numbers do not even characterize the homotopy type of a topological space.

We shall now describe some constructions that are useful for understanding the topological version of the Conley index as presented above in our treatment of simplicial homology. The essential idea there was to consider suitable pairs (P_0, P_1) where P_1 was some kind of exit set that represented whatever was not participating in the process going on in P_1, and which therefore should be considered as trivial and contracted to a point.

$(X, x_0)(x_0 \in X)$ is called a punctured space. (X, A) yields the punctured space X/A by identifying all $x \in A$ as a single point. More formally: $x \approx y$ if $x = y$ or $x, y \in A$ is an equivalence relation, and we let $[A]$ be the equivalence class of $x \in A$. We then put

$$X/A := (X/\approx, [A]).$$

Thus, we collapse the whole subset A to a single point. Moreover, we have the important special case

$$X/\emptyset = (X \amalg p, p), p \notin X.$$

(Here, $\amalg$ means the disjoint union, that is, here we are adding a point p to X that is not contained in X.)

A continuous map $f : (X, A) \to (Y, B)$ induces $[f] : X/A \to Y/B$ via $[f][x] := [f(x)]$.

If $(X, A) \sim (Y, B)$ then also $X/A \sim Y/B$.

We shall simply write X/A for the homotopy class $[X/A]$ of X/A.

Returning to our examples, we consider Z/A where A consists of the two boundary circles $x_3 = \pm 1$ of the cylinder Z. This space is homotopically equivalent to $S^2/\{p_1, p_2\}$, the sphere modulo two points, for example the north and the south pole. This in turn is homotopically equivalent to T^2/S' where the torus T^2 is obtained by rotating the circle $S' := \{(x_1, 0, x_3) : x_1 = 2 + \sin \theta, x_3 = \cos \theta\}$ about the x_3-axis.

We now come to the fundamental connection between homotopy and homology.

Theorem 6.5.1 *If $f \sim g : (X, A) \to (Y, B)$, then*

$$f_* = g_* : H_q(X, A; G) \to H_q(Y, B; G). \tag{6.5.5}$$

In particular, the corresponding homology groups of homotopy equivalent spaces are isomorphic. Their Betti and Euler numbers coincide.

This is the basic principle, that algebraic invariants attached to topological spaces are invariant under continuous deformations.

Proof For simplicity, and in order to bring out the basic geometric intuition more clearly, we shall provide a proof of this result only for the case where $A = \emptyset$. That is, we show that if $f, g : X \to Y$ are homotopic, then they induce the same homomorphism

$$f_* = g_* : H_q(X) \to H_q(Y), \qquad (6.5.6)$$

where we leave out the group G from our notation as it plays no role in the proof. References are [47] or [108].

Let $F : X \times [0, 1] \to Y$ be a homotopy between f and g, with $f(.) = F(., 0)$, $g(.) = F(., 1)$, and let $\gamma : \Delta_q \to X$ be a singular q-simplex. The following diagram describes the basic geometric situation (for $q = 1$).

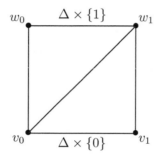

$$(6.5.7)$$

The square represents $\Delta \times [0, 1]$. The image under $F \circ \gamma$ of the bottom of this square is the image of $f \circ \gamma$, that of the top $g \circ \gamma$, whereas the left and the right boundary become images of $\partial \Delta \times [0, 1]$. These four pieces together constitute the boundary of $\Delta \times [0, 1]$. We thus see that, up to the image of a boundary, which will disappear when we shall consider cycles, and a boundary, that is, in homology, the images of $f \circ \gamma$ and $g \circ \gamma$ are the same, for every singular q-simplex. However, there is one technical issue, that $\Delta \times [0, 1]$ is not a simplex itself. But this is easily remedied by subdividing it into simplices. In the picture, this is achieved by the diagonal from v_0 to w_1. In higher dimensions, we need, of course, more simplices.

We now provide the details. We first subdivide $\Delta_q \times [0, 1]$ into $q + 1$ $(q + 1)$-dimensional simplices. We have the projection $\Delta_q \times [0, 1] \to \Delta_q$, and we arrange the vertices $v_0, \ldots, v_q$ of the simplex $\Delta_q \times \{0\}$ and the vertices $w_0, \ldots, w_q$ of $\Delta_q \times \{1\}$ in such a manner that v_i and w_i project to the same vertex of Δ_q, as in the above diagram. We consider maps $\phi_i : \Delta_q \to [0, 1]$ for $i = 0, \ldots, q$ with $\phi_0 = 0$ and

$$\phi_i(\lambda_0, \ldots, \lambda_1) = \sum_{\nu = q - i + 1}^{q} \lambda_\nu \text{ for } i = 1, \ldots, q$$

in barycentric coordinates (see (6.4.1)). The image of ϕ_i is the q-simplex $(v_0, \ldots, v_q) = \Delta \times \{0\}$ for $i = 0$ and the simplex $(v_0, \ldots, v_{q-i}, w_{q-i+1}, \ldots, w_q)$ for $i = 1, \ldots, q$. Also $\phi_i(x) \le \phi_{i+1}(x)$ for all $x \in \Delta_q$, and the region between these two graphs is the $(q + 1)$-dimensional simplex $(v_0, \ldots, v_{q-i}, w_{q-i}, \ldots, w_q)$. This yields the desired realization of $\Delta_q \times [0, 1]$ as a simplicial complex.

Let now $\gamma : \Delta_q \to X$ be a singular simplex. We extend it as $\gamma \times \mathrm{id} : \Delta_q \times [0, 1] \to X \times [0, 1]$ where of course id is the identity on $[0, 1]$. With the above subdivision, this then becomes a singular chain.

With these constructions, we can define

$$D : C_q(X) \to C_{q+1}(Y)$$

$$s_q \mapsto \sum_k (-1)^k F \circ (s_q \times \mathrm{id})|(v_0, \ldots, v_k, w_k, \ldots, w_q) \quad (6.5.8)$$

Our aim now is to verify the relation

$$\partial D = g_* - f_* - D\partial. \quad (6.5.9)$$

This formalizes what we have said when discussing the diagram (6.5.7). For showing (6.5.9), the essential point is to realize $D\partial$ as the image of $\partial D \times [0, 1]$. We compute

$$D\partial(\gamma_q) = \sum_{\kappa > k} (-1)^k (-1)^\kappa F \circ (\gamma_q \times \mathrm{id})|(v_0, \ldots, v_k, w_k, \ldots, \hat{w}_\kappa, \ldots, w_q)$$

$$+ \sum_{\kappa < k} (-1)^{k-1}(-1)^\kappa F \circ (\gamma_q \times \mathrm{id})|(v_0, \ldots, \hat{v}_\kappa, \ldots, v_k, w_k, \ldots, w_q).$$

$$(6.5.10)$$

We can now identify all the terms in

$$\partial D(\gamma_q) = \sum_{\kappa \leq k} (-1)^k (-1)^\kappa F \circ (\gamma_q \times \mathrm{id})|(v_0, \ldots, \hat{v}_\kappa, \ldots, v_k, w_k, \ldots, w_q)$$

$$+ \sum_{\kappa \geq k} (-1)^k (-1)^{\kappa+1} F \circ (\gamma_q \times \mathrm{id})|(v_0, \ldots, v_k, w_k, \ldots, \hat{w}_\kappa, \ldots, w_q).$$

The terms with $\kappa \neq k$ yield $-D\partial$ according to (6.5.10). The terms with $\kappa = k$ cancel except for the first term in the sum for $\kappa = k = 0$ which yields $g \circ \gamma_q$ and the second term in the sum for $\kappa = k = q$ which gives $-f \circ \gamma_q$. This proves (6.5.9).

The relation (6.5.9) then extends by linearity to singular q-chains s_q. When s_q is a cycle, then $\partial s_q = 0$, and (6.5.9) yields

$$\partial D(s_q) = g_*(s_q) - f_*(s_q). \quad (6.5.11)$$

Therefore, $g_*(s_q) = f_*(s_q)$ in homology, and we have verified (6.5.6).

The remaining claims of the Theorem are now easy. If X and Y are homotopy equivalent, then according to (6.5.4), there are

$$\phi : X \to Y, \ \psi : Y \to X \text{ with } \phi \circ \psi \sim \mathrm{id}_Y, \ \psi \circ \phi \sim \mathrm{id}_X,$$

and by what we have already shown, this implies that $\phi_* \circ \psi_*$ induces the identity on $H(Y)$, and $\psi_* \circ \phi_*$ induces the identity on $H(X)$. Therefore, for instance ϕ induces an isomorphism between $H(X)$ and $H(Y)$. $\qquad \square$

Theorems 6.4.1 (and its analogue for pairs of topological spaces which we did not state explicitly) and 6.5.1 yield

Theorem 6.5.2

$$H_q(.; G) : (X, A) \mapsto H_q(X, A; G) \qquad (6.5.12)$$

is a functor from the category of homotopy classes of pairs of topological spaces to the category of abelian groups.

We state some consequences of Theorem 6.5.1 which the reader will readily check.

Corollary 6.5.1 *1. If A is a deformation retract of X, then*

$$H_q(X, A) = 0 \text{ for all } q. \qquad (6.5.13)$$

2. If $B \subset A \subset X$ and A is a deformation retract of X, then the inclusion $i : A \to X$ induces isomorphisms

$$i_* : H_q(A, B) \to H_q(X, B). \qquad (6.5.14)$$

3. If $B \subset A \subset X$ and B is a deformation retract of A, then the inclusion $j : B \to A$ induces isomorphisms

$$j_* : H_q(X, B) \to H_q(X, A). \qquad (6.5.15)$$

Theorem 6.5.1 tells us that the homology groups yield topological invariants of a topological space, in the sense that they depend only on its homotopy equivalence class. Conversely, homology groups could then be used to distinguish between spaces that are *not* homotopy equivalent. In order to show that two spaces X and Y are not homotopy equivalent, it suffices to show that at least one of their homology groups is different. While this is very successful in many cases, unfortunately, the homology groups do not furnish a *complete* set of invariants. In other words, there do exist spaces X and Y that are not homotopy equivalent, but whose corresponding homology groups are all isomorphic.

6.6 Cohomology

We have briefly introduced cohomology already in Sect. 4.6. We now consider cohomology more systematically and in particular describe its connections with homology.

Let G be some fixed group. Subsequently, G will be assumed to be abelian, but we do not yet need that assumption. We then have the Hom-functor

$$\mathbf{Groups^{op}} \to \mathbf{Sets}$$
$$A \mapsto \mathrm{Hom}(A, G) \qquad (6.6.1)$$

which assigns to each group A the set of group homomorphisms from A to G and which maps each group homomorphism $f : A_1 \to A_2$ to

$$f^* : \mathrm{Hom}(A_2, G) \to \mathrm{Hom}(A_1, G)$$
$$\phi \mapsto \phi \circ f. \qquad (6.6.2)$$

Recalling the discussion in Sect. 8.3, the Hom-functor is contravariant, that is, the directions of homomorphisms, here f, are reversed, here f^*.

We now consider a simplicial chain group $C_q(X, \mathbb{Z})$ or a singular one, $S_q(X, \mathbb{Z})$. We then have the group $\mathrm{Hom}(C_q(X), G)$ of cochains with values in G and the coboundary operator

$$\delta^{q-1} := \partial_q^* : \mathrm{Hom}(C_{q-1}(X), G) \to \mathrm{Hom}(C_q(X), G)$$

$$\phi \mapsto \phi \circ \partial_q$$

i.e., $\quad \delta^{q-1}(\phi)(\sum g_i \sigma_q^i) = \phi(\sum g_i \partial_q \sigma_q^i).$ (6.6.3)

Lemma 6.6.1

$$\delta^q \circ \delta^{q-1} = 0 \tag{6.6.4}$$

for all q.

Proof This follows from the corresponding property of the boundary operators ∂_q, see (6.2.11)   □

Analogously to the definition (6.2.14) of the homology groups, we can then define cohomology groups, assuming from now that the group G is abelian.

Definition 6.6.1 The qth cohomology group of the simplicial complex (when we work with simplicial chains) or the topological space X (when we use singular chains) with coefficients in the abelian group G is

$$H^q(X, G) := \ker \delta^q / \mathrm{im}\, \delta^{q-1}. \tag{6.6.5}$$

Analogously, of course, one also defines the relative cohomology groups $H^q(X, A; G)$ of topological pairs (X, A).

For a chain $c_q \in C_q$ and a cochain $\phi^q \in \mathrm{Hom}(C_q, G)$, we write

$$\langle \phi^q, c_q \rangle := \phi^q(c_q) \in G. \tag{6.6.6}$$

Lemma 6.6.2 $\langle ., . \rangle$ *is bilinear and satisfies*

$$\langle \delta^{q-1} \phi^{q-1}, c_q \rangle = \langle \phi^{q-1}, \partial_q c_q \rangle. \tag{6.6.7}$$

In particular, the pairing of a cocycle ($\delta\phi = 0$) with a boundary vanishes, and so does the product of a cycle ($\partial c = 0$) with a coboundary ($\delta\phi$).

The *proof* is obvious.

Corollary 6.6.1 $\langle ., . \rangle$ *induces a bilinear pairing between the cohomology group $H^q(X, G)$ and the homology group $H_q(X, G)$, denoted by the same symbols $\langle ., . \rangle$.*

We shall now introduce products in cohomology. For that purpose, from now on, we need to assume that G be a ring, that is, in addition to the commutative group law written as $g + h$, we also have a multiplication, written as gh, that, for the sake of generality, need not be commutative. Of course the main example is the—commutative—ring $\mathbb{Z}$, and the fields $\mathbb{Z}_2$ and $\mathbb{R}$ are likewise important.

Definition 6.6.2 Let $\gamma : \Delta_n \to X$ be a singular n-simplex, and let $\phi \in \mathrm{Hom}(S_q(X), G)$ be a singular q-cochain with coefficients in the ring G, with $q \leq n$. The cap-product between γ and ϕ is defined by, for $g \in G$,

$$\phi \cap (g\gamma) := g\langle \phi, \gamma \circ [e_{n-q}, \ldots, e_n]\rangle(\gamma \circ [e_0, \ldots, e_{n-q}]) \in S_{n-q}(X) \otimes G. \tag{6.6.8}$$

The cap-product is linearly extended to singular chains.

Thus, we evaluate the q-cochain on the q-dimensional rear of the n-simplex γ and take this number as the coefficient of the $(n-q)$-dimensional front of γ.

Lemma 6.6.3 *For a singular n-chain c and a singular q-cochain ϕ,*

$$\partial(\phi \cap c) = (-1)^q (\delta\phi) \cap c + \phi \cap \partial c. \tag{6.6.9}$$

The *proof* proceeds by a somewhat tedious, but otherwise straightforward computation.

In particular, the cap-product of a cocycle (i.e., $\delta\phi = 0$) with a cycle is a cycle, and the product of a cocycle with a boundary is a boundary. Therefore, we can state

Definition 6.6.3 The cap-product between cohomology and homology is the product induced by (6.6.8)

$$H^q(X, G) \times H_n(X, G) \to H_{n-q}(X, G)$$
$$(\psi, c) \mapsto \psi \cap c. \tag{6.6.10}$$

By duality, we can now also define a product on cohomology.

Definition 6.6.4 Let $\phi \in H^p(X, G), \psi \in H^q(X, G)$. The cup-product between ϕ and ψ is defined by requiring that for all $c \in H_n(X, G)$ with $p + q = n$

$$\langle \phi \cup \psi, c\rangle = \langle \phi, \psi \cap c\rangle \in G. \tag{6.6.11}$$

Thus,

$$\phi \cup \psi \in H^n(X, G), \text{ with } p + q = n. \tag{6.6.12}$$

We could also entangle the definition of the cup-product in terms of the cap-product and define the former also at cocycle level (we shall utilize the same letters for cocycles and for cohomology classes, that is, equivalence classes of cocycles; this should cause no confusion): For $\phi \in \mathrm{Hom}(S_p(X), G), \psi \in \mathrm{Hom}(S_q(X), G)$ and a singular $(p+q)$-simplex $\gamma : \Delta_{p+q} \to X$ and $g \in G$, we have

$$\langle \phi \cup \psi, g\gamma\rangle = g\langle \phi, \gamma \circ [e_0, \ldots, e_p]\rangle\langle \psi, \gamma \circ [e_p, \ldots, e_{p+q}]\rangle, \tag{6.6.13}$$

that is, we multiply the evaluation of ϕ on the p-dimensional front by the evaluation of ψ on the q-dimensional back of the image of Δ_{p+q}.

Theorem 6.6.1 *Let G be a commutative ring with unit.*
Equipped with the cup-product,

$$H^*(X, G) := \bigoplus_{q \geq 0} H^q(X, G) \tag{6.6.14}$$

is an anticommutative ring with unit, called the **cohomology ring** *of X, and a module over the ring G.*
Anticommutativity here means

$$\phi \cup \psi = (-1)^{pq} \psi \cup \phi \text{ for } \phi \in H^p, \psi \in H^q. \tag{6.6.15}$$

Proof For the ring structure, we check the distribution law

$$(\phi^1 + \phi^2) \cup \psi = \phi^1 \cup \psi + \phi^2 \cup \psi \tag{6.6.16}$$

$$\phi \cup (\psi^1 + \psi^2) = \phi \cup \psi^1 + \phi \cup \psi^2 \tag{6.6.17}$$

and the associativity

$$(\phi \cup \psi) \cup \omega = \phi \cup (\psi \cup \omega). \tag{6.6.18}$$

For the module structure, we have the homogeneity

$$(g\phi) \cup \psi = \phi \cup (g\psi) = g(\phi \cup \psi). \tag{6.6.19}$$

The unit is the unit $1_X \in H^0(X, G)$ that satisfies

$$1_X \cup \phi = \phi \cup 1_X = \phi. \tag{6.6.20}$$

All these properties are readily checked.   □

Moreover, the cup-product is functorial in the sense of

Lemma 6.6.4 *For a continuous map $f : Y \to X$, we have*

$$f^*(\phi \cup \psi) = f^*\phi \cup f^*\psi. \tag{6.6.21}$$

Corollary 6.6.2

$$X \mapsto H^*(X, G) \tag{6.6.22}$$

is a functor from the category **Top**^{op} *of topological spaces (with reversed directions of morphisms) to the category of (anticommutative) rings and G-modules.*

It is important to note that we do not have an analogous ring structure on the homology of X. The cup-product on the cohomology of X carries information that is in general not contained in its homology. When X is a closed manifold, however, the cup-product on cohomology is dual to the intersection product on homology. This is a consequence of Poincaré duality for manifolds to which we shall now turn.

6.7   Poincaré Duality and Intersection Numbers

In this section, M is a connected manifold of dimension n, for the moment possibly with boundary ∂M. Here, ∂M is the boundary of M in the topological sense, the closure of M minus its interior, but for manifolds, this topological boundary is what underlies the homological boundary of M as an n-chain. We assume that (together with its boundary, if present) M is compact.

Definition 6.7.1 M is called orientable if

$$H_n(M, \partial M, \mathbb{Z}) \cong Z. \tag{6.7.1}$$

A choice of a generator of this group, $[M]$, which is called a fundamental class of M, is then called an orientation of M.

Remark In general, for a connected manifold, possibly with boundary, for all abelian groups G

$$H_n(M, \partial M; G) \text{ either } \cong G \text{ or } \cong \{g \in G : 2g = 0\}, \tag{6.7.2}$$

and in the first case, M is orientable. In particular, we always have

$$H_n(M, \partial M, \mathbb{Z}_2) \cong Z_2, \tag{6.7.3}$$

and we therefore always have a fundamental class $[M]_2 \in \mathbb{Z}_2$.

The geometric reason for (6.7.2), i.e., that H_n has only a single generator when G is $\mathbb{Z}$ or $\mathbb{Z}_2$, is that M itself is the only singular n-chain that is closed modulo ∂M. Any other n-chain has some boundary inside M and therefore does not represent a homology class modulo ∂M. The $\mathbb{Z}$-cohomology class corresponding to M is then $[M]$, and it either satisfies $2[M] = 0$, in which case M is not orientable (think of the Möbius band), or it freely generates $\mathbb{Z}$.

Möbius band

The **Poincaré duality theorem** is then

Theorem 6.7.1 *Let M be a compact orientable n-dimensional manifold without boundary. Then, for any $q \in \mathbb{Z}$, the cap-product with a fundamental class $[M]$ yields an isomorphism*

$$\cap [M] : H^q(M, \mathbb{Z}) \rightarrow H_{n-q}(M, \mathbb{Z})$$
$$\phi \mapsto \phi \cap [M]. \tag{6.7.4}$$

In any case, when M is not assumed to be orientable, we obtain an isomorphism

$$\cap [M]_2 : H^q(M, \mathbb{Z}_2) \rightarrow H_{n-q}(M, \mathbb{Z}_2)$$
$$\phi \mapsto \phi \cap [M]_2. \tag{6.7.5}$$

We do not prove this result here, as the proof is too long. The proof can be found in any good textbook on algebraic topology, for instance [47] or [108].

Corollary 6.7.1 *The   Betti   numbers   of   the   compact   orientable n-dimensional manifold M satisfy*

$$b_q = b_{n-q}. \tag{6.7.6}$$

In particular, when n is odd, the Euler characteristic of M vanishes:

$$\chi(M) = \sum_q (-1)^q b_q = 0. \tag{6.7.7}$$

Proof The universal coefficient theorem in cohomology (which we do not prove here) states that the cohomology groups always satisfy

$$H^q(X, G) \cong \mathrm{Hom}(H_q(X), G) \oplus \mathrm{Ext}(H_{q-1}(X), G), \tag{6.7.8}$$

where Ext stands for some finite group. In particular, the free parts of H^q and H_q agree, and so, the qth Betti number of X, defined as the dimension of the free part of $H_q(X, \mathbb{Z})$, agrees with the dimension of the free part of $H^q(X, \mathbb{Z})$. When X is a manifold, we may therefore apply Poincaré duality to get the result.   □

Theorem 6.7.1 allows us to define a product on homology for compact manifolds, the intersection product, which is dual to the cup product on cohomology.

Definition 6.7.2 Let M be a compact oriented manifold and let

$$P : H_q(M, \mathbb{Z}) \to H^{n-q}(M, \mathbb{Z}) \tag{6.7.9}$$

be the inverse of the Poincaré isomorphism (6.7.4). For $\alpha \in H_q(M, \mathbb{Z})$, $\beta \in H_{n-q}(M, \mathbb{Z})$, we define their intersection product as

$$\alpha \cdot \beta := \langle P(\beta) \cup P(\alpha), [M] \rangle \in \mathbb{Z}. \tag{6.7.10}$$

When M is not necessarily orientable, we can still define a $\mathbb{Z}_2$ intersection product by using $[M]_2$ in place of $[M]$ in (6.7.10).

Note that we take the factors on the right hand side of (6.7.10) in the opposite order. This leads to a sign normalization, in view of the formula

$$\alpha \cdot \beta = (-1)^{q(n-q)} \beta \cdot \alpha. \tag{6.7.11}$$

The geometric interpretation of (6.7.10) is that it counts the number of intersection points, with appropriate signs, of cycles representing the homology classes α, β. We shall now briefly sketch this geometric interpretation for the case where M is a *differentiable* manifold (see e.g. [29]). A differentiable manifold is oriented iff there exists an atlas as in Definition 5.3.3 for which all chart transitions have positive Jacobian determinant. This in turn is equivalent to the following condition. For a vector space V of dimension n over $\mathbb{R}$, we take some ordered basis $(e_1, \ldots, e_n)$ and declare it to be positive and then call any other basis that is obtained from this one by a linear transformation with positive determinant positive as well; otherwise it is called negative. Thus, the order of the basis vectors matters. When we permute an odd number of them, the resulting basis is no longer positive, but negative. In this way, we can orient every tangent space $T_x M$ of M. M

is then oriented if we can choose these orientations in a coherent manner. This means the following. In each coordinate chart, every tangent space is identified with $\mathbb{R}^n$, and an orientation on $\mathbb{R}^n$ then induces an orientation on each tangent space. Since for an oriented manifold, all chart transitions have positive Jacobian determinant, this orientation can therefore be consistently chosen independently of the coordinate chart.

We now let N_1, N_2 be compact oriented differentiable submanifolds, of dimensions q and $n - q$, resp. of the compact oriented differentiable manifold M of dimension n. We assume that N_1 and N_2 always intersect transversally. This means that at a point x of intersection, we can find oriented bases $(\epsilon_1, \ldots, \epsilon_q)$ of $T_x N_1$ and $(\eta_1, \ldots, \eta_{n-q})$ of $T_x N_2$ such that their combination $(\epsilon_1, \ldots, \epsilon_q, \eta_1, \ldots, \eta_{n-q})$ yields a basis of $T_x M$. In other words, the two tangent spaces $T_x N_1$, $T_x N_2$, considered as subspaces of $T_x M$, span the latter space. Because of the complementarity of the dimensions q and $n - q$, this means that there are no linear dependencies between tangent vectors of N_1 and tangent vectors of N_2. Differential topology (see e.g. [51]) tells us that we can always achieve such transversal intersections by slight perturbations of N_1 or N_2, in particular without affecting the homology classes that they define. Also, when such submanifolds of complementary dimensions intersect transversally, they intersect in at most finitely many points (the case where N_1 and N_2 are disjoint, that is, do not intersect at all, is also considered as transversal). We then assign $i(x) := +1$ to every intersection point x for which $(\epsilon_1, \ldots, \epsilon_q, \eta_1, \ldots, \eta_{n-q})$ as constructed above is a positive basis of $T_x M$, and $i(x) := -1$ when it is a negative basis. Letting $[N_1]$, $[N_2]$ be the homology classes defined by N_1, N_2, we have

$$[N_1] \cdot [N_2] = \sum_{x \in N_1 \cap N_2} i(x). \qquad (6.7.12)$$

Thus, we count the points in the intersection $N_1 \cap N_2$ with signs according to the orientations of the submanifolds and of $[M]$ itself. This is the desired geometric interpretation of intersection numbers.

The important point is that these intersection numbers depend only on the homology classes of the submanifolds N_1, N_2. In particular, they are invariant under perturbations of them.

The intersection numbers are particularly useful when n is even and $q = n - q = n/2$. In this case, we can define the self-intersection number of a compact oriented submanifold N of M as $[N] \cdot [N]$. Geometrically, this means that we perturb N into another (homologous) submanifold N' such that N and N' intersect transversally and then count their intersection points according to the above rule. In particular, when we can deform N in such a manner that N' is disjoint from N, then the self-intersection number of N vanishes. We point out that the self-intersection number of a submanifold can also be negative.

Let us also mention that there exists a Morse theoretic approach to the topology of differentiable manifolds, see for instance [100, 58]. In that approach, intersections play an important geometric role.

Structures

7

7.1 Generating Structures

Before returning to category theory in Chap. 8, we want to take up a somewhat different perspective on mathematical structures. This is embodied in the following

Definition 7.1.1 A *mathematical structure* consists of a set of *generators*, a (possibly empty) set of *relations* between them (*constraints*), and a set of *operations* with which the other elements of the structure are generated from the generators.

Examples:

1. A set is simply generated by its elements, without further relations or operations. This simply expresses the fact that a set as such does not carry any further structure.
2. For an equivalence relations on a set X, every $x \in X$ generates its equivalence class by the operation of equivalence.
3. A vector space is generated by a basis, the operation being taking linear combinations of vectors.
4. A group G is *finitely generated* if there exists a finite number of group elements $g_1, \ldots, g_n$ such that every $g \in G$ can be written as a product $g = \gamma_1\gamma_2 \ldots \gamma_N$ with $\gamma_i \in \{g_1, g_1^{-1}, \ldots, g_n, g_n^{-1}\}$ for $i = 1, \ldots, N$, that is, as a finite product of the generators and their inverses. These generators may satisfy certain relations between them, beyond the trivial relation $gg^{-1} = e$ for all $g \in G$. For instance, when the group is commutative, the generators also satisfy $ghg^{-1}h^{-1} = e$ for all $g, h \in G$. The generator 1 of the additive group $\mathbb{Z}_2$ satisfies $1 + 1 = 0$. The free abelian group $\mathbb{Z}$ is generated by 1 without any relation. When finitely many relations suffice to generate all the relations between finitely many generators, the group is said to be *finitely presented*. The operations in this example are the group laws, of course.

$\boxed{\mathbb{Z}_2}$

$\boxed{\mathbb{Z}}$

© Springer International Publishing Switzerland 2015
J. Jost, *Mathematical Concepts*, DOI 10.1007/978-3-319-20436-9_7

5. Let (X, d) be a metric space. We say that ξ is a midpoint of x_1, x_2 if

$$d(x_1, \xi) = d(x_2, \xi) = \frac{1}{2} d(x_1, x_2). \qquad (7.1.1)$$

When (X, d) is geodesic, such midpoints always exist. Simply take a shortest geodesic $\gamma : [0, 1] \to X$, parametrized proportionally to arc length (see (5.3.45)), with $\gamma(0) = x_1, \gamma(1) = x_2$. Then $\xi = \gamma(\frac{1}{2})$ is a midpoint. We also notice that midpoints need not be unique. For instance, when we take the north and the south pole on the sphere S^n, then all points on the equator are their midpoints.

Definition 7.1.2 Let X_0 be a subset of the geodesic metric space (X, d). Then the closure of the set of all points generated by iterated application of the midpoint operation starting with the elements of X_0 is called the *convex hull* of X_0.

That we iterated the midpoint operation simply means that when for instance ξ_1, ξ_2 are midpoints of elements of X_0, we can then in the next step take their midpoints, or midpoints of one of them with an element of X_0, and so on. Thus, here the operations are the midpoint construction and the topological closure.

For Euclidean space, we need infinitely many points to generate it in its entirety, because the convex hull of finitely many points is always bounded. In contrast, as the above example shows, the sphere S^n can already be generated by two points. We may then ask for any compact Riemannian manifold M how many points are needed to generate it.

6. When (X, d) is a metric space that is not geodesic, we can replace the midpoint construction by the betweenness relation of Definition 2.1.8, that is,

Definition 7.1.3 Let (X, d) be a metric space. For $x_1, x_2 \in X$, let $B(x_1, x_2)$ be the set of all points y that are between x_1 and x_2, i.e., that satisfy

$$d(x_1, y) + d(x_2, y) = d(x_1, x_2). \qquad (7.1.2)$$

The *convex hull* of $X_0 \subset X$ is then generated by the iterated application of the betweenness rule.

When (X, d) is geodesic, this leads to the same result as Definition 7.1.2.

7. The topology $\mathcal{O}(X)$ of a topological space is generated by a basis of the topology through the operations of possibly infinite unions and finite intersections. When (X, d) is a metric space, the balls $U(x, r) := \{y \in X : d(x, y) < r\}$ for $x \in X, r \geq 0$ generate a topology.
8. The Borel σ-algebra of a topological space is generated by its open sets through the operations of union, intersection, and complement.
9. The propositions of a logical system are generated from its axioms by the application of the operation of logical inference (modus ponens), see Sect. 9.3.

10. We consider a dynamical system (4.1.22) and (4.1.23), that is,

$$\dot{x}(t) = F(x(t)) \text{ for } x \in \mathbb{R}^d, t > 0 \qquad (7.1.3)$$

$$x(0) = x_0. \qquad (7.1.4)$$

We assume that we can find a solution of (7.1.3) with initial values (7.1.4) for $0 \leq t < T$ for some $T > 0$. For instance, when F is Lipschitz continuous, this follows from the Picard-Lindelöf theorem, see [59]. By this operation of the dynamical rule, $A \subset \mathbb{R}^d$ then generates the set of all such orbits $x(t)$ for $x_0 \in A$. In particular, each point x_0 generates its orbit $x(t)$.

When we supplement this by the operation of topological closure, we also get the asymptotic limit sets of such orbits, see [60].

11. In (4.1.31), we have defined the set of recombinations $R(x, y)$ of two binary strings, and we have then iterated the recombinations of strings from some given set in (4.1.33) and defined a corresponding closure operator.

Definition 7.1.4 A set G of generators of a mathematical structure is called *minimal* if no proper subset $G_0 \subsetneq G$ can generate all the elements of that structure.

Examples:

1. The groups $\mathbb{Z}, \mathbb{Z}_q$ have the single generator 1 which is then of course minimal.

    $\boxed{\mathbb{Z}, \mathbb{Z}_q}$

2. The symmetric group $\mathfrak{S}_n$ is minimally generated by the transpositions $(j, 1), j = 2, \dots, n$. Any other transposition $(k, j), k \neq j$ is a product $(j, 1) \circ (k, 1) \circ (j, 1)$, and any element of $\mathfrak{S}_n$ can be expressed as a product of transpositions (recall the discussion of the symmetric group in Sect. 2.1.6).

    $\boxed{\mathfrak{S}_n}$

3. For the Euclidean plane, there is no minimal set of generators for the operation of convex hull. As we had already observed, any set of generators for this space is necessarily infinite, and eliminating a finite number of points does not affect its generating ability.

    $\boxed{\text{Euclidean space}}$

4. When F is a dynamical system on $\mathbb{R}^d$ as in (7.1.3), a hypersurface $A \subset \mathbb{R}^d$ with the property that every point of $\mathbb{R}^d$ lies on a unique orbit $x(t)$ for some $x_0 = x(0) \in A$ is called a *Cauchy hypersurface*. Of course, such a hypersurface need not exist for a given F.

Definition 7.1.5 Let X_0 be a subset of the set of elements X of a mathematical structure. The set $\overline{X_0}$ of the elements of X generated by the elements of X_0 is called the *scope* of X_0. The scope of a single element x is also called the *orbit* of x.

An $x \notin \overline{X_0}$ is said to be *independent* of X_0.

The scope operator need not satisfy the axioms of the Kuratowski closure operator given in Theorem 4.1.2, because the scope of a union of two subsets could be larger than the union of their individual scopes.

Examples:

1. X_0 is a set of generators of X iff $\overline{X_0} = X$.
2. In a metric space, the scope of X_0 is its convex hull.
3. For a dynamical system (7.1.3), the scope of x_0 is the orbit $x(t)$ with $x(0) = x_0$.
4. For the recombination operator (4.1.31), we can then define a collection of binary strings of some given length n as independent if none of them can be obtained by the iterated application of the recombination operation from the others.

Genetic recombination

Definition 7.1.6 A subset X_0 of the set X of elements of a structure is called *nonpotent* if

$$\overline{X_0} = X_0, \tag{7.1.5}$$

that is, if its scope is not larger than itself.

Examples:

1. Trivially, X itself and $\emptyset$ are nonpotent.
2. The neutral element of a group is nonpotent.
3. In a metric space, any single point is nonpotent for the operation of convex hull.
4. More generally, any convex set is nonpotent because it only generates itself. This generalizes the previous example, as a single point always constitutes a convex set in a metric space.
5. In a topological space, any single open set is nonpotent, because it can only generate itself by taking a union or intersection. It can, however, generate a σ-algebra because taking its complement yields a set that is disjoint from itself.
6. A point x_0 is nonpotent for a dynamical system iff it is a *fixed point*, that is,

$$x(t) = x_0 \text{ for all } t \text{ for which the solution } x(t) \text{ exists.} \tag{7.1.6}$$

7.2   Complexity

With Definition 7.1.1 at hand, we can also relate the generation of structure to the concept of (algorithmic) complexity, as introduced by Kolmogorov [71], with independent related ideas by Solomonoff [103] and Chaitin [22]. The algorithmic complexity of a structure is defined as the length of the shortest computer program that can generate it. (While this length depends on the universal computer (Turing machine) used to implement it, translating it to another universal Turing machine only introduces an additive constant that can be controlled independently of the particular structure or program. This bound depends only on the Turing machines in question, because by their universality, they can simulate each other. We refer to [81] for details.)

Thus, the algorithmic complexity essentially measures the most compact or condensed description of a structure. What can be encoded in a short program is not very complex in this sense. Random structures, on the other hand, cannot be compressed, and hence possess high algorithmic complexity.[1]

The concept of algorithmic complexity is thus concerned with the efficiency of an encoding. Once one knows how to describe a structure in principle, one essentially understands it, according to this concept. Running the actual program is considered to be a trivial aspect. In Definition 7.1.1, we have emphasized a similar aspect. Once we know the generators, the relations or constraints between them and the generative rules, we understand the structure. Therefore, we can also measure the complexity of a mathematical structure in this sense when we compute or estimate the difficulty of such an encoding. We could, for instance, count the number of generators, the length of the description of the constraints, and an encoding of the generative rules. We would then consider this the algorithmic complexity of a mathematical structure. Again, concrete numbers would depend on the encoding scheme employed, but when that is universal, this dependence would again only lead to an additive bound independent of the particular structure under consideration.

A key aspect of the preceding is that symmetries reduce the algorithmic complexity of a structure. Rather than listing all elements of a structure, it is more efficient to describe it in terms of a minimal collection of elements together with the symmetry group which allows one to reconstruct the whole structure. Of course, symmetry groups by themselves can be complex in the sense that they require a long description. But the point here is that an important part of the complexity of a structure is represented by its symmetry group.

7.3 Independence

In this section, we shall refine the concepts of minimality and independence of Definitions 7.1.4 and 7.1.5. The idea is to represent sets of independent elements as the simplices of a simplicial complex.

Definition 7.3.1 A *matroid* consists of a set V and a nonempty simplicial complex Σ, called the *independence complex*, with vertex set V that satisfies the *exchange property*: Whenever σ, τ are simplices of Σ with $|\sigma| < |\tau|$, then there exists some vertex $v \in \tau \backslash \sigma$ for which $\sigma + v$ (the simplex spanned by the vertices of σ together with v) is also a simplex of Σ.

The vertex sets of the simplices of the independence complex Σ are called the *independent sets*.

---

[1] It is not always desirable to consider random structures as complex, simply because they are not very structured. Such aspects have been intensively discussed in the theory of complex systems, and other complexity measures have been proposed that avoid this consequence, but this is not our topic here, and I shall discuss this elsewhere.

In our subsequent notation, we shall be somewhat sloppy and often do not distinguish between subsets of V and the simplices they span in Σ.

We observe that the exchange property implies

Lemma 7.3.1 *All maximal simplices of a matroid have the same number of vertices.*   □

Definition 7.3.2 The number of vertices of the maximal simplices of a matroid is called its *rank*.

Examples:

1. Let V be a vector space. Letting every vector space basis span a simplex yields a matroid. The exchange property expresses the fact that any member of a basis can be exchanged against another one. When V is defined over the real or complex numbers and carries a Euclidean norm $\|.\|$, we could also take as vertex set the elements v with $\|v\| = 1$. We could also take the one-dimensional linear subspaces of V as vertices. Of course, the matroid in this example has an infinite vertex set when the vector space is defined over an infinite field. We could, however, apply the construction with some finite subset instead of all of V, in order to get a finite matroid. For instance, we can consider the vector

   | Vector space $\mathbb{Z}_2^n$ |

   space $\mathbb{Z}_2^n$ over the field $\mathbb{Z}_2$. For example, in $\mathbb{Z}_2^2$, we have the three basis sets $\{(1,0),(0,1)\}, \{(1,1),(0,1)\}, \{(1,0),(1,1)\}$. Thus, the corresponding matroid consists of the three vertices $(1,0),(0,1),(1,1)$ connected pairwise by edges. Thus, it is a triangle, that is, the complete

   | Complete graph K_3 |

   graph K_3

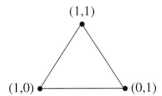

2. Let G be a finite graph with edge set E. We recall some concepts from Sect. 3.4.1. A cycle is a subgraph of G of the form $(v_0, v_1), (v_1 v_2), \ldots,$ $(v_m v_0)$ where $(v_{i-1} v_i), i = 1, \ldots, m$ are edges where all the vertices v_j involved are different, except that the last vertex coincides with the first. A *forest* is a subgraph of G that contains no cycles. A connected forest is a *tree*. We readily check a few observations. Assume that the graph is *connected*, that is, for any two vertices x, y, there exists an edge *path* joining them, that is, edges $(x_1 x_2), (x_2 x_3) \ldots (x_{k-1} x_k)$ with $x_1 = x, x_k = y$. When the graph then has m vertices, $|G| = m$, then a forest has at most $m - 1$ edges, and each forest with $m - 1$ edges is a tree. Conversely, every connected subgraph with $m - 1$ edges is a tree. In particular, any forest can be extended to a tree.

   We then have the matroid $M(G)$ with its vertex set being the *edge* set E of G and the simplices representing the forests in G. By the preceding observations, the dimension of the maximal simplices is $m - 2$, because they have $m - 1$ vertices (i.e., edges of G).

We leave it to the reader as an easy exercise to generalize this to non-connected graphs. We now discuss some examples of connected graphs.

The complete graph K_2 has a single edge, and therefore $M(K_2)$ consists of a single vertex,

$\boxed{K_2}$

where in this and the following diagrams, we represent a graph on the left and its matroid on the right.

K_3 has three edges, and any two of them span a tree; thus

$\boxed{K_3}$

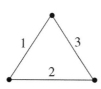
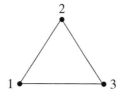

K_4 has six edges, each of which is contained in two triangles (which are not trees) and four trees of maximal size, that is, with three edges,

$\boxed{K_4}$

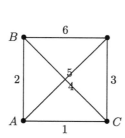
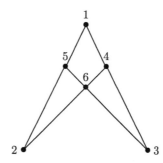

but here, we have not depicted the full matroid. The diagram needs to be read as follows. Any two vertices are connected because any two edges of the graph constitute either a tree or a forest. Thus, the edge graph of $M(K_4)$ is the complete graph K_6. Next, three vertices of our partial representation of $M(K_4)$ are on a straight line iff the corresponding edges in K_4 form a triangle; for example, the edges $1, 2, 5$ form the triangle A, B, C and therefore are represented by three vertices on a straight line. Conversely, any three edges not represented by vertices on a straight line constitute a tree in K_4. Thus, for example, the edges 1 and 4 form trees with the edges 2, 5 or 6. Thus, we need to fill a 2-simplex in $M(K_4)$ for every triple of vertices not lying on a straight line.

3. Let C be a convex set in a metric space. The points of C that are not between two other points are called the *extreme points* of C. For instance, when C is a triangle together with its interior in the Euclidean plane, then the three vertices are its extreme points. Similarly, a convex polygon in the Euclidean plane has only finitely many extreme points, its vertices.

Convex polygon

In contrast, the unit disk $\{x \in \mathbb{R}^2 : \|x\| \leq 1\}$ has the entire unit circle as extreme points.

The corresponding matroid then has at its vertex set the extreme points of C, and any subcollection of them defines a simplex. Thus, when C has only finitely many extreme points, say m of them, then this matroid is simply an $(m-1)$-dimensional simplex.

Genetic recombination

4. We return again to the recombination (4.1.31) of binary strings of some fixed length n. Here, however, the exchange property of matroids is not satisfied. For instance, as pointed out after (4.1.32), from the iterated recombination of $x = (0000)$ and $y = (1111)$, we can obtain any string of length 4. But the four strings $x_1 = (1000)$, $x_2 = (0100)$, $x_3 = (0010)$, $x_4 = (0001)$ are also independent generators of the collection of strings of length 4.

Definition 7.3.3 Let A be a subset of the vertex set V of a matroid Σ_V. The simplicial complex Σ_A containing those simplices of Σ whose vertices are all in A is then a submatroid of Σ_V. The rank of Σ_A is then denoted by $r(A)$.

Equivalently, the independent sets of $A \subset V$ are those subsets of A that are independent sets of V.

Of course, for this definition to be meaningful, it needs to be checked whether such a Σ_A is a matroid itself. But the crucial condition, the exchange property, is obviously satisfied because if σ, τ are simplices of Σ_A, then they are also simplices of Σ and hence satisfy the exchange property whenever $|\sigma| < |\tau|$.

Lemma 7.3.2 *The rank function of a matroid assigns to every $A \subset V$ a nonnegative integer or the value ∞; it satisfies*

$$r(\{v\}) = 1 \qquad \text{for every } v \in V, \qquad (7.3.1)$$

$$r(A) \leq |A| \qquad \text{for every } A \subset V, \qquad (7.3.2)$$

$$r(A) \leq r(A \cup \{v\}) \leq r(A) + 1 \quad \text{for every } v \in V, A \subset V, \qquad (7.3.3)$$

$$r(A \cup B) + r(A \cap B) \leq r(A) + r(B) \qquad \text{for every } A, B \subset V \qquad (7.3.4)$$

$$r(A) \leq r(B) \qquad \text{for every } A \subset B \subset V. \qquad (7.3.5)$$

Proof (7.3.1), (7.3.2) and (7.3.3) are clear, and (7.3.4) can then be derived by induction on the size $|B|$. (7.3.5) also follows from (7.3.3).

Definition 7.3.4 Let (V, Σ_V) be a matroid of finite rank, that is, $r(V) < \infty$. We define the *closure* of $A \subset V$ as

$$\overline{A} := \{v \in V : r(A \cup \{v\}) = r(A)\}. \qquad (7.3.6)$$

This closure operator satisfies the properties (i), (ii), and (iv) of Theorem 4.1.2, that is,

Lemma 7.3.3

$$A \subset \overline{A} \text{ for every} A \subset V, \tag{7.3.7}$$

$$A \subset B \Rightarrow \overline{A} \subset \overline{B} \text{ for every } A, B \subset V, \tag{7.3.8}$$

$$\overline{\overline{A}} = \overline{A} \text{ for every} A \subset V. \tag{7.3.9}$$

Proof (7.3.7) and (7.3.9) are easy. To see (7.3.8), and to understand the meaning of the definitions more generally, it is useful to disentangle what $v \in \overline{A}$ means. It says that the maximal simplices in $\Sigma_{A \cup \{v\}}$ have the same rank as those in Σ_A, or equivalently, that the maximal independent sets in A and $A \cup \{v\}$ are of the same size. Thus, when we add v to a maximal independent set of A, the result is no longer independent. Now, turning this the other way around when considering B, if v were not in $\overline{B}$ we would find a maximal independent set B_I in B which remains independent when we enlarge it by v. But taking subsets leaves independence unaffected. That is, both the intersection of B_I with A and the intersection of $B_I \cup \{v\}$ with $A \cup \{v\}$ would be independent as well. Also, the intersection of a maximal independent set like B_I with A yields a maximal independent subset A_I of A. But since the independent set $A_I \cup \{v\}$ is then larger than A_I, this would contradict $v \in \overline{A}$. This shows that when $v \in \overline{A}$ we must also have $v \in \overline{B}$ whenever $A \subset B$. Therefore $\overline{A} \subset \overline{B}$, that is (7.3.8). □

In general, however, property (iii) of Theorem 4.1.2 is not satisfied. In fact, none of the above three examples of matroids satisfies that property. As these examples show, typically, from a union of two sets of generators one can generate more than the union of what can be generated by the two sets individually.

Definition 7.3.5 The submatroid generated by a closed subspace $A = \overline{A}$ of V is called a *flat*.

Examples:

1. For a finite dimensional vector space V, the matroid whose independent sets are the vector space bases of V, the rank is equal to the dimension. For a subset $A \subset V$, $\overline{A}$ is the vector subspace generated by A, and a flat therefore is simply a linear subspace of V.
2. For the matroid of forests in a graph G with edge set E, the closure of $A \subset E$ contains all those $e \in E$ that when added to A create a cycle. That means that there exists a path consisting of edges $(x_1 x_2), (x_2 x_3), \ldots, (x_{k-1} x_k) \in A$ such that $e = (x_k x_1)$.
3. For the matroid of a convex set C in Euclidean space with finitely many extreme points, $v \in \overline{A}$ if v is in the convex hull of A. Since v is an extreme point, this is only possible if $v \in A$. Thus, every $A \subset V$ is closed.

Matroids were first introduced by H. Whitney [117]. References for matroids are [91, 113].

Categories

8

In this chapter, we shall take up the concepts introduced in Sect. 2.3.

Before starting this chapter, let me point out that much of the terminology of category theory has been chosen without any regard or respect for established terminology in other fields. Instances are the notions of "limit" or "topos" and even the notion of "category" itself which had been employed since the times of Aristotle with a meaning almost opposite to that of category theory.[1] However, I shall make no attempts to change that terminology here. (Let me also remark here that the physicists seem to be more responsible in this regard. They usually coin new terms for new concepts or phenomena, from entropy to the bewildering variety of atomic or subatomic particles, from electrons, protons and neutrons to quarks, leptons, hadrons, bosons, fermions,. . ., with certain exceptions like the very unfortunate term "atom" meaning "indivisible".)

Category theory embodies some important principles which I shall list here.

1. Objects should not be defined or characterized by intrinsic properties, but rather by their relations with other objects. Objects that are isomorphic in terms of relations therefore cannot be distinguished.
2. Constructions can be reflexively iterated, i.e., when constructions are performed with objects at a certain level, we can pass to a higher level, where the objects are the constructions of the previous level. Put differently, the relations between objects at one level can become the objects at the next level.

---

[1] Some history about how those names evolved can be found in [82], and of course, this development possesses some internal consistency that may serve as a justification for the choice of terminology.

© Springer International Publishing Switzerland 2015
J. Jost, *Mathematical Concepts*, DOI 10.1007/978-3-319-20436-9_8

8.1   Definitions

We recall the basic

Definition 8.1.1 A *category* **C** consists of *objects* A, B, C, ... and *arrows* or *morphisms*

$$f : A \to B \tag{8.1.1}$$

between objects, called the *domain* $A = \text{dom}(f)$ and *codomain* $B = \text{cod}(f)$ of f. Arrows can be composed, that is, given $f : A \to B$ and $g : B \to C$, there is an arrow

$$g \circ f : A \to C. \tag{8.1.2}$$

(The requirement for the composition is solely that $\text{cod}(f) = \text{dom}(g)$.) This composition is *associative*, that is,

$$h \circ (g \circ f) = (h \circ g) \circ f \tag{8.1.3}$$

for $f : A \to B, g : B \to C, h : C \to D$.
For each object A, we have the *identity arrow* ("doing nothing")

$$1_A : A \to A \tag{8.1.4}$$

which satisfies

$$f \circ 1_A = f = 1_B \circ f \tag{8.1.5}$$

for all $f : A \to B$.

We also recall the underlying idea that the objects of a category share some kind of structure, and that the morphisms then have to preserve that structure. A category thus consists of objects with structure and directed relations between them. A very useful aspect is that these relations can be considered as operations.

We also recall some general constructions from Sect. 2.3. For a category $\mathcal{C}$ of categories, the objects are categories **C**, and the morphisms $F : \mathbf{C} \to \mathbf{D}$ of $\mathcal{C}$, called *functors*, then preserve the category structure. This means that they map objects and arrows of **C** to objects and arrows of **D**, satisfying

$$F(f : A \to B) \text{ is given by } F(f) : F(A) \to F(B) \tag{8.1.6}$$

$$F(g \circ f) = F(g) \circ F(f) \tag{8.1.7}$$

$$F(1_A) = 1_{F(A)} \tag{8.1.8}$$

for all A, B, f, g. Thus, the image of an arrow under F is an arrow between the images of the corresponding objects (domain and codomain) under F, preserving compositions, and mapping identities to identities.

Definition 8.1.2 A category **C** is called *small* if the collection of objects and the collection of arrows of **C** are both sets (taken from a fixed universe U as described in Sect. 2.2). Otherwise, it is called *large*.

C is called *locally small* if for any two objects, the collection of arrows between them is a set.

We should note that many important categories are not small. For instance, **Sets** is large. It is locally small, however, because for any two sets X, Y, the collection of mappings from X to Y is a set.

The collection of arrows $\{f : A \to B\}$ between two objects A, B of $\mathbf{C}$ is written as

$$\mathrm{Hom}_{\mathbf{C}}(A, B) \tag{8.1.9}$$

and called a *hom-set* (when $\mathbf{C}$ is locally small).

For two categories $\mathbf{C}$, $\mathbf{D}$, we have the category $\mathbf{Fun}(\mathbf{C}, \mathbf{D}) =: \mathbf{D}^{\mathbf{C}}$ (the latter notation will be explained below in (8.2.63)) of all functors $F : \mathbf{C} \to \mathbf{D}$ whose morphisms are called *natural transformations*. Thus, a natural transformation

$$\theta : F \to G \tag{8.1.10}$$

maps a functor F to another functor G, preserving the structure of the category $\mathbf{Fun}(\mathbf{C}, \mathbf{D})$. A natural transformation $\theta : F \to G$ then, for each $C \in \mathbf{C}$, has to induce a morphism

$$\theta_C : FC \to GC \tag{8.1.11}$$

such that the diagram

$$
\begin{array}{ccc}
FC & \xrightarrow{\theta_C} & GC \\
{\scriptstyle Ff}\Big\downarrow & & \Big\downarrow{\scriptstyle Gf} \\
FC' & \xrightarrow{\theta_{C'}} & GC'
\end{array}
\tag{8.1.12}
$$

commutes.

Definition 8.1.3 A functor $F : \mathbf{C} \to \mathbf{D}$ is *faithful* if for all objects C, $D \in \mathbf{C}$,

$$F_{C,D} : \mathrm{Hom}_{\mathbf{C}}(C, D) \to \mathrm{Hom}_{\mathbf{D}}(FC, FD) \tag{8.1.13}$$

is injective. It is *full* if $F_{C,D}$ is always surjective.

F is an *embedding* if it is faithful, full, and injective on objects of $\mathbf{C}$.

Being faithful is weaker than being injective on arrows, because it requires only injectivity on arrows between the same objects.

Definition 8.1.4 Given a category $\mathbf{C}$, the opposite category $\mathbf{C}^{\mathrm{op}}$ is obtained from $\mathbf{C}$ by taking the same objects, but reversing the direction of all arrows.

Thus, each arrow $C \to D$ in $\mathbf{C}^{\mathrm{op}}$ corresponds to an arrow $D \to C$ in $\mathbf{C}$.

8.2 Universal Constructions

Definition 8.2.1 Let $\mathbf{C}$ be a category. $0 \in \mathbf{C}$ is an *initial object* if for any object $C \in \mathbf{C}$, there is a unique morphism

$$0 \to C. \tag{8.2.1}$$

1 ∈ **C** is a *terminal object* if for any object C ∈ **C**, there is a unique morphism

$$C \rightarrow 1. \tag{8.2.2}$$

This definition gives examples of a so-called *universal mapping property*. In diagrammatic language, if the category has an initial element 0, then for any morphism

$$C \longrightarrow D \tag{8.2.3}$$

we obtain a commuting diagram

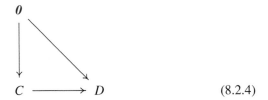

$$(8.2.4)$$

with the unique morphisms from *0* to C and D. This is an equivalent definition of an initial object.

This is an instance of a standard procedure in category theory to define certain objects or constructions, and we shall see several examples in the sequel. The general principle is that one wishes to define special objects without referring to their internal structure, but rather characterize them uniquely in terms of their relations to other objects. This is achieved by identifying universal properties that uniquely define that special object.

In this sense, it is clear, or at least easily checked, that initial (terminal) objects are unique up to isomorphism—if they exist. The latter need not be the case. In a set with more than one element, there is no initial or terminal element, as there are no arrows between different elements.

Here are some further examples of initial and terminal objects.

1. Ø and X are initial and terminal elements for the poset structure of $\mathcal{P}(X)$. That is, we have

$$\emptyset \subset A \quad \text{(and also } A \cap \emptyset = \emptyset) \tag{8.2.5}$$

and

$$A \subset X \quad \text{(and also } A \cup X = X) \tag{8.2.6}$$

for all $A \in \mathcal{P}(X)$.

2. In general, in a poset A, an element a is initial iff $a \leq b$ for all $b \in A$, that is, if a is a smallest element. Such an element need not exist. Likewise, a terminal object would be a largest object. For instance, the poset $(\mathbb{Z}, \leq)$ contains neither a smallest nor a largest element, and therefore, as a category, it does not possess an initial or terminal element.

Poset $(\mathbb{Z}, \leq)$

This, in fact, is an instance of a general aspect of the subsequent constructions. Specific objects in a category will be defined in terms of universal properties, but for a particular category, it then remains to check whether it possesses such an object.

3. In the category of sets, the empty set is initial whereas any one-element set is terminal. (Recall from (2.1.1) that there is also a map s from the one element set $I = \{1\}$ to any non-empty set S; simply take any $s \in S$ and put $s(1) = s$. Since the element $s \in S$ is arbitrary here, this map is not *unique* whenever S contains more than one element. In contrast, the map $\sigma : S \to I$ with $\sigma(s) = 1$ for all $s \in S$ is unique as required for a terminal object.)

4. In the category of groups, the trivial group with a single element is both initial and terminal.

5. In the category of commutative rings with identity 1, the trivial ring with $0 = 1$ is still terminal, but now the ring of integers $\mathbb{Z}$ is initial. To see this, let R be a commutative ring with identity 1. We denote the neutral element of the addition $+$ by 0, and the additive inverse of 1 by -1. We then map $0 \in \mathbb{Z}$ to $0 \in R$, and likewise $1 \in \mathbb{Z}$ to $1 \in R$. In order to have a ring homomorphism, we then have to map $n \in \mathbb{Z}$ to $1 + \cdots + 1$ (n times) for $n > 0$ and to $(-1) + \cdots (-1)$ ($-n$ times) for $n < 0$. For instance, when $R = \mathbb{Z}_2$, this means that an even $n \in \mathbb{Z}$ gets mapped to 1 and an odd one to 0.

   | Ring $\mathbb{Z}$ |

   | Ring $\mathbb{Z}_2$ |

   Thus, $\mathbb{Z}$ is indeed initial in the category of commutative rings with 1. Note that we cannot map the trivial ring with $0 = 1$ to any nontrivial ring R with 1 (in which case $0 \neq 1$ in R) because 0 has to be mapped to 0, but 1 has to be mapped to 1. However, we can map any ring R to this trivial ring. Thus, the trivial ring is the terminal object in this category.

6. Somewhat similarly, in the category of Boolean algebras, the Boolean algebra $B_0 = \{0, 1\}$ is initial, because for any Boolean algebra B, we map $0, 1 \in B_0$ to $0, 1 \in B$. The trivial Boolean algebra consisting only of 0 is again terminal.

   | Boolean algebra $\{0, 1\}$ |

7. In Sect. 2.3, we have introduced two different categories of metric spaces. The objects are the same. In the first category, morphisms are isometries, that is, mappings $f : (S_1, d_1) \to (S_2, d_2)$ with $d_2(f(x), f(y)) = d_1(x, y)$ for all $x, y \in S_1$. Of course, we can map the empty metric space isometrically to any other metric space, and so, this is an initial object. There is no terminal object in this category, however, because there is no metric space (S_∞, d_∞) to which we can map any other metric space (S, d) isometrically (you may want to reflect a little about this example, as it involves some not entirely trivial aspect). In the second category of metric spaces, morphisms had to satisfy $d_2(f(x), f(y)) \leq d_1(x, y)$ for all $x, y \in S_1$, that is, be distance nonincreasing. Again, the empty metric space is initial. This time, however, we have the distance nonincreasing map $g : (S, d) \to (S_0, d_0)$, the trivial metric space (S_0, d_0) consisting of a single element x_0, with $g(y) = x_0$ for every $y \in S$, and so, the trivial metric space becomes a terminal object.

8. Somewhat similarly, in the category of graphs, the initial object is the empty graph and the terminal object is the graph with a single vertex v_0 and an edge from v_0 to itself.

Perhaps, we can discern a rough pattern from these examples. Whenever the empty set is an object of the category under consideration, it can be taken as the initial object. When, in contrast, an object of a category needs to possess certain distinguished elements, then any initial object also has to contain those, and when, as in the case of commutative rings with 1, such an element generates a series of other elements, then an initial object also has to allow for such a generative process. The terminal object has to support the relations or operations of the objects of the category in question in a minimal manner. Therefore, it typically contains a single element.

We had already observed in (2.1.1) that we can consider any element s of a set S as the image of a morphism $s_I : I \to S$ by simply putting $s_I(1) = s$. More generally, whenever the category $\mathbf{C}$ possesses a terminal object I, an arrow $I \to C$ for an object C is called a *global element* of C.

Definition 8.2.2 Let $\mathbf{C}$ be a category, and $\mathbf{I}$ another category, called an *index set*, whose objects will be denoted by $i, j, \ldots$. A *diagram* of type $\mathbf{I}$ in $\mathbf{C}$ is then a functor

$$D : \mathbf{I} \to \mathbf{C}, \qquad (8.2.7)$$

and we write D_i in place of $D(i)$, and $D_{i \to j}$ for the morphism $D(i \to j) : D_i \to D_j$ that is the image of a morphism $i \to j$.

Thus, we select certain objects of $\mathbf{C}$ by equipping them with an index $i \in \mathbf{I}$, and we require that these index elements then carry the arrows from $\mathbf{I}$.

We then have the category $\mathbf{C}^{\mathbf{I}} = Fun(\mathbf{I}, \mathbf{C})$ (see (8.2.63) below for an explanation of the notation) of diagrams in $\mathbf{C}$ of type $\mathbf{I}$.

When $\mathbf{I} = \{1\} =: \mathbf{1}$ has only a single object[2] and only the identity arrow of that object, then a diagram is simply an object in $\mathbf{C}$. In particular, any object of $\mathbf{C}$ can be identified with a functor

$$\mathbf{1} \to \mathbf{C}, \qquad (8.2.8)$$

up to isomorphism, as always. For instance, the elements of a set S may be represented by arrows from a one-element set into S, as we had already observed in (2.1.1). Thus, (8.2.8) tells us that in general, the objects of a category can be considered as functors or diagrams.

Similarly, when $\mathbf{I} = \{1, 2\} =: \mathbf{2}$ has only two objects and only their identity arrows and one non-identity arrow $1 \to 2$, then each functor

$$\mathbf{2} \to \mathbf{C} \qquad (8.2.9)$$

corresponds to an arrow of $\mathbf{C}$.

We now consider the category **Sets** of sets and take as the index set the category $\mathbf{\Gamma} = \{1, 0\}$ with two arrows

$$\alpha, \omega : 1 \rightrightarrows 0. \qquad (8.2.10)$$

A corresponding diagram G is then given by a pair of sets with a pair of arrows

$$g_\alpha, g_\omega : G_1 \rightrightarrows G_0, \qquad (8.2.11)$$

---

[2] Denoting that object by 1 is not meant to carry any implications; in particular, this neither means a terminal object nor an identity element. It is simply an arbitrary label, like others used below, including $0, 2, 3, \ldots$

besides the identity arrows, of course, but those will often be omitted from our descriptions of categories because they always have to be there and thus do not deserve explicit mention.

Thus, to each element e of G_1, we associate two elements $\alpha(e)$, $\omega(e)$ of G_0. This, however, is nothing but a directed graph with edge set G_1 and vertex set G_0, with $\alpha(e)$ and $\omega(e)$ being the initial and terminal vertex of an edge. Compare Sect. 3.2.

Definition 8.2.3 A *cone* over a diagram D consists of an object C of **C** and a family of arrows

$$c_i : C \to D_i \text{ for } i \in \mathbf{I}, \tag{8.2.12}$$

with the property that for each arrow $i \to j$ in **I**, the diagram

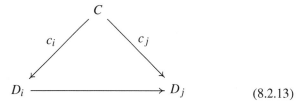

$$\tag{8.2.13}$$

commutes.

Such a C has to have arrows to all elements of the indexed family, respecting morphisms.

Let us compare this with the notion of an initial element. When we take $\mathbf{I} = \mathbf{C}$ and D as the identity functor, then a cone for which all the morphisms are unique is an initial element. In general, however, the morphisms from C to the D_i are not unique. In fact, whenever there is a nontrivial morphism $d_i : D_i \to D_i$ of some D_i to itself, induced by a morphism from i to itself in **I**, then $d_i \circ c_i$ is also sa morphism from C to D_i.

We then obtain the category **Cone**(D) of cones over D, with morphisms

$$\gamma : (C, c_i) \to (C', c_i') \tag{8.2.14}$$

having to satisfy

$$c_i = c_i' \circ \gamma \text{ for all } i \in \mathbf{I}, \tag{8.2.15}$$

that is, the diagram

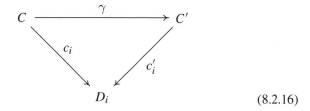

$$\tag{8.2.16}$$

commutes.

Definition 8.2.4 A *limit* $p_i : \varprojlim_{cone(D)} C_{cone(D)} \to D_i, i \in \mathbf{I}$, for a diagram $D : \mathbf{I} \to \mathbf{C}$ is a terminal object in the category **Cone**(D).

Thus, for any cone (C, c_i) over D, we have a unique $\gamma : C \to \lim_{\overleftarrow{cone(D)}} C_{cone(D)}$ with

$$p_i \circ \gamma = c_i \text{ for all } i. \tag{8.2.17}$$

Of course, the terminology "limit" here is rather unfortunate, for two reasons. First, it corresponds to the notion of an inverse or projective limit in topology, whereas a direct or inductive limit in topology becomes a colimit in category theory, as defined below. Secondly, a limit in category theory has little to do with the concept of a limit as used in analysis. In fact, while certain constructions of limits of groups or classifying spaces were indeed originally obtained as limits in the analytical sense of results of convergence, it was then found that the criterion of convergence can be replaced by that of universality. For the latter reason, it might be better to call it a "universal" instead, but it seems that we cannot change the terminology established in the field. In any case, it is a universal construction that includes many important constructions as special cases. For instance, to start with the most trivial case, when $\mathbf{I}$ is the empty category, there is only one diagram $D : \mathbf{I} \to \mathbf{C}$, and a limit is then simply a terminal element in $\mathbf{C}$. The general concept of a limit then generalizes this example in the sense that one not only looks at morphisms to objects, but at morphisms to diagrams, that is, specific arrangements of several objects connected by morphisms.

When $\mathbf{I} = \{1\} =: \mathbf{1}$ has only a single object and only the identity arrow of that object, then, as we have already observed, a diagram is an object in $\mathbf{C}$. A cone is then simply an arrow $C \to D$. Likewise, when $\mathbf{I} = \{1, 2\} =: \mathbf{2}$ has only two objects and only their identity arrows and one non-identity arrow $1 \to 2$, then each functor

$$\mathbf{2} \to \mathbf{C} \tag{8.2.18}$$

corresponds to an arrow of $\mathbf{C}$. In this case, a cone is a commutative diagram

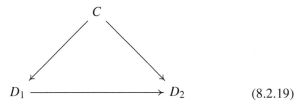

$$\tag{8.2.19}$$

In the case of the category $\mathbf{1}$ as the index category, the limit of a diagram (8.2.8) is that object itself, that is, in that case, a diagram is its own limit. This example is not very helpful, however, for understanding the concept of a limit. Likewise, for the index category $\mathbf{2}$, a terminal cone is simply an arrow $D_1 \to D_2$. In order to come to a less trivial case, let now $\mathbf{I} = \{1, 2\}$ be a two-element set considered as a category, i.e., the only arrows are the identity arrows of the elements. A diagram $D : \mathbf{I} \to \mathbf{C}$ is then simply a pair D_1, D_2 of objects of $\mathbf{C}$, and a cone is then an object C with arrows

$$D_1 \xleftarrow{\ c_1\ } C \xrightarrow{\ c_2\ } D_2, \tag{8.2.20}$$

A terminal such cone is called a *product* $D_1 \times D_2$. Thus, a product $D_1 \times D_2$ of the two objects D_1, D_2 is an object with morphisms $d_1 : D_1 \times D_2 \to D_1, d_2 : D_1 \times D_2 \to D_2$ (which are also called projections) and the universal property that whenever C is an object with morphisms $c_1 : C \to D_1, c_2 : C \to D_2$, then there is a unique morphism $c : C \to D_1 \times D_2$ with $c_1 = d_1 \circ c, c_2 = d_2 \circ c$. In particular, a product, like other such universal objects, is unique up to (unique) isomorphism. The corresponding commutative diagram is

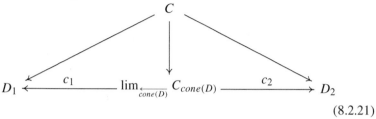

$$(8.2.21)$$

for any cone C over the diagram. Thus, again, for any object C in the same constellation as $\lim_{\overleftarrow{cone(D)}} C_{cone(D)}$, there has to exist a unique arrow from C to $\lim_{\overleftarrow{cone(D)}} C_{cone(D)}$, respecting the constellation. In this sense, this generalizes the notion of a terminal object in a category, which simply corresponds to the case where the constellation in question is trivial. So, this allows us to understand the general concept of a limit as a natural generalization of that of a terminal object. Similarly, below we shall introduce colimits as such a generalization of initial objects.

Let us return to the specific example of products. The standard example motivating the name and the construction is of course the Cartesian product of sets. Thus, let S, T be sets, and put

$$S \times T = \{(s, t) : s \in S, t \in T\} \tag{8.2.22}$$
$$\pi_S(s, t) = s, \qquad \pi_T(s, t) = t.$$

For any set X with maps $x_S : X \to S, x_T : X \to T$, we then have the required commutative diagram

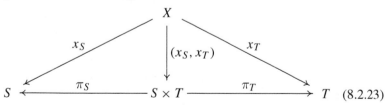

$$(8.2.23)$$

When we have sets equipped with algebraic or other structures, like monoids, groups, rings, vector spaces, we can then simply equip their Cartesian products with the corresponding product operations or structures. For instance, when G, H are groups, their Cartesian product $G \times H$ becomes a group with the group law defined in (2.1.138), that is,

$$(g_1, h_1)(g_2, h_2) = (g_1g_2, h_1h_2). \tag{8.2.24}$$

$G \times H$ equipped with (8.2.24) is then the product of G and H in the category of groups. (Note that in (2.1.138), we had written the product simply as GH in place of $G \times H$.)

Similarly, we had already defined the product $(X \times Y, \mathcal{O}(X \times Y))$ of two topological spaces $(X, \mathcal{O}(X)), (Y, \mathcal{O}(Y))$ in Sect. 4.1. Again, this is the product within the category of topological spaces, as the reader will readily check.

Again, products need not exist. Let us consider once more the example where our category is given by a poset. A product $a \times b$ of two elements a, b then has to have arrows to a and b, that is, has to satisfy

$$a \times b \leq a \text{ and } a \times b \leq b, \tag{8.2.25}$$

and since it has to be limit, whenever c satisfies $c \leq a, c \leq b$, then also $c \leq a \times b$. That is, $a \times b$ is the greatest lower bound of a and b, and this need not exist. When the poset is the powerset $\mathcal{P}(X)$ of a set X, that is, the set of its subsets, with partial order given by inclusion, then we have

$$a \times b = a \cap b, \tag{8.2.26}$$

that is, the product is given by the intersection, because $a \cap b$ is the largest common subset of a and b.

We also observe that when $\mathbf{C}$ possesses a terminal object I and products, then for any object C, the product $I \times C$ is isomorphic to C itself; this is easily inferred from the commutative diagram

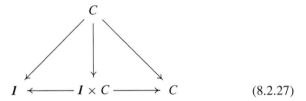

$$\tag{8.2.27}$$

with the obvious morphisms.

Let us now recall the category $\boldsymbol{\Gamma} = \{1, 0\}$ (8.2.10) with two arrows

$$1 \underset{\omega}{\overset{\alpha}{\longrightarrow}} 0 \tag{8.2.28}$$

A diagram D for $\boldsymbol{\Gamma} = \{1, 0\}$ is given by a pair of arrows

$$D_1 \underset{d_\omega}{\overset{d_\alpha}{\longrightarrow}} D_0 \tag{8.2.29}$$

(identity arrows not shown), and a cone over D then is given by a commutative diagram

$$\tag{8.2.30}$$

with

$$d_\alpha \circ c_1 = c_0 \text{ and } d_\omega \circ c_1 = c_0, \text{ that is, } d_\alpha \circ c_1 = d_\omega \circ c_1. \tag{8.2.31}$$

A limit for this diagram is called an *equalizer* of the morphisms d_α and d_ω. In the category of sets, the equalizer of $f : X \to Y$ and $g : X \to Y$ consists of the set of those $x \in X$ with $f(x) = g(x)$, with the inclusion map into X. In the category of groups, the equalizer of two homomorphisms $\chi, \rho : G \to H$ is given by the kernel of $\chi(\rho)^{-1}$, that is, by those elements $g \in G$ that satisfy $\chi(g)(\rho(g))^{-1} = e$.

As another example, let us consider the following index category $\mathbf{I} = \{1, 2, 3\}$

$$
\begin{array}{ccc}
& & 1 \\
& & \downarrow \\
2 & \longrightarrow & 3
\end{array}
\qquad (8.2.32)
$$

A limit of a corresponding diagram

$$
\begin{array}{ccc}
& & D_1 \\
& & \downarrow{\scriptstyle d_\alpha} \\
D_2 & \xrightarrow{\ d_\beta\ } & D_3
\end{array}
\qquad (8.2.33)
$$

is then called a *pullback* of d_α, d_β; it is universal for the diagram

$$
\begin{array}{ccc}
\varprojlim_{cone(D)} C_{cone(D)} & \longrightarrow & D_1 \\
\downarrow & & \downarrow{\scriptstyle d_\alpha} \\
D_2 & \xrightarrow{\ d_\beta\ } & D_3
\end{array}
\qquad (8.2.34)
$$

In the category of sets, the pullback P of $f : X \to Z, g : Y \to Z$ consists of the pairs (x, y) with $x \in X, y \in Y, f(x) = g(y)$ and the projection morphisms π_X, π_Y to X and Y. Thus, we have the diagram

$$
\begin{array}{ccc}
P & \xrightarrow{\ \pi_X\ } & X \\
\downarrow{\scriptstyle \pi_Y} & & \downarrow{\scriptstyle f} \\
Y & \xrightarrow{\ g\ } & Z
\end{array}
\qquad (8.2.35)
$$

and the condition $f(x) = g(y)$ becomes

$$
f \circ \pi_X = g \circ \pi_Y,
\qquad (8.2.36)
$$

that is, the minimal requirement for the diagram (8.2.35) to commute. In particular, when we have any other commuting diagram

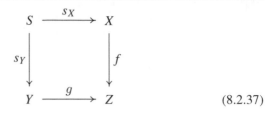

$$(8.2.37)$$

we get the map $s_P : S \to P$ with $s_P(s) = s_X(s), s_Y(s)$ which is well defined because of the commutativity of (8.2.37), that is, $f \circ s_X(s) = g \circ s_Y(s)$ for all $s \in S$. In particular, when f is the inclusion $i : Z' \to Z$ of a subset Z' of Z, then the pullback is the subset $g^{-1}(Z')$ of Y. The diagram is

$$
\begin{array}{ccc}
g^{-1}(Z') & \xrightarrow{\ g\ } & Z' \\
\big\downarrow{\scriptstyle i} & & \big\downarrow{\scriptstyle i} \\
Y & \xrightarrow{\ g\ } & Z
\end{array}
\qquad (8.2.38)
$$

If g is an inclusion $i'' : Z'' \to Z$ as well, then the pullback is simply the intersection $Z' \cap Z''$.

Analogously for sets with structures, like groups or topological spaces.

Again, for our poset example, if we have $a \le c, b \le c$, then a pullback would be some d with $d \le a, d \le b$ and whenever $e \le a, e \le b$, then also $e \le d$. Such a d would be the greatest lower bound of a and b, or their meet in the sense of Definition 2.1.6, see (2.1.36). Again, such a d need not exist. Thus, the notion of a meet in a poset can be given a diagrammatical formulation. Similarly, for our join, see (2.1.37), one could give a diagrammatical formulation as a pushforward. We leave it to the reader to develop this definition and to then apply it to this and other examples.

We also note that when the category **C** possesses a terminal object I, then with $D_3 = I$, the pullback of (8.2.34) is simply the product $D_1 \times D_2$. This follows because since there exist unique morphisms from D_1 and D_2 to I, the diagrams for pullbacks reduce to the corresponding ones for products, see (8.2.20), (8.2.21). In particular, a category with a terminal object and pullbacks also possesses products.

We also observe that equalizers can be seen as pullbacks. We simply play around with the diagrams (8.2.30) and (8.2.34) and present them as

$$
\begin{array}{ccccc}
 & & C & & \\
 & {\scriptstyle c_1}\swarrow & \big\downarrow{\scriptstyle c_0} & \searrow{\scriptstyle c_1} & \\
D_1 & \xrightarrow{\ d_\alpha\ } & D_0 & \xleftarrow{\ d_\omega\ } & D_1
\end{array}
\qquad (8.2.39)
$$

and

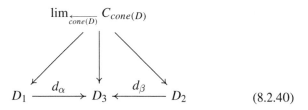

$$D_1 \xrightarrow{\ d_\alpha\ } D_3 \xleftarrow{\ d_\beta\ } D_2 \qquad\qquad (8.2.40)$$

(where we have added the vertical morphism in (8.2.40) whose existence follows from the commutativity of the diagram) and see that (8.2.39) is a special case of (8.2.40), with $D_1 = D_2$.

As an illustration of pullbacks, let us consider the following

Lemma 8.2.1 *The arrow $f : A \to B$ is a monomorphism iff the pullback of f along itself is an isomorphism, that is, iff the diagram*

$$
\begin{array}{ccc}
A & \xrightarrow{\ 1_A\ } & A \\
{\scriptstyle 1_A}\big\downarrow & & \big\downarrow{\scriptstyle f} \\
A & \xrightarrow{\ f\ } & B
\end{array}
\qquad\qquad (8.2.41)
$$

is a pullback.

Proof By definition, f is a monomorphism iff for any two arrows $g_1, g_2 : C \to A$, $f \circ g_1 = f \circ g_2$ implies $g_1 = g_2$. We consider the diagram

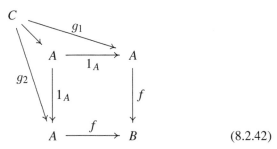

$$(8.2.42)$$

When f is a monomorphism, this diagram commutes, with the unlabelled arrow $C \to A$ being given by $g_1 = g_2$. Conversely, when f is a pullback along itself, this implies that there exists an arrow, which we now call $g : C \to A$, that makes the diagram commutative. But then $g_1 = 1_A \circ g = g_2$. $\qquad\square$

Also, pullbacks satisfy functorial properties. For instance, we have the useful

Lemma 8.2.2 *Consider a commutative diagram of the form*

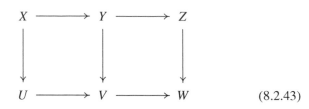

$$(8.2.43)$$

If either the two subdiagrams $\begin{array}{ccc} X & \to & Y \\ \downarrow & & \downarrow \\ U & \to & V \end{array}$, $\begin{array}{ccc} Y & \to & Z \\ \downarrow & & \downarrow \\ V & \to & W \end{array}$ *(that is, the left and the right subdiagram)*

or the two subdiagrams $\begin{array}{ccc} X & \to & Z \\ \downarrow & & \downarrow \\ U & \to & W \end{array}$, $\begin{array}{ccc} Y & \to & Z \\ \downarrow & & \downarrow \\ V & \to & W \end{array}$ *(that is, the outer and the right diagram) are pullbacks, then so is the third.*

Proof Let us prove the first claim and leave the second one to the reader. Thus, assume that we have the following commutative diagram of black arrows (in this diagram, the steps of the proof will already be indicated by colors)

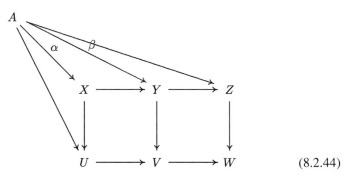

$$(8.2.44)$$

We then need to prove that there exists a morphism $\alpha : A \to X$ making the diagram commute. Since the right diagram is a pullback, and since we have morphisms from A to Z and to V (composing the morphism from A to U with that from U to V), we get a morphism $\beta : A \to Y$ so that the resulting diagram commutes. Thus, we now have arrows from A to Y and to U, and we can indeed use the pullback property of the left diagram to get the desired morphism $\alpha : A \to X$. □

We say that *a category possesses limits, products, equalizers, pullbacks, ...,* whenever the corresponding limit objects always exist in that category. For instance, it possesses (binary) products whenever for any two objects A, B, their product $A \times B$ exists. In fact, whenever such binary products exists, then products of any finite number of objects also exist in that category. This is what is meant by the existence of products.

Similarly, *a functor F between categories preserves limits, products, . . . ,* if it maps the relevant diagrams in one category to the corresponding diagrams in the other one. Thus, for instance, F preserves products when we canonically have $F(A \times B) = F(A) \times F(B)$ for all objects A, B.

Analogously to limits, we can also define colimits (also called direct or inductive limits). In detail, a cocone for the diagram $D : \mathbf{I} \to \mathbf{C}$ is given by an object $B \in \mathbf{C}$ and morphisms $b_i : D_i \to B$ with

$$b_j \circ D_{i \to j} = b_i \text{ for all morphisms } i \to j \text{ in } \mathbf{I}. \tag{8.2.45}$$

A *colimit* $q_i : D_i \to \varinjlim_{cone(D)} C_{cone(D)}$ is then an initial object in the category of cocones. Thus, for every cocone (B, b_i) for D, we have a unique $\lambda : \varinjlim_{cone(D)} C_{cone(D)} \to B$ with

$$b_i = \lambda \circ q_i \text{ for all } i. \tag{8.2.46}$$

For the two-element set $\mathbf{I}$ as a category as in the definition of products, an initial such cone is called a *coproduct* $D_1 + D_2$. The corresponding commutative diagram is

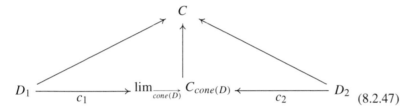

$$\tag{8.2.47}$$

for any cone C over the diagram. Of course, a coproduct in a category $\mathbf{C}$ corresponds to a product in the opposite category $\mathbf{C}^{op}$, as we are reversing all the arrows. In the same manner, we can dualize other universal constructions.

In the categories of sets or topological spaces, the coproduct is simply the disjoint union. That is, when we have two sets X and Y, we consider every element of X as being different from any element of Y and form the corresponding union. That is, when for instance $X, Y \subset Z$ with $X \cap Y \neq \emptyset$, every element of that intersection occurs twice in the disjoint union, once as an element of X and once as an element of Y. In a more formal manner, we can identify X with $X \times \{0\}$ and Y with $Y \times \{1\}$ in $Z \times \{0, 1\}$ and then consider the union $X \times \{0\} \cup Y \times \{1\}$. The index 0 or 1 then distinguishes the elements in X from those in Y. One also writes $\dot{\cup}$ for the disjoint union. We leave it to the reader to check that this disjoint union is indeed the coproduct in **Sets**. Of course, the essential point is that a map on $X \dot{\cup} Y$ is uniquely specified through its values on X and Y.

In the same manner that the definition of a pullback generalizes that of a product, we can also generalize the definition of a coproduct. The resulting structure is called a pushforward. We shall now elaborate on this issue, although perhaps the construction is already clear to the reader. Thus, we consider the index category $\mathbf{I}^\star = \{1, 2, 3\}$

$$3 \longrightarrow 1$$
$$\downarrow$$
$$2 \qquad\qquad (8.2.48)$$

A limit of a corresponding diagram

$$D_3 \xrightarrow{\ d_\alpha\ } D_1$$
$$\Big\downarrow {\scriptstyle d_\beta}$$
$$D_2 \qquad\qquad (8.2.49)$$

is then called a *pushforward* of d_α, d_β; it is universal for the diagram

$$
\begin{array}{ccc}
D_3 & \xrightarrow{\quad d_\alpha \quad} & D_1 \\
\Big\downarrow {\scriptstyle d_\beta} & & \Big\downarrow \\
D_2 & \xrightarrow{\qquad} & \varinjlim_{cocone(D)} C_{cocone(D)}
\end{array}
\qquad (8.2.50)
$$

In the category of sets, for the pushforward $P^\star$ of $\phi : Z \to X, \gamma : Z \to Y$, we take the disjoint union of X and Y and identify $\phi(z) \in X$ with $\gamma(z) \in Y$ for all $z \in Z$ and the inclusion morphisms i_X, i_Y from X and Y. Thus, we have the diagram

$$
\begin{array}{ccc}
Z & \xrightarrow{\ \phi\ } & X \\
\Big\downarrow {\scriptstyle \gamma} & & \Big\downarrow {\scriptstyle i_X} \\
Y & \xrightarrow{\ i_Y\ } & P^\star
\end{array}
\qquad (8.2.51)
$$

and since we identify $\phi(z)$ and $\gamma(z)$, we have

$$i_X \circ \phi = i_Y \circ \gamma, \qquad (8.2.52)$$

so that the diagram (8.2.51) commutes. As for pullbacks, the diagram (8.2.51) is universal.

In particular, when we have inclusions $X \cap Y \to X, X \cap Y \to Y, X \to X \cup Y, Y \to X \cup Y$, we get the diagram

$$
\begin{array}{ccc}
X \cap Y & \longrightarrow & X \\
\Big\downarrow & & \Big\downarrow \\
Y & \longrightarrow & X \cup Y
\end{array}
\qquad (8.2.53)
$$

in which $X \cap Y$ is a pullback and $X \cup Y$ is a pushforward.

Analogously for meet and join in a poset.

We now describe a construction that is not given by limits or colimits.

Definition 8.2.5 Let the category $\mathbf{C}$ possess products. The *exponential*[3] of objects B and C is

$$\text{an object } C^B \text{ and an arrow } \epsilon : B \times C^B \to C, \tag{8.2.54}$$

called the *evaluation*, with the property that for any object P and arrow $f : B \times P \to C$, there exists a unique arrow

$$F : P \to C^B \tag{8.2.55}$$

with

$$\epsilon \circ (1_B \times F) = f. \tag{8.2.56}$$

In terms of diagrams,

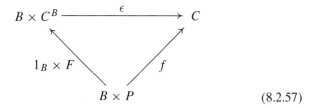

$$\tag{8.2.57}$$

Again, the basic example we should have in mind here is the category **Sets**. When B and C are sets, an element of C^B is simply a map $f : B \to C$, and the evaluation simply applies such an f to an element of B,

$$\epsilon(x, f) = f(x). \tag{8.2.58}$$

In terms of Hom-sets (8.1.9), we have

$$\text{Hom}_{\mathbf{Sets}}(A \times B, C) = \text{Hom}_{\mathbf{Sets}}(A, C^B). \tag{8.2.59}$$

This means that we view a map from $A \times B$ to C as a family of maps, parametrized by B, from A to C. Of course, by exchanging the roles of A and B, we obtain from (8.2.59)

$$\text{Hom}_{\mathbf{Sets}}(A, C^B) = \text{Hom}_{\mathbf{Sets}}(B, C^A). \tag{8.2.60}$$

On the basis of this example, we should view C^B as the set of morphisms from B to C, and the evaluation simply applies such a morphism to an element of B. $f : B \times P \to C$ can be considered as a family of morphisms from B to C parametrized by P. C^B is the universal parameter space for morphisms from B to C, and any other parameter space P can then be mapped into that one while keeping the element of B where the morphisms are to be evaluated.

---

[3] The name "exponential" seems to derive from the notation employed. It has little to do with the exponential function in analysis or the exponential map in Riemannian geometry and Lie groups. Perhaps a small justification is the following. When B and C are finite sets with n and m elements, resp., then the set C^B has m^n elements.

A simple, but important example is the following. In the category **Sets**, we consider the object $2 := \{0, 1\}$. Then for any $X \in$ **Sets**, the exponential 2^X is simply the power set $\mathcal{P}(X)$ of X, that is, the set of its subsets. In fact, any subset $A \subset X$ is characterized by the morphism

$$\chi_A : X \to 2 \text{ with } \chi_A(x) = \begin{cases} 1 & \text{if } x \in A \\ 0 & \text{if } x \notin A. \end{cases} \tag{8.2.61}$$

In turn, the implication (4.1.4),

$$(A, B) \mapsto A \Rightarrow B := (X \setminus A) \cup B \tag{8.2.62}$$

in the power set $\mathcal{P}(X)$ is an exponential B^A. This follows from the observation that $C \cap A \subset B$ implies $C \subset (A \Rightarrow B)$. (Note that the product in the poset $\mathcal{P}(X)$ is given by the intersection $\cap$, see (8.2.26).)

Definition 8.2.6 A category is *Cartesian closed* if it possesses products and exponentials.

One can then show that the category **Cat** of small categories and functors is Cartesian closed. The exponential of two categories **C**, **D** is given by the functors between them,

$$\mathbf{D}^{\mathbf{C}} = \text{Fun}(\mathbf{C}, \mathbf{D}). \tag{8.2.63}$$

We omit the proof which is not difficult.

8.3   Categories of Diagrams

Generalizing the category **Sets**$^\Gamma$ of graphs considered above, we now look at functor categories of the form

$$\mathbf{Sets}^{\mathbf{C}} \tag{8.3.1}$$

for some locally small category **C**. Its objects are set-valued diagrams on **C**, that is, functors

$$F, G : \mathbf{C} \to \mathbf{Sets}, \tag{8.3.2}$$

and its arrows are natural transformations

$$\phi, \psi : F \to G, \tag{8.3.3}$$

see (8.1.10)–(8.1.12).

For any object $C \in \mathbf{C}$, we then have the evaluation

$$\epsilon_C : \mathbf{Sets}^{\mathbf{C}} \to \mathbf{Sets}, \tag{8.3.4}$$

given by

$$\epsilon_C(F) = FC \text{ and } \epsilon_C(\phi)(F) = (\phi \circ F)I. \tag{8.3.5}$$

For an object $C \in \mathbf{C}$, we have the functor

$$\text{Hom}_{\mathbf{C}}(C, -) : \mathbf{C} \to \mathbf{Sets} \tag{8.3.6}$$

that takes an object $D \in \mathbf{C}$ to the set of morphisms $\mathrm{Hom}_{\mathbf{C}}(C, D)$ from C to D in the category $\mathbf{C}$, and a morphism $g : D \rightarrow D'$ to the morphism $\mathrm{Hom}_{\mathbf{C}}(C, D) \rightarrow \mathrm{Hom}_{\mathbf{C}}(C, D')$ in the category **Sets** that maps $h : C \rightarrow D$ to $g \circ h : C \rightarrow D'$. Thus,

$$\mathrm{Hom}_{\mathbf{C}}(C, -) \in \mathbf{Sets}^{\mathbf{C}}. \qquad (8.3.7)$$

Such a functor is called *representable*.

Now, however, when $f : C \rightarrow C'$ is a morphism in $\mathbf{C}$, we obtain the natural transformation

$$\mathrm{Hom}_{\mathbf{C}}(f, -) : \mathrm{Hom}_{\mathbf{C}}(C', -) \rightarrow \mathrm{Hom}_{\mathbf{C}}(C, -) \qquad (8.3.8)$$

which goes in the opposite direction. Here, a morphism $h' : C' \rightarrow D$ goes to the morphism $h' \circ f : C \rightarrow D$. We call such a behavior *contravariant*. Thus

$$xC := \mathrm{Hom}_{\mathbf{C}}(C, -) \qquad (8.3.9)$$

defines a contravariant functor

$$
\begin{aligned}
x : \quad & \mathbf{C}^{\mathrm{op}} \rightarrow \mathbf{Sets}^{\mathbf{C}} \\
& C \mapsto \mathrm{Hom}_{\mathbf{C}}(C, -) \qquad (8.3.10)
\end{aligned}
$$

from the category $\mathbf{C}^{\mathrm{op}}$ obtained from $\mathbf{C}$ by taking the same objects, but reversing the direction of all arrows, see Definition 8.1.4. Since we are reversing the arrows, x now maps arrows to arrows, instead of to arrows in the opposite direction. Thus, x is a functor.

Let us analyze once more what we have done here. For objects C, D in a locally small category $\mathbf{C}$, we can look at the set $\mathrm{Hom}_{\mathbf{C}}(C, D)$ of all morphisms $f : C \rightarrow D$. When we then let D vary, but keep C fixed, we are probing C by its relations with other objects of the category, and since morphisms encode the relations in a category, somehow the collection of these sets for varying D should yield a good characterization of C. Also, when we consider this as a function of the varying D, we obtain a functor from $\mathbf{C}$ to **Sets**, mapping each object D to the set $\mathrm{Hom}_{\mathbf{C}}(C, D)$, and mapping morphisms $g : D \rightarrow D'$ to mappings between those sets, simply by composing morphisms $f : C \rightarrow D$ with g to obtain morphisms $f' := g \circ f : C \rightarrow D'$. When we then also let C vary, then we obtain a family of such functors from $\mathbf{C}$ to **Sets**, that is, a family of elements of $\mathbf{Sets}^{\mathbf{C}}$. Since now morphisms $f : C \rightarrow C'$ induce mappings in the opposite direction, altogether, we obtain a morphism $\mathbf{C}^{\mathrm{op}} \rightarrow \mathbf{Sets}^{\mathbf{C}}$.

Similarly, we have

$$\mathrm{Hom}_{\mathbf{C}}(-, C) : \mathbf{C}^{\mathrm{op}} \rightarrow \mathbf{Sets} \qquad (8.3.11)$$

from $\mathbf{C}^{\mathrm{op}}$. Thus, $D \in \mathbf{C}^{\mathrm{op}}$ is now taken to the set of morphisms from $D \rightarrow C$ in $\mathbf{C}$, and a morphism $g : D \rightarrow D'$ in $\mathbf{C}^{\mathrm{op}}$, that is, a morphism $D' \rightarrow D$ in $\mathbf{C}$, is now taken to a morphism in **Sets**.

This time, a morphism $f : C \rightarrow C'$ in $\mathbf{C}$ is taken to a morphism in the same direction,

$$\mathrm{Hom}_{\mathbf{C}}(-, f) : \mathrm{Hom}_{\mathbf{C}}(-, C) \rightarrow \mathrm{Hom}_{\mathbf{C}}(-, C'). \qquad (8.3.12)$$

This behavior is called *covariant*.

Altogether, we obtain a functor

$$y : \quad \mathbf{C} \to \mathbf{Sets}^{\mathbf{C}^{op}}$$
$$C \mapsto \mathrm{Hom}_{\mathbf{C}}(-, C), \qquad\qquad (8.3.13)$$

called the *Yoneda functor*. Thus, y sends $C \in \mathbf{C}$ to the functor that sends $D \in \mathbf{C}$ to the set $\mathrm{Hom}_{\mathbf{C}}(D, C)$.

Thus, y maps the category $\mathbf{C}$ to the functor category $\mathbf{Sets}^{\mathbf{C}^{op}}$. The image of each object C of $\mathbf{C}$ thus is a functor. The Yoneda lemma below will then compare the functor yC with other functors F in this category; more precisely, it will characterize the morphisms between the functors yC and F. Such a functor $F \in \mathbf{Sets}^{\mathbf{C}^{op}}$ will be called a presheaf, see Sect. 8.4.

Let us consider the trivial example (which you may well wish to skip). When $\mathbf{C} = \{0\}$ consists of a single element 0 with the identity 1_0 as the only morphism, then y sends 0 to a single element set. In particular, when F is a presheaf (to be defined below in general) on $\{0\}$, that is, a single set $F0$, then we can send 0 to any element of this set. This is then a natural transformation from $\mathrm{Hom}_{\{0\}}(-, 0) = \{1_0\}$ to F. The construction of Yoneda will now generalize this in a functorial manner, that is, respecting morphisms of a general category $\mathbf{C}$.

Definition 8.3.1 A functor $F \in \mathbf{Sets}^{\mathbf{C}^{op}}$ that is of the form $\mathrm{Hom}_{\mathbf{C}}(-, C)$ for some $C \in \mathbf{C}$ is called *representable*.

The main result here is the Yoneda Lemma:

Theorem 8.3.1 *Let $\mathbf{C}$ be a locally small category. Then, for any functor $F \in \mathbf{Sets}^{\mathbf{C}^{op}}$,*

$$\mathrm{Hom}_{\mathbf{Sets}^{\mathbf{C}^{op}}}(yC, F) \cong FC, \qquad\qquad (8.3.14)$$

that is, the natural transformations from $\mathrm{Hom}_{\mathbf{C}}(-, C)$ to F are in natural correspondence with the elements of the set FC.

Moreover, the Yoneda functor is an embedding in the sense of Definition 8.1.3.

This result has the following important corollary, which follows from the definition of an embedding.

Corollary 8.3.1 *If the functors $\mathrm{Hom}_{\mathbf{C}}(-, C)$ and $\mathrm{Hom}_{\mathbf{C}}(-, C')$ for two objects C, C' of $\mathbf{C}$ are isomorphic, then C and C' are isomorphic themselves. More generally, for $C, C' \in \mathbf{C}$, the morphisms between the functors $\mathrm{Hom}_{\mathbf{C}}(-, C)$ and $\mathrm{Hom}_{\mathbf{C}}(-, C')$ correspond to the morphisms between C and C'.*

This corollary furnishes us with a useful strategy for checking the isomorphisms of two objects of some (locally small) category.

The principle of the proof of Theorem 8.3.1 is not difficult. An element $\theta \in \mathrm{Hom}_{\mathbf{Sets}^{\mathbf{C}^{op}}}(yC, F)$ is a morphism $\theta : \mathrm{Hom}_{\mathbf{C}}(-, C) \to F$. We can apply this to $D \in \mathbf{C}$ to get a map $\mathrm{Hom}_{\mathbf{C}}(D, C) \to FD$. Thus, an element of $\mathrm{Hom}_{\mathbf{C}}(D, C)$, that is, a morphism from D to C gets mapped to an element of the set FD. We thus obtain one direction of the correspondence

(8.3.14) by sending θ to $\theta(1_C)$ where $1_C \in \mathrm{Hom}_\mathbf{C}(C, C)$ is the identity morphism of C. For the other direction, we send $\xi \in FC$ to the morphism $\theta_\xi \in \mathrm{Hom}_{\mathbf{Sets}^{\mathbf{C}^{\mathrm{op}}}}(yC, F)$ that sends $f \in \mathrm{Hom}_\mathbf{C}(D, C)$ to $Ff(\xi) \in FD$ (note that, by contravariance, $Ff : FC \to FD$ for $f : D \to C$). For the embedding, we simply apply the first part to $F = \mathrm{Hom}_\mathbf{C}(-, C')$. In slightly more detail, we have

$$\mathrm{Hom}_{\mathbf{Sets}^{\mathbf{C}^{\mathrm{op}}}}(yC, yC') \cong yC'(C) \text{ by the first part}$$
$$\cong \mathrm{Hom}_\mathbf{C}(C, C') \text{ by definition of } y \quad (8.3.15)$$

whence all the embedding properties follow. In fact, (8.3.15) is the key identity underlying the Yoneda lemma, and it directly explains the result of the corollary.

In order to check the details of this proof, we make the following constructions.[4] Essentially, this will consist in entangling the hierarchical or iterated constructions underlying the Yoneda lemma (a functor is a morphism between categories, a natural transformation is a morphism between functors, y is a functor that maps each object of $\mathbf{C}$ to a functor, etc.) Thus, perhaps it may be more insightful to reflect on the above principle of the proof than to go through all the subsequent details.

For a functor $F \in \mathbf{Sets}^{\mathbf{C}^{\mathrm{op}}}$, considering $\mathrm{Hom}_{\mathbf{Sets}^{\mathbf{C}^{\mathrm{op}}}}(yC, F)$, the morphisms from the functor yC to F, we note that both yC and F are functors from $\mathbf{C}^{\mathrm{op}}$ to $\mathbf{Sets}$. yC is considered to be a special functor, because its image is a Hom-set. Given an element of $\mathrm{Hom}(yC, F)$ (note that we leave out the subscript $\mathbf{Sets}^{\mathbf{C}^{\mathrm{op}}}$ indicating the category for simplicity of notation), that is some morphisms

$$\theta : \mathrm{Hom}_\mathbf{C}(-, C) \to F, \quad (8.3.16)$$

we can apply it to any $D \in \mathbf{C}$ to get

$$\theta_D : \mathrm{Hom}_\mathbf{C}(D, C) \to FD. \quad (8.3.17)$$

Here, both $\mathrm{Hom}_\mathbf{C}(D, C)$ and FD are simply sets.

In particular, we can form

$$x_\theta := \eta_{C,F}(\theta) := \theta_C(1_C) \in FC. \quad (8.3.18)$$

Thus, we have constructed a morphism

$$\eta_{C,F} : \mathrm{Hom}(yC, F) \to FC. \quad (8.3.19)$$

We want to show that this is an isomorphism.

For that purpose, we shall now construct a morphism in the opposite direction which we shall subsequently identify as the inverse of $\eta_{C,F}$. Thus, for $\xi \in FC$ we want to construct a morphism $\theta_\xi : yC \to F$. Of course, we need to define this on objects $D \in \mathbf{C}^{\mathrm{op}}$, that is, we need to define

$$(\theta_\xi)_D : \mathrm{Hom}_\mathbf{C}(D, C) \to FD, \quad (8.3.20)$$

---

[4] In order to facilitate the reading for those who are not accustomed to the abstract language employed in category theory, we try to spell out all the details. Therefore, more advanced readers can skip much of what follows below.

and we do this by setting

$$(\theta_\xi)_D(f) := Ff(\xi). \qquad (8.3.21)$$

Note that $Ff : FC \to FD$ for $f : D \to C$ as F maps the category $\mathbf{C}^{op}$ to **Sets**. In particular, given such a morphism $f : D \to C$ in the category $\mathbf{C}$, we obtain a map $Ff : FC \to FD$ between sets. This is how F is defined. Thus, we construct for any $\xi \in FC$ a natural transformation (a property to be verified below) from the functor $yC = \mathrm{Hom}_{\mathbf{C}}(-, C)$ to the functor F by evaluating $Ff : FC \to FD$ for any morphism $f : D \to C$ at the element $\xi \in FC$.

In order to show that θ_ξ is a morphism between the functors yC and F, we need to verify that it is natural. This means that for a morphism $g : D' \to D$,

$$
\begin{array}{ccc}
\mathrm{Hom}(D, C) & \xrightarrow{(\theta_\xi)_D} & FD \\
{\scriptstyle \mathrm{Hom}(g,C)}\big\downarrow & & \big\downarrow {\scriptstyle Fg} \\
\mathrm{Hom}(D', C) & \xrightarrow{(\theta_\xi)_{D'}} & FD'
\end{array}
\qquad (8.3.22)
$$

commutes. This follows from the following computation that only uses the functoriality of F, for any $h \in yC(D)$, i.e., $h : D \to C$,

$$
\begin{aligned}
(\theta_\xi)_{D'} \circ \mathrm{Hom}(g, C)(h) &= (\theta_\xi)_{D'}(h \circ g) \\
&= F(h \circ g)(\xi) \text{ by } (8.3.21) \\
&= F(h) \circ F(g)(\xi) \text{ since } F \text{ is a functor} \\
&= F(h)(\theta_\xi)_D(g) \text{ by } (8.3.21) \text{ again.}
\end{aligned}
$$

We can now show that the two morphisms are the inverses of each other. For that purpose, we compute θ_{x_θ}, see (8.3.21) and (8.3.18). Thus, for $f : D \to C$, we have

$$(\theta_{x_\theta})_D(f) = Ff(\theta_C(1_C)). \qquad (8.3.23)$$

Since θ is natural, see (8.3.22),

$$
\begin{array}{ccc}
yC(C) & \xrightarrow{\theta_C} & FC \\
{\scriptstyle yC(f)}\big\downarrow & & \big\downarrow {\scriptstyle Ff} \\
yC(D) & \xrightarrow{\theta_D} & FD
\end{array}
\qquad (8.3.24)
$$

commutes. Thus,

$$
\begin{aligned}
(\theta_{x_\theta})_D(f) &= Ff(\theta_C(1_C)) \text{ by } (8.3.23) \\
&= \theta_D \circ yC(f)(1_C) \text{ by } (8.3.24) \\
&= \theta_D(f) \text{ by definition of } yC.
\end{aligned}
$$

This implies

$$\theta_{x_\theta} = \theta. \qquad (8.3.25)$$

For the other direction, for $\xi \in FC$,

$$x_{\theta_\xi} = (\theta_\xi)_C(1_C) \text{ by (8.3.18) again}$$
$$= F(1_C)(\xi) \text{ by (8.3.21) again}$$
$$= 1_{FC}(\xi) \text{ since } F \text{ maps identities to identities}$$
$$= \xi.$$

Thus,

$$x_{\theta_\xi} = \xi, \tag{8.3.26}$$

and we have verified that

$$\mathrm{Hom}(yC, F) \cong FC. \tag{8.3.27}$$

This is the proof of the first part, and in fact the key part of the proof. Before we come to the second statement, we observe that the constructed isomorphism is natural in the sense that it yields commutative diagrams from any arrow of the objects and morphisms involved. First of all, consider a morphism

$$\theta : \mathrm{Hom}_{\mathbf{C}}(-, C) \to F \tag{8.3.28}$$

between functors as in (8.3.16). Such a morphism is then natural by definition in the sense that for any morphism $f : D \to C$, the diagram (8.3.24),

$$
\begin{array}{ccc}
(yC)C & \xrightarrow{\ \theta_C\ } & FC \\
{\scriptstyle (yC)f}\big\downarrow & & \big\downarrow{\scriptstyle Ff} \\
(yC)D & \xrightarrow{\ \theta_D\ } & FD
\end{array}
\tag{8.3.29}
$$

commutes (note that contravariance here leads to a reversal of the direction of arrows). Leaving the checking of the details to the reader, from this, we obtain the commutative diagram

$$
\begin{array}{ccc}
\mathrm{Hom}(yC, F) & \xrightarrow{\ \eta_{C,F}\ } & FC \\
{\scriptstyle \mathrm{Hom}(yf,F)}\big\downarrow & & \big\downarrow{\scriptstyle Ff} \\
\mathrm{Hom}(yD, F) & \xrightarrow{\ \eta_{D,F}\ } & FD.
\end{array}
\tag{8.3.30}
$$

Similarly, given a morphism $\chi : F \to G$, the diagram

$$
\begin{array}{ccc}
\mathrm{Hom}(yC, F) & \xrightarrow{\ \eta_{C,F}\ } & FC \\
{\scriptstyle \mathrm{Hom}(yC,\chi)}\big\downarrow & & \big\downarrow{\scriptstyle \chi_C} \\
\mathrm{Hom}(yC, G) & \xrightarrow{\ \eta_{C,G}\ } & GC
\end{array}
\tag{8.3.31}
$$

commutes.

The embedding part is now straightforward to prove. For objects $C, C' \in \mathbf{C}$, by the first part, we obtain an isomorphism

$$\mathrm{Hom}_{\mathbf{C}}(C, C') = yC'(C) \cong \mathrm{Hom}(yC, yC'). \tag{8.3.32}$$

This isomorphism is, of course, induced by y. In fact, $f : C \rightarrow C'$ is mapped to the natural transformation $\theta_f : yC \rightarrow yC'$ which operates on $h : D \rightarrow C$ via

$$(\theta_f)_D h = (yC')h(f) = \text{Hom}_{\mathbf{C}}(h, C')(f) = f \circ h = (yf)_D h, \quad (8.3.33)$$

that is, $\theta_f = yf$. Thus, y is faithful and full. Finally, for the injectivity, if $yC = yC'$, then

$$\text{Hom}_{\mathbf{C}}(C, C) = (yC)C = (yC')C = \text{Hom}_{\mathbf{C}}(C, C'), \quad (8.3.34)$$

and since 1_C is in this Hom-set, we have an isomorphism between C and C'.

This completes the proof of Theorem 8.3.1.

8.4 Presheaves

In Sect. 8.3, we have studied the category $\mathbf{Sets}^{\mathbf{C}^{op}}$ of contravariant set-valued functors on some fixed small category $\mathbf{C}$. In fact, this category has already appeared in Sect. 4.5, and we recall

Definition 8.4.1 An element P of $\mathbf{Sets}^{\mathbf{C}^{op}}$ is called a *presheaf* on $\mathbf{C}$.

For an arrow $f : V \rightarrow U$ in $\mathbf{C}$, and $x \in PU$, the value $Pf(x)$, where $Pf : PU \rightarrow PV$ is the image of f under P, is called the *restriction* of x along f.

A presheaf thus assigns to each object in $\mathbf{C}$ some set, and this is functorial in the contravariant sense, that is, those sets can be pulled back along arrows in $\mathbf{C}$. We can therefore view a presheaf as a collection of sets indexed by the objects of $\mathbf{C}$ in a functorial manner. For instance, when the category $\mathbf{C}$ is simply some set X (the objects thus being the elements of X, and with only the identity arrows and no arrows between different elements), we just have a collection of sets indexed by the elements of X.

In fact, for a fixed set X, we have an isomorphism of categories

$$\mathbf{Sets}^X \cong \mathbf{Sets}/X, \quad (8.4.1)$$

where $\mathbf{Sets}/X$ is the slice category introduced in Sect. 2.3, see (2.3.30). On the left of (8.4.1), we have a family $\{Z_x : x \in X\}$ of sets indexed by X whereas on the right, we have the function

$$\xi : \ Z := \coprod Z_x \rightarrow X$$
$$Z_x \rightarrow x \quad (8.4.2)$$

from the disjoint union of the sets Z_x to the base X. In other words, a map from the set X into the category $\mathbf{Sets}$, which assigns a set Z_x to every $x \in X$ can be equivalently described as the collection of the preimages under the projection from Z to X, where Z is the disjoint union of the Z_x.

Emphasizing in contrast the functorial aspect, a presheaf formalizes the possibility of restricting collections of objects, that is, the—possibly structured—sets assigned to objects of $\mathbf{C}$, to preimages of arrows.

We recall from Sect. 8.3 that each object $U \in \mathbf{C}$ yields the presheaf yU on $\mathbf{C}$, the contravariant Hom-functor that is defined on an object V by

$$yU(V) = \mathrm{Hom}_{\mathbf{C}}(V, U) \tag{8.4.3}$$

and on a morphism $f : W \to V$ by

$$
\begin{aligned}
yU(f) : \quad \mathrm{Hom}_{\mathbf{C}}(V, U) &\to \mathrm{Hom}_{\mathbf{C}}(W, U) \\
h &\mapsto h \circ f.
\end{aligned}
\tag{8.4.4}
$$

Definition 8.4.2 A presheaf of the form yU for some object $U \in \mathbf{C}$ is called a *representable functor*.

We also recall that when $f : U_1 \to U_2$ is a morphism of $\mathbf{C}$, we obtain a natural transformation $yU_1 \to yU_2$ by composition with f, so that we get the Yoneda embedding (Theorem 8.3.1)

$$y : \mathbf{C} \to \mathbf{Sets}^{\mathbf{C}^{\mathrm{op}}}. \tag{8.4.5}$$

8.5   Adjunctions and Pairings

While the concept of an adjoint is fundamental to category theory, we shall only briefly consider it here. In fact, this will not be seriously used in later sections, so the present section might be skipped.

Definition 8.5.1 An *adjunction* between the categories $\mathbf{C}, \mathbf{D}$ consists of functors

$$L : \mathbf{C} \leftrightarrows \mathbf{D} : R \tag{8.5.1}$$

with the property that for any objects $C \in \mathbf{C}, D \in \mathbf{D}$, there is an isomorphism

$$\lambda : \mathrm{Hom}_{\mathbf{D}}(LC, D) \cong \mathrm{Hom}_{\mathbf{C}}(C, RD) \tag{8.5.2}$$

that is natural in C and D.

We then also call L the *left adjoint* of R, and R the *right adjoint* of L.

Let us consider the following important example. This example concerns power set categories, that is, the categories $\mathcal{P}(Y)$, the subsets of some set Y. When we have another set X, we can consider the projection

$$\pi : X \times Y \to Y \tag{8.5.3}$$

and the induced pullback

$$\pi^* : \mathcal{P}(Y) \to \mathcal{P}(X \times Y). \tag{8.5.4}$$

Let now $A \subset Y, B \subset X \times Y$. Then

$$\pi^*(A) = \{(x, y) : x \in X, y \in A\} \subset B \text{ iff for all } x \in X, A \subset \{y : (x, y) \in B\}. \tag{8.5.5}$$

We put

$$\forall_\pi B := \{y \in Y : (x, y) \in B \text{ for all } x \in X\}. \tag{8.5.6}$$

Then (8.5.5) becomes

$$\pi^*(A) \subset B \text{ iff } A \subset \forall_\pi B. \tag{8.5.7}$$

Similarly, defining

$$\exists_\pi B := \{y \in Y : \text{there exists some } x \in X \text{ with } (x, y) \in B\}, \tag{8.5.8}$$

we have

$$B \subset \pi^*(A) \text{ iff } \exists_\pi B \subset A. \tag{8.5.9}$$

We now recall that in a category $\mathcal{P}(Z)$, $\mathrm{Hom}(Z_1, Z_2)$ has a single element iff $Z_1 \subset Z_2$ and is empty otherwise. We thus conclude

Theorem 8.5.1 *The functor* $\forall_\pi : \mathcal{P}(X \times Y) \to \mathcal{P}(Y)$ *is the right adjoint of* $\pi^* : \mathcal{P}(Y) \to \mathcal{P}(X \times Y)$, *and* $\exists_\pi : \mathcal{P}(X \times Y) \to \mathcal{P}(Y)$ *is the left adjoint of* π^*.

Of course, this construction can be generalized to arbitrary maps between sets in place of projections.

Here is another example. We have the *forgetful functor*

$$U : \textbf{Groups} \to \textbf{Sets} \tag{8.5.10}$$

that assigns to every group G the set $U(G)$ of its elements. Its left adjoint is the functor

$$F : \quad \textbf{Sets} \to \textbf{Groups}$$
$$X \mapsto G_X \tag{8.5.11}$$

where G_X is the free group generated by X. The elements of G_X are all monomials $x_1^{n_1} x_2^{n_2} \ldots$ with $x_i \in X$, $n_i \in \mathbb{Z}$ and only finitely many $n_i \neq 0$. The only group laws are $x^n x^m = x^{n+m}$ for $x \in X, n, m \in \mathbb{Z}$, and $x^0 = 1$, the unit of the group, for every $x \in X$. Then the morphisms

$$FX \to G \text{ in } \textbf{Groups} \tag{8.5.12}$$

correspond to the morphisms

$$X \to U(G) \text{ in } \textbf{Sets} \tag{8.5.13}$$

because a morphism from a free group F to another group G is determined by the images of the generators of F, and conversely, assigning images to all generators of F determines a morphism of F.

Analogously, we can consider the category of abelian groups and assign to every set X the free abelian group A_X whose elements are all formal sums $n_1 x_1 + n_2 x_2 + \cdots$, again with only finitely many nonzero $n_i \in \mathbb{Z}$, and where now $n_1 x_1 + n_2 x_2 = n_2 x_2 + n_1 x_1$ to make the group abelian.

Also, the above construction of a free group generated by a set X can be modified to yield the free monoid $X^* := M_X$ over the set X, consisting again of formal products $x_1 x_2 \ldots$ of elements of X. Here, X is considered as an "alphabet" from which the "words" in X^* are formed. (This is indeed the same as the construction of the free group G_X, as we see when we write

xx for x^2 and so on.) Again, to make X^* free, we do not allow for any nontrivial relations among the monomials.

In a somewhat similar vein, we can consider the *forgetful functor*

$$U : \textbf{Top} \to \textbf{Sets} \tag{8.5.14}$$

from the category of topological spaces to the category of sets that simply assigns to a topological space its underlying set. The left adjoint F gives a set its discrete topology, denoted by $\mathcal{O}_d(X)$, where every subset is open, see Sect. 4.1. Then any morphism from a set X to $U(\mathcal{Y})$ for some topological space $\mathcal{Y}$ corresponds to a continuous map from $(X, \mathcal{O}_d(X))$ to $\mathcal{Y}$ because any map from a space with the discrete topology is continuous. Similarly, U possesses a right adjoint G that assigns to every set X its indiscrete topology $\mathcal{O}_i(X)$, see Sect. 4.1 again. The right adjointness now follows from the fact that any map from any topological space into a space with the indiscrete topology is continuous.

Limits can likewise be written as adjoints. A cone over a diagram $D_{\textbf{I}}$ of type $\textbf{I}$ in a category $\textbf{C}$ is a morphism

$$c : C \to D_{\textbf{I}} \tag{8.5.15}$$

from an object C of $\textbf{C}$ to the diagram $D_{\textbf{I}}$. In particular, we have the diagonal morphism c_Δ with $c_\Delta(C)_i = C$ for all indices i in $\textbf{I}$. Thus, a diagram is a morphism in

$$\text{Hom}(c_\Delta(C), D_{\textbf{I}}) \tag{8.5.16}$$

and when a diagram has a limit $\lim_{\overleftarrow{cone(D)}} C_{cone(D)}$, then for each C there is a unique morphism $C \to \lim_{\overleftarrow{cone(D)}} C_{cone(D)}$. This means that

$$\text{Hom}(c_\Delta(C), D_{\textbf{I}}) = \text{Hom}(C, \lim_{\overleftarrow{cone(D)}} C_{cone(D)}) \tag{8.5.17}$$

and the limit thus is a right adjoint of the diagonal morphism. Similarly, colimits are left adjoints of diagonals.

From an adjunction, we obtain a natural transformation

$$\eta : 1_{\textbf{C}} \to R \circ L \tag{8.5.18}$$

by

$$\eta_C = \lambda(1_{LC}), \tag{8.5.19}$$

and conversely, given such an η, we can set

$$\lambda(f) = Rf \circ \eta_C. \tag{8.5.20}$$

We omit the details of the verification of this claim.

Lemma 8.5.1 *Right adjoints preserve limits, and therefore by duality, left adjoints preserve colimits.*

Proof We consider a diagram $\Delta : \mathbf{I} \rightarrow \mathbf{D}$ and a cone K with morphisms $K \rightarrow D_i$ over that diagram. For any $D \in \mathbf{D}$, we then get an induced diagram $\Delta_D : \mathbf{I} \rightarrow \mathrm{Hom}_{\mathbf{D}}(D, .)$ and a cone $\mathrm{Hom}(D, K)$ with morphisms $\mathrm{Hom}(D, K) \rightarrow \mathrm{Hom}(D, D_i)$. We then have

$$\lim_{\overleftarrow{cone(D)}} \mathrm{Hom}(D, C_{cone(D)}) = \mathrm{Hom}(D, \lim_{\overleftarrow{cone(D)}} C_{cone(D)}) \qquad (8.5.21)$$

whenever $\lim_{\overleftarrow{cone(D)}} C_{cone(D)}$ exists in $\mathbf{D}$. In that situation, we therefore find

$$\mathrm{Hom}_{\mathbf{C}}(C, R(\lim_{\overleftarrow{cone(D)}} C_{cone(D)})) \cong \mathrm{Hom}_{\mathbf{D}}(LC, \lim_{\overleftarrow{cone(D)}} C_{cone(D)})$$

$$\cong \lim_{\overleftarrow{cone(D)}} \mathrm{Hom}_{\mathbf{D}}(LC, C_{cone(D)})$$

$$\cong \lim_{\overleftarrow{cone(D)}} \mathrm{Hom}_{\mathbf{C}}(C, RC_{cone(D)})$$

$$\cong \mathrm{Hom}_{\mathbf{C}}(C, \lim_{\overleftarrow{cone(D)}} RC_{cone(D)}).$$

The Yoneda Theorem 8.3.1 then implies the isomorphism

$$R(\lim_{\overleftarrow{cone(D)}} C_{cone(D)}) \cong \lim_{\overleftarrow{cone(D)}} RC_{cone(D)}. \qquad (8.5.22)$$

$\square$

The concept of an adjoint allows for the unified treatment of many mathematical constructions, see [5, 82]. At this moment, we only demonstrate how this can be applied to the case of adjoint operators between Hilbert spaces.

We start with some vector space V, over the reals $\mathbb{R}$ (or some other ground field $\mathbb{K}$ which would then replace $\mathbb{R}$ in the subsequent constructions) and consider the category $\mathbb{P}^1 V$ of oriented one-dimensional subspaces of V, with morphisms

$$\mathrm{Hom}(l_1, l_2) \cong \mathbb{R}. \qquad (8.5.23)$$

| Hilbert space |

When we have a Hilbert space H with scalar product $\langle ., . \rangle$,[5] we consider the same objects, that is, the elements of $\mathbb{P}^1 H$, but the only morphism between l_1 and l_2 now being

$$\langle e_1, e_2 \rangle \qquad (8.5.24)$$

where e_i is a positive generator of l_i with $\langle e_i, e_i \rangle = 1$.

With these definitions, the standard concept of the adjoint $L^\star$ of an operator $L : H \rightarrow H'$, i.e.,

$$\langle Lx, y \rangle_{H'} = \langle x, L^\star y \rangle_H \qquad (8.5.25)$$

---

[5] The scalar product $\langle ., . \rangle : H \times H \rightarrow \mathbb{R}$ is symmetric, bilinear, and positive definite, that is,

$$\langle v, w \rangle = \langle w, v \rangle \quad \text{for all } v, w \in H$$

$$\langle \alpha v_1 + \beta v_2, w \rangle = \alpha \langle v_1, w \rangle + \beta \langle v_2, w \rangle \quad \text{for all } \alpha, \beta \in \mathbb{R}, v_1, v_2, w \in H$$

$$\langle v, v \rangle > 0 \quad \text{for all } v \neq 0 \in H.$$

now becomes

$$\text{Hom}_{H'}(Lx, y) = \text{Hom}_H(x, L^\star y) \tag{8.5.26}$$

in hopefully obvious notation, when we identify a vector in a Hilbert space with the linear subspace it spans.

We now put

$$\mathbf{C}^* := \mathbf{Fun}(\mathbf{C}, \mathbb{R}). \tag{8.5.27}$$

For $\alpha \in \mathbf{C}^*$ and $C \in \mathbf{C}$, we then have

$$(\mathbf{C}, \alpha) := \alpha(C) \in \mathbb{R}. \tag{8.5.28}$$

When we then have some

$$i : \mathbf{C} \to \mathbf{C}^* \tag{8.5.29}$$

we obtain a generalized scalar product via

$$\langle C, D \rangle := (C, i(D)) = i(D)(C). \tag{8.5.30}$$

When $\mathbf{C}$ is a group, we may require

$$i(C)(C) > 0 \text{ for } C \neq 0, \tag{8.5.31}$$

that is, positive definiteness of $\langle ., . \rangle$.

Category theory was invented by Eilenberg and Mac Lane [32] in order to provide a foundation and a formal framework for algebraic topology [31]. Important contributions to the theory were the Yoneda lemma [120] and the notion of adjunctions, discovered by Kan [65].

In this chapter, I have often used the treatment in [5]. A general reference for category theory is [82]. A more elementary exposition of the basic ideas and constructions is [79].

Topoi

9

In this (almost) final chapter, we describe and analyze a concept, that of a topos, that emerged from the works of Grothendieck in algebraic geometry, see in particular [4], and Lawvere in logic, see for instance [77, 78] and also the contribution of [110], and that provides a general framework for the mathematical structures of geometry and logic. As always in mathematics, when a concept unifies hitherto separate mathematical domains, it leads to substantial and fundamental insights in both of them. A reason for this is, of course, that it makes the concepts and methods developed in each of the fields concerned available in the other.

In an intuitive sense—to be refined and corrected in this chapter—, a topos is a category of structured families of sets where the type of structure characterizes the topos in question. In the approach of Lawvere, the concept of a topos emerges from abstracting certain fundamental properties of the category of sets. One of those properties is that, for a set X, there is a correspondence between subsets A of X and characteristic functions $\chi :$ $X \to \{0, 1\}$. The subset A corresponds to χ_A with $\chi_A(x) = 1$ if $x \in A$ and $\chi_A(x) = 0$ if $x \notin A$. One can interpret 1 here as the truth value "true" and 0 as "false". Thus $\chi_A(x) = 1$ iff it is true that $x \in A$. In a structured family X_i of sets, we can then ask whether $x \in X_i$ for every i, and we then get the corresponding truth values for each i. This is the idea of a subobject classifier that will be developed in Sect. 9.1.

Also, in the category of sets, whenever $x \in X$ and $F : X \to X$ is a mapping, there exists a mapping $\eta : \mathbb{N} \to X$ defined by $\eta(n) = F^n(x)$ for $n \in \mathbb{N}$. (This was described as a dynamical system in Sect. 2.5.) In that sense, the natural numbers can be made to operate on X. That latter property, albeit important as well, will not be systematically developed in this chapter.

This chapter heavily depends on the material developed in the previous one. Good references for topoi are [42, 84] from which much of the material has been taken. The references [10, 43] provide more details on geometric modality. Other references for topos theory and logic are [74, 86]. Finally, we mention the compendium [55].

© Springer International Publishing Switzerland 2015
J. Jost, *Mathematical Concepts*, DOI 10.1007/978-3-319-20436-9_9

9.1   Subobject Classifiers

For the sequel, we shall have to work with sets of morphisms, and therefore, to be on the safe side, we shall assume henceforth that all categories involved are small, even though there will be several instances where this assumption will not be necessary.

Definition 9.1.1 A *subobject* A of an object B of a category $\mathbf{C}$ is a monomorphism

$$i : A \rightarrowtail B. \tag{9.1.1}$$

A subobject $i' : A' \rightarrowtail B$ is *included* in the subobject A if there exists a morphism $j : A' \to A$ with

$$i' = i \circ j, \tag{9.1.2}$$

i.e., the diagram

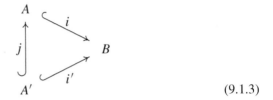

$$\tag{9.1.3}$$

commutes (note that j is automatically monic).

The subobjects $A_1, A_2 \rightarrowtail B$ are equivalent when each is included in the other.

$$\mathrm{Sub}_{\mathbf{C}}(B) \tag{9.1.4}$$

is the set of equivalence classes of subobjects of B in $\mathbf{C}$.

$\mathrm{Sub}_{\mathbf{C}}(B)$ is a poset with the ordering induced from the inclusion of subobjects. In the category **Sets**,

$$\mathrm{Sub}_{\mathbf{Sets}}(X) = \mathcal{P}(X), \tag{9.1.5}$$

the power set of X. (We have to be careful here. We do not admit arbitrary injective maps between subsets of X, but only set theoretical inclusions. That is, we consider the set X as a collection of specific elements, as above when a set has been considered as a category itself, and all morphisms have to preserve these elements.)

Equivalently, a subset A of X can be characterized by its characteristic function

$$\chi_A(x) := \begin{cases} 1 & \text{if } x \in A \\ 0 & \text{if } x \notin A. \end{cases} \tag{9.1.6}$$

Thus, χ_A takes its values in $\mathbf{2} := \{0, 1\}$. We consider $\mathbf{2}$ as a set of truth values, with 1 corresponding to the value "true". Thus, we have the monomorphism

$$\mathrm{true} : \mathbf{1} := \{1\} \rightarrowtail \mathbf{2}, \quad 1 \mapsto 1. \tag{9.1.7}$$

Thus, the subset A, given by the equivalence class of the monomorphism $i : A \rightarrowtail X$, is obtained as the pullback of true along χ_A:

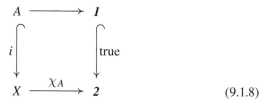

$$(9.1.8)$$

We now generalize this.

Definition 9.1.2 In a category $\mathbf{C}$ with finite limits, a *subobject classifier* consists of an object Ω and a monomorphism

$$\text{true} : \mathbf{1} \rightarrowtail \Omega \qquad (9.1.9)$$

with the property that for every monomorphism $A \rightarrowtail X$ in $\mathbf{C}$, there is a unique morphism $\chi : X \to \Omega$ with a commutative pullback diagram

$$
\begin{array}{ccc}
A & \longrightarrow & \mathbf{1} \\
\downarrow & & \downarrow{\scriptstyle\text{true}} \\
X & \xrightarrow{\ \chi\ } & \Omega
\end{array}
\qquad (9.1.10)
$$

In order to demonstrate that the existence of a subobject classifier has nontrivial consequences, we present

Lemma 9.1.1 *In a category $\mathbf{C}$ that possesses a subobject classifier, every monomorphism is an equalizer, and a morphism that is both a mono- and an epimorphism has to be an isomorphism.*

Or shorter: In a category with a subobject classifier, monic epics are iso.

Proof We consider the diagram (9.1.10) in the following form

$$
\begin{array}{ccc}
A & \longrightarrow & \mathbf{1} \\
{\scriptstyle f}\downarrow & \nearrow & \downarrow{\scriptstyle\text{true}} \\
X & \xrightarrow{\ \chi_f\ } & \Omega
\end{array}
\qquad (9.1.11)
$$

where the diagonal is the unique morphism $t_X : X \to \mathbf{1}$. A monic $f : A \to X$ thus equalizes χ_f and true $\circ\, t_X$.

Now, when f equalizes morphisms $h_1, h_2 : X \to Y$, i.e., $h_1 \circ f = h_2 \circ f$ and is an epimorphism, then $h_1 = h_2 =: h$. Thus, f equalizes h with itself. Now, an equalizer of a morphism with itself has to be an isomorphism, because also 1_X is such an equalizer. We see this from the universal property of (8.2.30), that is, from

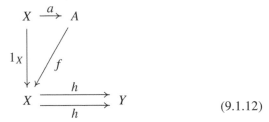

$$(9.1.12)$$

where the existence of the morphism a comes from the universality property of equalizers. This then implies that $f \circ a = 1_X$, $a \circ f = 1_A$, i.e., f is an isomorphism. $\square$

While we have seen the subobject classifier in **Sets**, let us now consider an example of a category without a subobject classifier, the category **Ab** of (small) abelian groups A. The terminal object $\mathbf{1}$ in **Ab** is the trivial group consisting only of e.[1] Thus, a homomorphism $t : \mathbf{1} \rightarrowtail \Omega$ into a putative subobject classifier, an abelian group Ω, has to send e to $e \in \Omega$. Therefore, when we pull t back along a group homomorphism $\phi : A \to \Omega$, we obtain a diagram

$$
\begin{array}{ccc}
S & \longrightarrow & \mathbf{1} \\
\downarrow{\scriptstyle i} & & \downarrow{\scriptstyle t} \\
A & \xrightarrow{\;\phi\;} & \Omega
\end{array}
\qquad (9.1.13)
$$

with $S = \ker \phi = \phi^{-1}(e)$. Since by commutativity of the diagram, e has to be the image of S under the group homomorphism $\phi \circ i$, this homomorphism then has to map A to the quotient group A/S which then has to be a subgroup of Ω. Thus, Ω would have to contain every such quotient group A/S as a subgroup. There is obviously no such small abelian group Ω. This shows that the category **Ab** cannot possess a subobject classifier. Intuitively, the structure of this category is too constrained, in the sense that it poses severe restrictions on morphisms, in order to permit a subobject classifier.

An even simpler counterexample is a poset category like $\mathcal{P}(Y)$, the set of subsets of some fixed set Y, with the only morphisms being the inclusions $A \subset B$. It does not possess a subobject classifier. The terminal object $\mathbf{1}$ in $\mathcal{P}(Y)$ is Y itself because that is the only one for which there exists a unique morphism $A \subset Y$ for every A. But then, a putative Ω would also have to be Y itself, because no other object can receive a morphism from Y as required for the arrow true : $\mathbf{1} \to \Omega$. But then, (9.1.10) can be a pullback diagram only for X itself, as χ then necessarily has to be the inclusion $X \subset Y$, and so, X itself is the only pullback of the corresponding diagram, and we cannot satisfy the pullback condition for any nontrivial subobjects A of X in $\mathcal{P}(Y)$. The only exception where this does not create a problem for the

---

[1] Here we write e instead of 0 for the neutral element of an abelian group, to avoid conflict with our convention that when the terminal object $\mathbf{1}$ of a category has a single element, we denote that element by 1, and not by 0.

pullback condition are sets Y with at most one element. We can also explain the situation by the following diagram

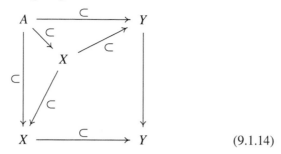

$$(9.1.14)$$

This simply expresses the fact that every diagram with such inclusions between A, X, Y can always be factored through X. The reason is essentially that the bottom arrow $X \subset Y$ does not depend on A. Note the contrast with (9.1.8) where the arrow χ_A depends on the subset A and where consequently we cannot factor through X when $A \neq X$.

We now turn to the more positive issue of describing categories that do possess a subobject classifier. All these categories will be derived from the basic category **Sets**, and they will somehow correspond to categories of parametrized families of sets, with suitable functorial properties, of course. This is, in fact, no coincidence because the concept of a topos, to be described below and one of whose essential features is the existence of a subobject classifier, has been developed precisely as a formal framework for such categories of parametrized families of sets.

In any case, in a category other than **Sets**, such a subobject classifier Ω, when it exists at all, can be more complicated. Let us recall the category of directed graphs (8.2.10) and (8.2.11), that is, the category of diagrams G consisting of a pair of sets (G_0, G_1) (vertices and directed edges) with a pair of arrows

$$g_\alpha, g_\omega : G_1 \rightrightarrows G_0, \qquad (9.1.15)$$

(besides the identity arrows).

Thus, to each element e of G_1, that is, to each directed edge, we associate two elements $\alpha(e), \omega(e)$ of G_0, its initial and terminal vertices. Since we are in the category of graphs, a subobject classifier has to be a graph itself. This graph Ω is depicted in the diagram (9.1.16). The morphism true : $\mathbf{1} \rightarrowtail \Omega$ maps the point 1 of $\mathbf{1}$ to $1 \in \Omega$ (the element of Ω corresponding to the presence of a vertex), and the identity arrow of $\mathbf{1}$ to the arrow labelled $+ + +$. This corresponds to the situation when an edge of G_1 is also an edge in a subgraph (Γ_0, Γ_1). The arrow $+ + -$ in Ω corresponds to the case where the initial and terminal vertices of an edge are contained in Γ_0, but the edge itself is missing from Γ_1. When only the initial vertex is in Γ_0, we have the arrow $+ - -$, while when only the terminal vertex is present, we have $- + -$. Finally, when neither of those vertices is present, we have the arrow $- - -$ from 0 (corresponding to absence of a vertex) to itself.

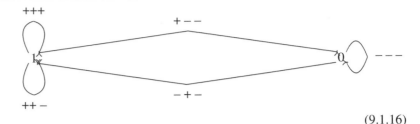

$$\text{(9.1.16)}$$

We next consider the category $\mathbf{Sets}^{\circlearrowleft}$ of sets with endomorphisms (equivalently: automata or discrete dynamical systems, see Sect. 2.5). An object of $\mathbf{Sets}^{\circlearrowleft}$ thus is a set A equipped with an endomorphism, that is, a self-map,

$$\alpha : A \to A, \qquad\qquad\qquad (9.1.17)$$

and a morphism between two such objects (A, α), (B, β) is given by a map

$$f : A \to B \text{ with } f \circ \alpha = \beta \circ f. \qquad\qquad (9.1.18)$$

Since such an endomorphism α can be iterated, that is, repeatedly applied, this indeed yields an automaton or a discrete dynamical system.

A subautomaton of such an automaton is then given by a subset $A' \subset A$ and $\alpha' = $ restriction of α to A' with the property that A' be closed under the dynamics, that is, whenever $a \in A'$, then also $\alpha(a) \in A'$. On the other hand, when $a \notin A'$, there may exist some $n \in \mathbb{N}$ with $\alpha^n(a) \in A'$. This leads to the following subobject classifier, a particular set with an endomorphism.

$$\bullet_1 \longleftarrow \bullet_{\frac{1}{2}} \longleftarrow \bullet_{\frac{1}{3}} \longleftarrow \cdots \qquad \bullet_0$$
$$\circlearrowleft \qquad\qquad\qquad\qquad\qquad\qquad \circlearrowleft \qquad (9.1.19)$$

Here, the bullet labelled $\frac{1}{n+1} \in \mathbb{N}$ corresponds to the smallest $n \in \mathbb{N}$ for which $\alpha^n(a) \in A'$, and the bullet labelled 0 corresponds to those $a \in A$ that will never be mapped to A' by any iteration of α. The self-arrow at 1 expresses the fact that when $a \in A'$, then also $\alpha(a) \in A'$. More colloquially, n is the "time till truth", that is, how many steps it will take to get the element a into the subset A'.

More generally, we consider the category of morphisms of sets, $\mathbf{Sets}^{\to}$. The objects of this category thus are maps $f : A \to B$, and the morphisms are commutative diagrams

$$
\begin{array}{ccc}
C & \xrightarrow{\ \ g\ \ } & D \\
\downarrow & & \downarrow \\
A & \xrightarrow{\ \ f\ \ } & B
\end{array}
\qquad (9.1.20)
$$

In particular, a subobject $f' : A' \to B'$ of $f : A \to B$ requires monomorphisms $i : A' \to A$, $j : B' \to B$ with a commutative diagram

$$A' \xrightarrow{f'} B'$$

(9.1.21)

For $x \in A$, we then have three possibilities:

(1) $x \in A'$ ("it is there")

($\frac{1}{2}$) $x \notin A'$, but $f(x) \in B'$ ("it will get there")

(0) $x \notin A'$ and $f(x) \notin B'$ ("it will never get there").

The subobject classifier is then given by the sets $3 := \{1, \frac{1}{2}, 0\}$, $2 = \{1, 0\}$ and the map $t : 3 \to 2$ with $t(1) = t(\frac{1}{2}) = 1$, $t(0) = 0$. Here, the elements of 3 correspond to the above three cases, and those of 2 correspond to $y \in B'$ and $y \notin B'$. The commutative diagram for subobject classification is then, with $1 = \{1\}$ and $i_t(1) = 1$, $j_t(1) = 1$,

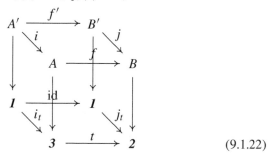

(9.1.22)

We next consider the category $M - \mathbf{Sets}$ of representations of the monoid M, see (2.3.10). In this category, the terminal object is the set $1 = \{e\}$ with a single object and the trivial operation of M, i.e., $\mu_e(m, e) = e$ for all $m \in M$. We consider $\Omega = (\Lambda_M, \omega)$ where Λ_M is the set of all left ideals of M with the operation

$$\omega : M \times \Lambda_M \to \Lambda_M$$
$$(m, L) \mapsto \{n : nm \in L\} =: L_m. \qquad (9.1.23)$$

L_m is indeed an ideal because $kL_m = \{kn : nm \in L\} \subset L_m$ since if $nm \in L$ then also $k(nm) \in L$ as L is an ideal. We also observe that the largest ideal, M itself, is fixed by ω, $\omega(m, M) = M$ for all $m \in M$.

Thus, Ω is an M-set. We shall now verify that Ω is the subobject classifier in the category $M - \mathbf{Sets}$. First, $t : 1 \to \Omega$ is given by $t(e) = M$. We note that the equivariance condition (2.3.12) is satisfied: $t(\mu_e(m, e)) = \omega(m, t(e)) = \omega(m, M) = M$.

Let now $i : (X, \mu) \to (Y, \lambda)$ be an inclusion (by equivariance, see (2.3.12), we have, with $i(x) = x$, $\lambda(m, x) = \mu(m, x)$ for all $x \in X, m \in M$). We put

$$\chi_i : (Y, \lambda) \to \Omega = (\Lambda_M, \omega)$$
$$y \mapsto \{m : \lambda(m, y) \in X\} \qquad (9.1.24)$$

Again, $\chi_i(y)$ is an ideal because $k\chi_i(y) = \{km : \lambda(m, y) \in X\} \subset \chi_i(y)$ since $\lambda(km, y) = \lambda(k, \lambda(m, y))$ is in X as the operation of M on X maps X to itself and therefore $k \in M$ maps the element $\lambda(m, y) \in X$ to another element of X. Finally, we have $\chi_i(y) = M$ iff for all $m \in M$, $\lambda(m, y) \in X$ iff $\lambda(e, y) \in X$ iff $y \in X$. Thus, Ω is indeed the subobject classifier.

To put this example into perspective, let us observe that when M is the trivial monoid $\{e\}$, then the operation of M on any set is trivial. Hence, in this case $M - \textbf{Sets} = \textbf{Sets}$. Since in this case M possesses only two ideals, $\emptyset$ and M itself, Ω has two elements and reduces to the subobject classifier $\mathbf{2}$ in **Sets**. More generally, when the monoid M is a group G, then we have only two ideals, $\emptyset$ and G itself, and so Ω reduces again to $\mathbf{2}$, with the trivial G-action. In fact, the reason why the subobject classifier here is the same as in **Sets** is that whenever a subset X of Y is invariant under the action of G, then so is its complement: for $y \in Y \setminus X$, if we had $gy \in X$ for some $g \in G$, then also $y = g^{-1}gy \in X$, a contradiction. So, we can indeed proceed as in **Sets**. For a general monoid action on Y that leaves $X \subset Y$ invariant, in general $Y \setminus X$ is not invariant, and so, we send $y \in Y$ to the set, in fact ideal, L_y of all those $m \in M$ with $my \in X$. L_y is the maximal ideal M iff $y \in X$, and this then gives the subobject classifier. Conversely, $L_y = \emptyset$ if y is never moved into X by the action of M.

The above category $\textbf{Sets}^{\circlearrowright}$ of sets with endomorphisms represents a special case of $M - \textbf{Sets}$, with $M = \mathbb{N}_0$, the monoid of nonnegative integers with addition.

We now look at the category $\textbf{Sets}^{\mathbf{C}^{op}}$ of presheaves on $\mathbf{C}$, see Sect. 8.4. We need a

Definition 9.1.3 A *sieve* on an object C of $\mathbf{C}$ is a set S of arrows $f : . \to C$ (from an arbitrary object of $\mathbf{C}$ to C) with the following property. If $f \in S$, $f : D \to C$, then for any arrow $g : D' \to D$, also $f \circ g \in S$.

Thus a sieve is closed under precomposition. For instance, when the category is the power set $\mathcal{P}(B)$ of a set B, that is, the collection of subsets of B, with the morphisms being the inclusions $A' \subset A$, then a sieve S for A can be considered as a collection of subsets of A with the property that whenever $A_2 \subset A_1 \in S$, then also $A_2 \in S$. More generally, when the category is a poset $(\mathbf{P}, \leq)$, then a sieve S on $p \in \mathbf{P}$ can be identified with a collection of elements $q \leq p$ such that $r \leq q \in S$ implies also $r \in S$. —When the category is a monoid, then a sieve is simply a right ideal.

We have the following important functorial property. If S is a sieve on C, and $\phi : D \to C$ is a morphism, then $\phi^* S = \{g : D' \to D : \phi \circ g \in S\}$ is a sieve on D. That is, sieves can be pulled back under morphisms. Thus, a sieve satisfies the functorial property of a presheaf.

When we assign to each object C of $\mathbf{C}$ the so-called *total sieve* $\bar{S}(C) := \{f : . \to C\}$ of all morphisms into C, we obtain a presheaf on $\mathbf{C}$. In fact, this is the terminal object $\mathbf{1}$ in the category $\textbf{Sets}^{\mathbf{C}^{op}}$ of presheaves on $\mathbf{C}$. The reason is simply that any presheaf on $\mathbf{C}$ has to cover any arrow of $\mathbf{C}$. We also obtain a presheaf Ω on $\mathbf{C}$ by

$$\Omega(C) := \{S \text{ a sieve on } C\}, \tag{9.1.25}$$

and we then have a natural monomorphism

$$1 \rightarrowtail \Omega. \tag{9.1.26}$$

This, in fact, yields the subobject classifier for the category $\mathbf{Sets}^{\mathbf{C}^{op}}$. This is seen as follows. For any presheaf F on $\mathbf{C}$ and any subobject $U \rightarrowtail F$, we need to define $u : F \to \Omega$. For $C \in \mathbf{C}$ and $x \in FC$, we put

$$u_C(x) := \{f : D \to C : Uf(x) \in UD\}. \tag{9.1.27}$$

Thus, $u_C(x)$ is the sieve of arrows into C that pull back the element x of the set FC to an element of the subset UD of FD. In other words, we consider the collection of all pullbacks $Uf(x)$ and check for which arrows f they are in the set defined by the subpresheaf. When this happens for all f, then x, or more precisely, the collection of the $Uf(x)$, belongs to the subpresheaf U. This yields the condition for the subobject classifier. In other words, for each arrow $f : D \to C$, we ask whether $Uf(x)$ is contained in the subset UD of FD. Thus, for every arrow f, we have two truth values, "yes" or "no", "true" or "false". These values for different arrows are not independent, because when $y \in UD$, then for every arrow $g : E \to D$, necessarily $g^*(y) \in UE$, by the presheaf condition for U. The notion of a sieve allows us to keep track of these dependencies. The total sieve corresponds to the situation where the answer is "always true". The empty sieve corresponds to "always false". Other sieves mean "sometimes, but not always true".

Actually, the preceding examples are special cases of this one. For instance, the category $\mathbf{Sets}^{\to}$ is the category of presheaves over the category $\to$ consisting of two objects, called i_0 and i_1, with the morphisms $i_0 \to i_0, i_0 \to i_1, i_1 \to i_1$. A presheaf F is then described by the sets $F(i_0)$, $F(i_1)$ and a morphism $F(i_1) \to F(i_0)$. On the object i_0, we then have two sieves, the total sieve with the arrow $i_0 \to i_0$, and the empty sieve. On the object i_1, we have three sieves, the total sieve containing the arrows $i_1 \to i_1, i_0 \to i_1$, the sieve with the arrow $i_0 \to i_1$, and the empty sieve. These correspond to the cases (1), $(\frac{1}{2})$, (0), resp., in the above discussion of that example. The case of $\mathbf{Sets}$ is trivial. It is the presheaf category over the category 1 with the single object i_0 with a single morphism $i_0 \to i_0$ where we have two sieves over i_0, the one with the morphism $i_0 \to i_0$ and the empty sieve.

Of course, the preceding construction is an instantiation of the Yoneda lemma, and this allows us to repeat in a slightly different way what we have just said (if you do not like repetitions, simply jump ahead to the next section). By that result, elements of $\Omega(C)$ for the presheaf Ω correspond to morphisms (natural transformations) from $\mathrm{Hom}_{\mathbf{C}}(-, C)$ to Ω. Since Ω is supposed to be a subobject classifier, they thus have to correspond to subobjects of $\mathrm{Hom}_{\mathbf{C}}(-, C)$. A subfunctor F of $\mathrm{Hom}_{\mathbf{C}}(-, C)$ is given by a set of the form

$$S = \{f \in \mathrm{Hom}_{\mathbf{C}}(D, C) : f \in FD\} \text{ (where } D \text{ varies)}. \tag{9.1.28}$$

By the presheaf property, S is a sieve. This then leads to (9.1.25). The natural transformation from $\mathrm{Hom}_{\mathbf{C}}(-, C)$ to Ω then assigns the sieve $\phi^* S$ to a $\phi \in \mathrm{Hom}_{\mathbf{C}}(D, C)$.

Similarly, elements of $\boldsymbol{1}(C)$ correspond to morphisms (natural transformations) from $\mathrm{Hom}_{\mathbf{C}}(-, C)$ to $\boldsymbol{1}$. By the structure of $\boldsymbol{1}(C)$, this means that we associate to $\phi \in \mathrm{Hom}_{\mathbf{C}}(D, C)$ the pull-back of the total sieve $\bar{S}(C) = \mathrm{Hom}_{\mathbf{C}}(., C)$ under ϕ which is then the total sieve on the domain D of ϕ. The assignment of C to the total sieve $\mathrm{Hom}_{\mathbf{C}}(., C)$ is of course just what the Yoneda functor y does, in the sense that this yields the presheaf $D \mapsto \mathrm{Hom}_{\mathbf{C}}(D, C)$. Moreover, for any presheaf U on $\mathbf{C}$, there is a unique morphism $U \to \boldsymbol{1}$, assigning to each $C \in \mathbf{C}$ the total sieve $\boldsymbol{1}(C)$. Also, for a morphism $m : F \to \Omega$, the preimage of $\boldsymbol{1}$ (as a subpresheaf of Ω) at C consists of those $x \in FC$ for which $Uf(x) \in m^{-1}\boldsymbol{1}(D)$ for every morphism $f : D \to C$, that is, those $x \in FC$ that can be pulled back under any arrow. This is then a subpresheaf of F, as described in (9.1.27).

9.2   Topoi

We now come to the key concept of this chapter. This concept encodes the fundamental properties of categories of indexed families of sets, like presheaves. The concept of a topos turns out to be very important in geometry and logic, and we shall try to indicate its power. In fact, we shall give two – equivalent – definitions. The first one is general and lazy.

Definition 9.2.1 A *topos*[2] is a category $\mathbf{E}$ that possesses

1. all finite limits,
2. all finite colimits,
3. a subobject classifier, and
4. all exponentials.

It turns out, however, that this definition is somewhat redundant. For instance, it can be shown that the existence of colimits already follows from the other properties. We therefore now provide an alternative definition that lists the minimal requirements.

Definition 9.2.2 A *topos* is a category $\mathbf{E}$ that possesses

1. a terminal object $\boldsymbol{1}$,
2. a pullback for every diagram $A \to C \leftarrow B$,
3. a subobject classifier, that is, an object Ω with a monic arrow true : $\boldsymbol{1} \rightarrowtail \Omega$, with the property that for every monomorphism $A \rightarrowtail X$ in $\mathbf{E}$, there is a unique morphism $\chi_A : X \to \Omega$ with a commutative pullback diagram

---

[2] The origin of this word is the Greek $\tau \acute{o} \pi o \varsigma$, meaning "place, position", whose plural is $\tau \acute{o} \pi o \iota$, topoi.

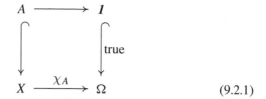

$$(9.2.1)$$

4. for each object X another object PX, called a *power object*, and a morphism $\epsilon_X : X \times PX \to \Omega$ with the property that for every morphism $f : X \times B \to \Omega$, there is a unique morphism $F : B \to PX$ making the following diagram commutative

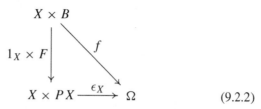

$$(9.2.2)$$

We first of all note that because of conditions 1 and 2, a topos in particular admits products, see the discussion after (8.2.34) Therefore, the last condition 4 is meaningful.

We recall the structural property Lemma 9.1.1 which then holds in any topos. Moreover, in a topos, any arrow can be factored into the composition of an epic and a monic arrow. In order to explain this, we say that an arrow $f : X \to Y$ has as image a monic $m = \operatorname{im} f : Z \to Y$ if f factors through $m : Z \to Y$ if $f = m \circ e$ for some $e : X \to Z$, and if m is universal in the sense that whenever f factors through some m', then so does m. As a diagram, this becomes

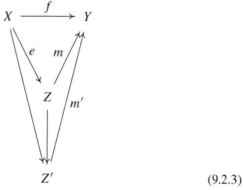

$$(9.2.3)$$

Lemma 9.2.1 *In a topos, every arrow f possesses an image m and factors as $f = m \circ e$ with an epic e.*

Condition 4, of course, is a special case of an exponential, that is,

$$PX = \Omega^X. \qquad (9.2.4)$$

In particular, we have

$$P\mathbf{1} = \Omega. \qquad (9.2.5)$$

Let us consider this last condition in the category **Sets**, recalling the above discussion of exponentials, see (8.2.61). We recall that in this category, $\Omega = \mathbf{2} = \{0, 1\}$. We put

$$\epsilon_X(x, A) := \begin{cases} 1 & \text{if } x \in A \\ 0 & \text{else,} \end{cases} \tag{9.2.6}$$

and for $f : X \times B \to \Omega$, we put

$$F(b) := \{x \in X : f(x, b) = 1\}. \tag{9.2.7}$$

Then $\epsilon_X \circ (1_X \times F)(x, b) = 1$ iff $x \in F(b) = \{\xi \in X : f(\xi, b) = 1\}$ iff $f(x, b) = 1$ so that the diagram (9.2.2) indeed commutes.

We can also express the last two requirements by

$$\mathrm{Sub}_\mathbf{E} X \cong \mathrm{Hom}_\mathbf{E}(X, \Omega) \tag{9.2.8}$$

and

$$\mathrm{Hom}_\mathbf{E}(X \times B, \Omega) \cong \mathrm{Hom}_\mathbf{E}(B, PX). \tag{9.2.9}$$

Thus, the subobject functor and the functor $\mathrm{Hom}_\mathbf{E}(X \times \cdot, \Omega)$ are both representable. In particular, combining these two isomorphisms, we have for $B = \mathbf{1}$

$$\mathrm{Sub}_\mathbf{E} X \cong \mathrm{Hom}_\mathbf{E}(X, \Omega) \cong \mathrm{Hom}_\mathbf{E}(\mathbf{1}, PX). \tag{9.2.10}$$

Therefore, a subobject A of X can be alternatively described as

a monic arrow $\quad A \rightarrowtail X \tag{9.2.11}$

a morphism $\quad \chi : X \to \Omega \tag{9.2.12}$

a morphism $\quad a : \mathbf{1} \to PX. \tag{9.2.13}$

In the category **Sets**, we have

$$\mathrm{Hom}_{\mathbf{Sets}}(\mathbf{1}, \Omega) = \Omega^{\mathbf{1}} = P\mathbf{1} \tag{9.2.14}$$

according to (9.2.5). This relationship, however, is not true in general. We consider the monoid M_2 and the category of $M_2 - $**Sets**. We have seen that the subobject classifier Ω of this category is the set of left ideals of M_2, that is,

| Monoid M_2 |

$$\Lambda_{M_2} = \{\emptyset, \{0\}, \{0, 1\}\} \tag{9.2.15}$$

with its natural M_2-action ω (9.1.23). Likewise, the terminal object $\mathbf{1}$ is the set $\{0\}$ with the trivial M_2-action λ_0.[3] Thus, when $h : \mathbf{1} \to \Omega$ is a morphism, then $h : \{0\} \to \Lambda_{M_2}$ is equivariant w.r.t. λ_0 and ω. Thus $\omega(0, h(0)) = h(\lambda_0(0), 0) = h(0)$. The key point now is that the value $h(0) = \{0\}$ is therefore not possible, because $\omega(0, \{0\}) = \{0, 1\} \neq \{0\}$. Thus, we have only two possible values, $h(0) = \{0, 1\}$ or $h(0) = \emptyset$. Thus,

$$\mathrm{Hom}_{M_2 - \mathbf{Sets}}(\mathbf{1}, \Omega) = \{\emptyset, \{0, 1\}\} \neq \Omega = \{\emptyset, \{0\}, \{0, 1\}\}. \tag{9.2.16}$$

---

[3] Here, we cannot avoid denoting the single element of $\mathbf{1}$ by 0, as the latter is the neutral element of M_2.

The reason for this discrepancy is that in general, the set of subobjects of X, $\mathrm{Sub}_{\mathbf{C}}(X) = \mathrm{Hom}_{\mathbf{C}}(X, \Omega)$, is only a set, but not an object of $\mathbf{C}$ itself, in contrast to $PX = \Omega^X$. One expresses this by saying that $\mathrm{Hom}_{\mathbf{C}}(X, \Omega)$ is external, while PX is internal to the category $\mathbf{C}$. Since these two things need not coincide, the external and the internal point of view of a category may lead to different results.

Often, one uses the notation

$$\top := \text{true},\tag{9.2.17}$$

as shall we do frequently. As remarked above, a topos possesses finite colimits, and so, in particular, it has an initial object $\mathbf{0}$. There is then a unique morphism

$$! : \mathbf{0} \to \mathbf{1}.\tag{9.2.18}$$

We can then define the arrow

$$\bot := \text{false} : \mathbf{1} \to \Omega\tag{9.2.19}$$

as the unique arrow for which

$$\tag{9.2.20}$$

is a pullback, that is, $\bot$ is the characteristic arrow of $!$, i.e., $\bot = \chi_!$. In the topos **Sets**, we have $\Omega = \{0, 1\}$, $\mathbf{0} = \emptyset$, $\mathbf{1} = \{1\}$, and

$$\top 1 = 1, \ \bot 1 = 0.\tag{9.2.21}$$

The reason is that in this case for no other object A of **Sets** than $\mathbf{0}$ is there such a commutative diagram

$$\tag{9.2.22}$$

and so, (9.2.19) trivially satisfies the pullback condition. Had we assigned the value 1 to $\bot 1$, however, then there would have been such diagrams, and the pullback property would not have been satisfied, because there are no morphisms $A \to \mathbf{0} = \emptyset$.

We return to the general case where we have the following important result.

Theorem 9.2.1 *(a) For any object X in a topos $\mathbf{E}$, the power object PX carries the structure of a Heyting algebra. In particular, the subobject classifier $\Omega = P1$ of a topos is a Heyting algebra.*

(b) For any object X of a topos **E***, the set of subobjects,* $\text{Sub}_{\mathbf{E}}(X) = \text{Hom}_{\mathbf{E}}(X, \Omega)$ *carries the structure of a Heyting algebra. In particular,* $\text{Hom}_{\mathbf{E}}(1, \Omega)$*, where 1 is the terminal object of the topos* **E***, is a Heyting algebra.*

Thus, whether we work internally (a) or externally (b), we obtain a Heyting algebra, but these two Heyting algebras may be different from each other as the preceding example shows. Also, either of them may or may not be a Boolean algebra. This will be important in Sect. 9.3 below. In any case, the key point of Theorem 9.2.1 is that it provides us with an algebraic structure—that of a Heyting algebra—on the subobject classifier and on the set of subobjects.

Let us sketch the relevant constructions for the proof of Theorem 9.2.1. The key point is to define the operations of meet, join, and implication in terms of appropriate pullback diagrams. We shall do this for the subobject classifier $\top : 1 \to \Omega$. First of all, $\neg = \chi_{\perp}$, the characteristic map of "false". That is,

$$
\begin{array}{ccc}
1 & \longrightarrow & 1 \\
{\scriptstyle \perp}\downarrow & & \downarrow{\scriptstyle \top} \\
\Omega & \xrightarrow{\ \neg = \chi_{\perp}\ } & \Omega
\end{array}
\tag{9.2.23}
$$

is a pullback diagram. Next, $\cap = \chi_{(\top,\top)}$ is the character of the product arrow $(\top, \top) : 1 \to \Omega \times \Omega$; the corresponding diagram is

$$
\begin{array}{ccc}
1 & \longrightarrow & 1 \\
{\scriptstyle (\top, \top)}\downarrow & & \downarrow{\scriptstyle \top} \\
\Omega \times \Omega & \xrightarrow{\ \cap = \chi_{(\top,\top)}\ } & \Omega
\end{array}
\tag{9.2.24}
$$

For $\cup$, we need the coproduct $\Omega + \Omega$ of Ω with itself (in **Sets**, the coproduct of two sets is simply their disjoint union; for the general definition, recall (8.2.47)). The two arrows $(\top, 1_{\Omega})$ and $(1_{\Omega}, \top)$ from Ω to $\Omega \times \Omega$ then define an arrow $u : \Omega + \Omega \to \Omega \times \Omega$, as in

$$
\begin{array}{ccccc}
\Omega & \longrightarrow & \Omega + \Omega & \longleftarrow & \Omega \\
& {\scriptstyle (\top, 1_{\Omega})}\searrow & \downarrow{\scriptstyle u} & \swarrow{\scriptstyle (1_{\Omega}, \top)} & \\
& & \Omega \times \Omega & &
\end{array}
\tag{9.2.25}
$$

Then $\cup = \chi_u$, as in

$$
\begin{array}{ccc}
\Omega + \Omega & \longrightarrow & 1 \\
{\scriptstyle u}\downarrow & & \downarrow{\scriptstyle \top} \\
\Omega \times \Omega & \xrightarrow{\ \cup = \chi_u\ } & \Omega
\end{array}
\tag{9.2.26}
$$

Finally, one needs to define an implication $\Rightarrow$ in terms of the other operations. One then needs to check that these operations satisfy the properties required for a Heyting algebra, see (2.1.67), (2.1.75)–(2.1.78) and Lemma (2.1.7).

Having constructed the Heyting algebra operations on Ω, we can then extend them to $\mathrm{Hom}_{\mathbf{E}}(X, \Omega)$: When we have arrows $f, g : X \to \Omega$, they define an arrow $(f, g) : X \to \Omega \times \Omega$ by the universal property of $\Omega \times \Omega$ (see the definition and discussion of products around (8.2.21) in Sect. 8.2), and we can then compose this with $\cap$, $\cup$ or $\Rightarrow$. And, of course, an arrow f can be composed with $\neg$ to obtain the arrow $\neg f : X \to \Omega$.

More specifically, we recall from the end of Sect. 9.1 that the subobject classifier for a presheaf category $\mathbf{Sets}^{\mathbf{C}^{\mathrm{op}}}$ is the presheaf Ω of sieves on the objects of $\mathbf{C}$. We can then check directly

Lemma 9.2.2 *The set $\Omega(C)$ of sieves on C is a Heyting algebra for each object C of the category $\mathbf{C}$, and hence so is the sieve presheaf Ω itself, by performing the Heyting algebra operations on each object C.*

Proof Each sieve is a set (of morphisms), and so, we can define the union and intersection of sieves. It is clear that unions and intersections of sieves are again sieves. The implication operator $S \Rightarrow S'$ is defined by

$$(f : D \to C) \in (S \Rightarrow S')(C) \qquad (9.2.27)$$

iff whenever, for $g : D' \to D$, $f \circ g \in S(C)$, then also $f \circ g \in S'(C)$.

To show that this is indeed a sieve we have to verify that if $f \in S \Rightarrow S'$ and $h : D'' \to D$, then also $f \circ h \in S \Rightarrow S'$. To check this: If in this situation $(f \circ h) \circ k \in S(C)$ for $k : D' \to D''$, then (9.2.27) with $g = h \circ k$ yields $(f \circ h) \circ k \in S'(C)$. This implication operator then satisfies

$$S_0 \subset (S \Rightarrow S') \text{ iff } S_0 \cap S \subset S' \qquad (9.2.28)$$

for all sieves S_0, as required for the implication operator by (2.1.59). $\qquad \square$

That there are many topoi is guaranteed by the following result which says that slice categories (introduced in Sect. 2.3, see (2.3.30)) over objects of topoi are again topoi.

Theorem 9.2.2 *For a topos $\mathbf{E}$ and an object E of $\mathbf{E}$, the slice category $\mathbf{E}/E$ of objects over E is also a topos.*

This result can be considered as a generalization of (8.4.1). We do not provide the complete proof (which is not difficult, but the construction of power objects is somewhat lengthy), but note only that the subobject classifier in $\mathbf{E}/E$ is simply given by $\Omega \times E \to E$ where Ω is, of course, the subobject classifier of $\mathbf{E}$ and the arrow is the projection onto the second factor.

This result also gives us an opportunity to reflect again the essential idea behind the concept of a topos. We do this very heuristically, but the preceding formal considerations will hopefully enable the reader to both understand and be able to make this precise. When X is a set, and $A \subset X$ is a subset, then for $x \in X$, we can check whether $x \in A$ or not. In other words, we can check whether the property A is true for x or not. Formally, this is what the subobject classifier Ω that consists of the two values 1 for true and 0 for false allows us to do. When, more generally, we look at a family X_e of sets, indexed by some E, with subsets $A_e \subset X_e$, and a section $x_e \in X_e$, we can then check for which $e \in E$, we have $x_e \in A_e$, that is, for which e the property A is true. The corresponding subobject classifier Ω_E which would consist of the individual subobject classifiers Ω_e then allows us to formalize this aspect that a property may hold for some, but perhaps not for all e. And when E then carries some additional relation structure, as expressed by a collection of morphisms, then the corresponding subobject classifier will also incorporate that structure. For that purpose, we have used concepts like sieves. And this is the reason why topoi are appropriate tools for versions of logic where assertions may be true only sometimes, under certain conditions or in certain situations. This is what we shall now turn to.

9.3   Topoi and Logic

We wish to first describe the main ideas of the present and the subsequent sections. The operators of classical propositional logic, $\wedge$ (and), $\vee$ (or), $\sim$ (not) and $\supset$ (implies),[4] obey the rules of a Boolean algebra, which are the same as for the set theoretical operations $\cap$ (intersection), $\cup$ (union), $\neg$ (complement) and $\Rightarrow$ (implies) in the power set $\mathcal{P}(X)$ of some set X. In particular, we may consider the simplest Boolean algebra $\{0, 1\}$, which we can identify with the power set of a 1-element set. Therefore, when one has a collection of logical propositions, formed from letters by the above logical operations, one considers so-called valuations that map sentences to elements of such a Boolean algebra, preserving the above correspondences between logical and Boolean operations. When the logical operations are also considered as constituting a Boolean algebra, such a valuation would then be a homomorphism of Boolean algebras in the sense of Sect. 2.1.4. The realm of logical formulas, however, is typically infinite whereas the range of a valuation usually is a finite Boolean algebra, like $\{0, 1\}$. Nevertheless, the issue of interest for classical propositional logic is whether a formula is true (i.e., can be derived from the axioms) or not, that is, a simple binary distinction. The sentences that can be derived within classical propositional logic then correspond to those that are mapped to 1 in the Boolean algebra by such a valuation. In fact, one may take any Boolean algebra here, for instance $\{0, 1\}$ or the power set $\mathcal{P}(X)$ of some set X. In particular, the law of the excluded middle $\alpha \vee \sim \alpha$ for any α in the latter case then corresponds to $A \cup X \backslash A = X$ for any subset A of a set X. Likewise, the equivalent

---

[4] The symbol $\supset$ (implies) is *not* the reverse of the symbol $\subset$ (subset). It rather goes in the same direction. If $A \subset B$, then $(x \in A) \supset (x \in B)$. I hope that this will not cause much confusion.

formulation $\sim\sim \alpha = \alpha$ corresponds to the fact that each $A \subset X$ equals the complement of its complement, i.e., $X \backslash (X \backslash A) = A$.

Now in intuitionistic logic, the law of the excluded middle is no longer accepted. Therefore, the logical operations of intuitionistic logic only correspond to those of a Heyting algebra. Again, there is a topological version of this, the collection $\mathcal{O}(X)$ of open subsets of a topological space X. In a Heyting algebra, we only have a pseudocomplement instead of a complement, and in $\mathcal{O}(X)$, this is the interior $(X \backslash A)^\circ$ of the set theoretical complement $X \backslash A$. In particular, in general $A \cup (X \backslash A)^\circ$ may not be all of X, but a proper open subset of it. Thus, for intuitionistic logic, one naturally considers Heyting algebra valuations. In contrast to the classical Boolean case, here it no longer suffices to take a single Heyting algebra, but one rather needs to consider all of them when one wants to check via valuations whether a sentence can be derived in intuitionistic logic. Following Kripke (who essentially revived an old idea of Leibniz), one then considers a poset $\mathbf{P}$ of so-called possible worlds, where $p \leq q$ is interpreted as the world q being accessible from p, or a possible successor of p. One then requires for a statement to hold at p, it has to continue to hold at all q with $p \leq q$. Likewise, for $\sim \alpha$ to hold at p, α must not hold at any q with $p \leq q$. In particular, the law of the excluded middle does not hold here, and we are in the realm of intuitionistic logic.

The poset here could again be a power set $\mathcal{P}(X)$, or $\mathcal{O}(X)$ for a topological space X, but the role will now be different. We have the natural association of $p \in \mathbf{P}$ to the set $A_p := \{r \in \mathbf{P} : r \leq p\}$, but this does not define a presheaf on $\mathbf{P}$, because it is not contravariant, but rather covariant, that is, $A_p \subset A_q$ when $p \leq q$. Thus, it rather defines a presheaf on $\mathbf{P}^{\mathrm{op}}$. This is easily remedied, however, by taking instead $F_p := \{q \in \mathbf{P} : p \leq q\}$. This does indeed yield a presheaf on $\mathbf{P}$. (There may be a danger of confusion here with the Yoneda lemma. Yoneda would tell us to construct a presheaf from the A_p via $r \mapsto \mathrm{Hom}(r, p)$, that is, in this special case of a poset, assigning to r the unique element $r \rightarrow p$ of $A_p(r)$ in case $r \leq p$, and $\emptyset$ else. Conversely, F_p arises as $p \rightarrow \bigcup_q \mathrm{Hom}(p, q)$ from the Yoneda functors $p \mapsto \mathrm{Hom}(p, q)$ for each q.)

Here, now, comes the main point. We consider variables whose range of values is indexed by the members of our poset $\mathbf{P}$. That is, the variable ranges, called types, are presheaves (in fact, sheaves, but we'll come to that point) on $\mathbf{P}$. That is, the range of possible values of a variable depends on the possible world p in a functorial manner. In particular, for formulas, the type consists of truth values, that is, it is the subobject classifier presheaf Ω. In particular, a formula can be true in some possible world p, and then has to remain true in all worlds subsequent to p, but may be false in others. Again, in this setting, the law of the excluded middle does not need to hold, and we are in the realm of intuitionistic logic. Operations like inserting a variable x into a formula $\phi(.)$ then become morphisms of presheaves $X \rightarrow \Omega$ where X is the type of x. Likewise substituting a particular value a for a variable x is described by a morphism $\mathbf{1} \rightarrow X$. The point, again, is that this has to be done over each possible world p, in a manner that respects the morphisms $p \leq q$.

In this section, we shall only give an outline and omit most proofs. A more detailed treatment can be found [42] which we shall also follow here in various places. A good reference for logic is [26].

We start with classical propositional logic (PL). There, we have an alphabet $\Pi_0 = \{\pi_0, \pi_1, \dots\}$ consisting of propositional variables (letters), the symbols $\sim, \wedge, \vee, \supset$ and the brackets (,). From these, we can form sentences by taking letters and applying the above operations to them. The collection of sentences is denoted by Π. Classical logic (CL) has a certain collection of axioms, that is, sentences that are considered to be universally valid, and one rule of inference. The axioms may be different in different treatises, but those different collections are all equivalent, hopefully. We list here the following collection:

1. $\beta \supset (\alpha \supset \beta)$
2. $(\alpha \supset (\beta \supset \gamma)) \supset ((\alpha \supset \beta) \supset (\alpha \supset \gamma))$
3. $(\alpha \wedge \beta) \supset \alpha$
4. $(\alpha \wedge \beta) \supset \beta$
5. $(\alpha \supset \beta) \supset ((\alpha \supset \gamma) \supset (\alpha \supset (\beta \wedge \gamma)))$
6. $\alpha \supset (\alpha \vee \beta)$
7. $\beta \supset (\alpha \vee \beta)$
8. $(\alpha \supset \gamma) \supset ((\beta \supset \gamma) \supset ((\alpha \vee \beta) \supset \gamma))$
9. $(\alpha \supset \beta) \supset (\sim \beta \supset \sim \alpha)$
10. $\alpha \supset \sim\sim \alpha$
11. $\sim\sim \alpha \supset \alpha$

The inference rule is modus ponendo ponens (usually simply called *modus ponens*):

From the sentences α and $\alpha \supset \beta$, we can derive the sentence β.

One should note that this inference rule does not have the same content as the axiom $(\alpha \wedge (\alpha \supset \beta)) \supset \beta$ would have. The inference rule only says that when we can *derive* α and $\alpha \supset \beta$ within our formalism, then we get β for free. In other words, the inference rule has a status different from an axiom.

The CL–theorems, in symbols

$$\vdash_{\mathrm{CL}} \alpha, \tag{9.3.1}$$

are those sentences α that can be derived from the axioms via applications of the inference rule and substitutions.[5] A substitution or insertion consists in replacing in a sentence containing a letter α that letter by some sentence. Thus, the inference rule enables us to derive theorems from the axioms with the help of substitutions.

With this notation, the modus ponens says that

$$\vdash_{\mathrm{CL}} \alpha \text{ and } \vdash_{\mathrm{CL}} (\alpha \supset \beta) \text{ imply } \vdash_{\mathrm{CL}} \beta$$

---

[5] The symbol $\vdash$ is often pronounced as "entail"; its shape also suggests calling it "turnstile".

which is different from the axiom

$$\vdash_{CL} (\alpha \wedge (\alpha \supset \beta)) \supset \beta).$$

Intuitionistic logic (IL) accepts the first 10 axioms and the rule of inference, but not the 11th axiom, i.e., it rejects the law of the excluded middle $\alpha \vee \sim \alpha$. The IL-theorems are expressed by

$$\vdash_{IL} \alpha. \tag{9.3.2}$$

Semantics then consists in assigning truth values to sentences. In topos logic, these truth values will be elements of the subobject classifier Ω of a topos $\mathbf{E}$, or in $H = \operatorname{Hom}_{\mathbf{E}}(1, \Omega)$. The simplest case is **Sets** where Ω is the Boolean algebra $2 = \{0, 1\}$. The essential point for the general case is given by Theorem 9.2.1 which says that every such subobject classifier carries the structure of a Heyting algebra. We thus let $(H, \sqsubset)$ be a Heyting algebra, with the operations of meet, join, implication, and pseudo-complement denoted by $\sqcap, \sqcup, \Rightarrow, \neg$, to distinguish them from the above logical operations.

For the version of topos logic that we shall explore in the sequel, the relevant Heyting algebra is

$$H = \operatorname{Hom}_{\mathbf{E}}(\mathbf{1}, \Omega), \tag{9.3.3}$$

see Theorem 9.2.1.

Definition 9.3.1 An *H-valuation*, where H is a Heyting algebra, is given by a function $V : \Pi_0 \to H$ that is extended to a function $V : \Pi \to H$ by the rules

1. $V(\sim \alpha) = \neg V(\alpha)$
2. $V(\alpha \wedge \beta) = V(\alpha) \sqcap V(\beta)$
3. $V(\alpha \vee \beta) = V(\alpha) \sqcup V(\beta)$
4. $V(\alpha \supset \beta) = \neg V(\alpha) \sqcup V(\beta) = V(\alpha) \Rightarrow V(\beta)$.

A sentence $\alpha \in \Pi$ is called *H-valid*, or an *H-tautology*, in symbols

$$H \models \alpha \tag{9.3.4}$$

if

$$V(\alpha) = 1 \text{ for every } H\text{-valuation.} \tag{9.3.5}$$

In particular, α is called *classically valid* if it is valid for the Boolean algebra $\{0, 1\}$.

We then have the following fundamental result.

Theorem 9.3.1 *The following four statements are equivalent.*

1.

$$\vdash_{CL} \alpha. \tag{9.3.6}$$

2. α is classically valid.
3. α is B-valid for some Boolean algebra $(B, \sqsubset)$.
4. α is B-valid for any Boolean algebra.

That CL-theorems are classically valid, i.e., the implication $1 \Rightarrow 2$ is called *soundness*, while the reverse implication $2 \Rightarrow 1$ is called *completeness* of the above classical system of axioms. Soundness, in contrast to completeness, is rather easy to prove. In turn, completeness is the more useful part of the result. In order to check whether a formula can be derived, one has to come up with a derivation of it which may require some ingenuity. In contrast, classical validity can be verified with the help of truth tables which, in principle, is a mechanical (although in practice typically lengthy) procedure.

Similarly, concerning *intuitionistic logic*, we have

Theorem 9.3.2 *The following statements are equivalent.*

1.

$$\vdash_{\text{IL}} \alpha. \tag{9.3.7}$$

2. α *is H-valid for any Heyting algebra H.*

We observe that Theorem 9.3.2 is not as strong as Theorem 9.3.1 insofar as here it does not suffice to check validity for a single algebra, like the classical Boolean algebra $\{0, 1\}$, but that one rather has to check validity for all Heyting algebras. In fact, this is a key insight into intuitionistic logic that emerged from the work of Kripke.

When the quantifiers $\exists, \forall$ are included, one is led to *first-order logic*. We shall not go into the details, but rather move on to the connection between topoi and logic. For this purpose, it will be insightful to discuss the approach of Kripke to intuitionistic and modal logic. One starts with a category **P** of so-called *possible worlds*. $p \to q$ for objects p, q of **P** means that the world q can be reached from the world p, i.e., that it is a possible successor of p. (For this interpretation, it may be helpful to assume that **P** is a poset, or at least to assume that between any two objects, there exists at most one morphism.) We then consider a presheaf F on $\mathbf{P}^{\text{op}}$, that is, an element of $\mathbf{Sets}^{\mathbf{P}}$. The interpretation is that we assign to every world p a set F_p of possible ranges of variables, or of possible states. Thus, when $p \to q$, we get an arrow $F_{pq} : F_p \to F_q$, with $F_{pp} = 1_{F_p}$ and $F_{pr} = F_{qr} \circ F_{pq}$ for $p \to q \to r$. Moreover, we assume that for each p, we have a set of possible relations between variables that are preserved under arrows. That is, if R_p is, for instance, a binary relation, then

$$x R_p y \text{ implies } F_{pq}(x) R_q F_{pq}(y) \text{ for } p \to q. \tag{9.3.8}$$

We also assume that for each p, we can verify for which values $v^1, \ldots, v^m \in F_p$ of variables $x^1, \ldots, x^m$ a formula $\phi(x^1, \ldots, x^m)$ is true. We'll have to return to this issue, but for the moment, we simply write this as

$$\mathcal{M}_p \models \phi(v^1, \ldots, v^m). \tag{9.3.9}$$

Such a formula could, for instance, express a relation $v^1 R_p v^2$. In fact, such relations and identities $v \approx u$ comprise the so-called atomic formulae. In other words, atomic formulae must not contain the logical symbols $\wedge, \vee, \supset, \sim$ or the quantifiers $\exists, \forall$.

We then define the *validity* at p in **P** of a formula in several steps, denoting it by

$$\mathcal{M} \models_p \phi(v^1, \ldots, v^m), \tag{9.3.10}$$

in terms of validities $\mathcal{M}_p \models \phi(v^1, \ldots, v^m)$ (note the different position of the subscript p here and in (9.3.10). We use the abbreviation $v_q := F_{pq}v$, and we adopt the convention that a variable with a superscript i can only be inserted into the ith slot of a formula.

1. For an atomic formula, $\mathcal{M} \models_p \phi(v^1, \ldots, v^m)$ iff $\mathcal{M}_p \models \phi(v^1, \ldots, v^m)$.
2. $\mathcal{M} \models_p \phi \wedge \psi(v^1, \ldots, v^m)$ iff $\mathcal{M} \models_p \phi(v^1, \ldots, v^m)$ and $\mathcal{M} \models_p \psi(v^1, \ldots, v^m)$.
3. $\mathcal{M} \models_p \phi \vee \psi(v^1, \ldots, v^m)$ iff $\mathcal{M} \models_p \phi(v^1, \ldots, v^m)$ or $\mathcal{M} \models_p \psi(v^1, \ldots, v^m)$.
4. $\mathcal{M} \models_p \sim \phi(v^1, \ldots, v^m)$ iff whenever $p \to q$, not $\mathcal{M} \models_q \phi(v_q^1, \ldots, v_q^m)$.
5. $\mathcal{M} \models_p \phi \supset \psi(v^1, \ldots, v^m)$ iff whenever $p \to q$, if $\mathcal{M} \models_q \phi(v_q^1, \ldots, v_q^m)$, then $\mathcal{M} \models_q \psi(v_q^1, \ldots, v_q^m)$.
6. $\mathcal{M} \models_p \exists w^i \phi(v^1, \ldots, v^{i-1}, w^i, v^{i+1}, \ldots, v^m)$ iff for some $v^i \in F_p$, $\mathcal{M} \models_p \phi(v^1, \ldots, v^i \ldots, v^m)$.
7. $\mathcal{M} \models_p \forall w^i \phi(v^1, \ldots w^i, \ldots, v^m)$ iff whenever $p \to q$, for all $v \in F_q$, $\mathcal{M} \models_q \phi(v_q^1, \ldots, v_q^{i-1}, v, v_q^{i+1}, \ldots, v_q^m)$.

The important items here are the negation, the implication and the all quantifier. For those, we not only require validity at the instance (world) p, but also at all later instances (accessible worlds) q.

We now return to the validities $\mathcal{M}_p$ that were employed in (9.3.10). We consider the set $\mathbf{P}^+$ of hereditary subsets of **P**; here a hereditary subset S has to satisfy

$$\text{if } p \in S \text{ and } p \to q, \text{ then also } q \in S. \tag{9.3.11}$$

We observe the analogy with the Definition 9.1.3 of a sieve—such a hereditary S is simply a sieve on $\mathbf{P}^{\mathrm{op}}$. As in the proof of Lemma 9.2.2, $\mathbf{P}^+$ is a Heyting algebra, and so it can receive a valuation $V : \Pi \to \mathbf{P}^+$. For such a valuation, for a sentence $\phi \in \Pi$, $V(\phi)$ is then considered as the set of those worlds where ϕ is valid, and since the elements of the Heyting algebra $\mathbf{P}^+$ satisfy (9.3.11), whenever ϕ is valid at p, it then remains valid at all q with $p \to q$, that is, in all worlds that can be reached from the world p. This is then the key point of Kripke's possible world semantics of intuitionistic logic. A model $\mathcal{M}$ is then defined to be such a pair $(\mathbf{P}, V)$ with a valuation $V : \Pi \to \mathbf{P}^+$.

Let us consider a simple example: **P** is the arrow category $\to$ with two objects i_0, i_1 and three morphisms $i_0 \to i_0, i_0 \to i_1, i_1 \to i_1$. Assume that for some sentence ϕ, we have $V(\phi) = \{i_1\}$, that is, ϕ is valid at i_1, but not at i_0. The latter is expressed in symbols as $\mathcal{M} \not\models_{i_0} \phi$. Since $\mathcal{M} \models_{i_1} \phi$ and we have the arrow $i_0 \to i_1$, then also $\mathcal{M} \not\models_{i_0} \sim \phi$. Therefore

$$\mathcal{M} \not\models_{i_0} (\phi \vee \sim \phi), \tag{9.3.12}$$

that is, the *law of the excluded middle* does not hold here – we are in the realm of intuitionistic logic. We have, however, $\mathcal{M} \models_{i_0} \sim\sim \phi$ since $\mathcal{M} \not\models_{i_0} \sim \phi$ and $\mathcal{M} \not\models_{i_1} \sim \phi$, and hence also

$$\mathcal{M} \not\models_{i_0} (\sim\sim \phi \Rightarrow \phi). \tag{9.3.13}$$

Definition 9.3.2 The sentence $\phi \in \Pi$ is *valid* for the valuation $V : \Pi \to \mathbf{P}^+$ if it is valid for every $p \in \mathbf{P}$, that is, if $V(\phi)$ is the total hereditary set $\mathbf{P}$ itself.

ϕ is valid in the sense of Kripke if it is valid for all such valuations V.

In the terminology of Definition 9.1.3, thus the sentence $\phi \in \Pi$ is valid for the valuation $V : \Pi \to \mathbf{P}^+$ if $V(\phi)$ is the total sieve of all elements of $\mathbf{P}$.

For propositional sentences (i.e., ones that do not involve the quantifiers $\exists, \forall$), we have

Theorem 9.3.3 *Let $\mathbf{P}$ be a poset. Then for any propositional sentence $\alpha \in \Pi$,*

$$\mathbf{Sets}^{\mathbf{P}} \models \alpha \text{ iff } \mathbf{P} \models \alpha, \tag{9.3.14}$$

where on the left-hand side, we have validity in the sense of the topos $\mathbf{Sets}^{\mathbf{P}}$ (see 9.3.3) whereas on the right hand side, we have validity in the sense of Kripke.

Later on, in 9.6, we shall put this result into a more general context, and we shall shift the perspective. Topos validity will be the main concern, and the Kripke rules will then emerge as an example of how to describe the validity in a special topos, $\mathbf{Sets}^{\mathbf{P}}$ in the present case. We shall consider more general topoi, in particular those of sheaves on a site (as defined in the sections to follow), that allow for local amalgamations of formulae.

9.4   Topoi Topologies and Modal Logic

Definition 9.4.1 A *(Lawvere-Tierney) topology*, also called a *local operator*, on a topos $\mathbf{E}$ with subobject classifier Ω is given by a morphism $j : \Omega \to \Omega$ satisfying

(i)
$$j \circ \text{true} = \text{true} \tag{9.4.1}$$

(ii)
$$j \circ j = j \tag{9.4.2}$$

(iii)
$$j \circ \wedge = \wedge \circ (j \times j) \tag{9.4.3}$$

where $\wedge$ is the meet operator in the Heyting algebra Ω (see Theorem 9.2.1).

The morphism $j : \Omega \to \Omega$ then classifies a subobject J of Ω, that is, we get a commutative diagram

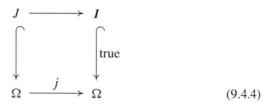

$$(9.4.4)$$

A good example of such a j is the double negation operator $\neg\neg$. Of course, there also exist trivial examples.

In order to understand this definition, let us start by considering a topological space $(X, \mathcal{O}(X))$. Let $\mathcal{U} := (U_i)_{i \in I}$ be a collection of open subsets of X, and let $j_0(\mathcal{U})$ be the collection of all open subsets of X contained in $\bigcup_i U_i$, that is, all open sets covered by $\mathcal{U}$. We then have

$$j_0(\mathcal{O}(X)) = \mathcal{O}(X) \tag{9.4.5}$$

and

$$j_0(j_0(\mathcal{U})) = j_0(\mathcal{U}). \tag{9.4.6}$$

Also,

$$j_0(\mathcal{U}_1 \cap \mathcal{U}_2) \subset j_0(\mathcal{U}_1) \cap j_0(\mathcal{U}_2) \tag{9.4.7}$$

for two such collections. Here, $\mathcal{U}_1 \cap \mathcal{U}_2$ consists of those open sets that are contained in both $\mathcal{U}_1$ and $\mathcal{U}_2$, and these may be few; in fact, $\mathcal{U}_1 \cap \mathcal{U}_2$ may well be empty even though, for instance, $\bigcup_{V \in \mathcal{U}_1} V$ might equal $\bigcup_{V \in \mathcal{U}_2} V$, i.e., the two collections cover the same open set, but possibly utilizing different sets. In particular, in general, the inclusion in (9.4.7) may be strict. However, if $\mathcal{U}_\alpha, \alpha = 1, 2$, are sieves (we recall that $\mathcal{U}$ is a sieve if whenever $V \subset U$ for some $V \in \mathcal{O}(X)$ and $U \in \mathcal{U}$, then also $V \in \mathcal{U}$), then we have equality in (9.4.7).

After this preparation, we look at the topos $\mathbf{Sets}^{\mathcal{O}(X)^{\mathrm{op}}}$. The subobject classifier in this topos is the presheaf that assigns to each $U \in \mathcal{O}(X)$ the set $\Omega(U)$ of all sieves on U, see (9.1.26). The terminal object $\mathbf{1}$ assigns to U the total sieve $\bar{S}(U)$ of all open subsets of U, and true $: \mathbf{1} \rightarrowtail \Omega$ is then the inclusion, see the construction leading to (9.1.27). We now define a subobject J of Ω by

$$J(U) := \{S \text{ a sieve on } U \text{ with } U = \bigcup_{V \in S} V\}. \tag{9.4.8}$$

Also, when U_0 is an open subset of U, we then have $U_0 = \bigcup_{V \in S_0} V$ where S_0 is the sieve consisting of all $V \cap U_0$, $V \in S$. Thus, the covering property of (9.4.8) is preserved under restrictions to open subsets, and J therefore is a subfunctor of Ω. The classifying morphism $j : \Omega \to \Omega$ of this subfunctor J then assigns to a sieve S_U on an open subset U the total sieve

$$j(S_U) := \bar{S}(U') \text{ with } U' = \bigcup_{V \in S_U} V \tag{9.4.9}$$

of all the open sets covered by S_U. Since j maps total sieves to total sieves, property (9.4.1) is satisfied, and (9.4.2) holds for the same reason. (9.4.3) is also easily verified, recalling the discussion of equality in (9.4.7). Thus, j defines a topology on the topos $\mathbf{Sets}^{\mathcal{O}(X)^{\mathrm{op}}}$.

Given such a topology j on a topos $\mathbf{E}$, and a subobject A of an object X with characteristic morphism χ_A, we then define the closure $\overline{A}^j$ (we use the superscript j here not only to indicate the specific topology used, but more importantly to distinguish this closure from the closure operator in topology (see Definition 4.1.9 and Theorem 4.1.2), as the two in general are different—and, in fact, are defined in different circumstances) as the subobject of X with characteristic morphism $j \circ \chi_A$, that is,

$$\chi_{\overline{A}^j} = j \circ \chi_A. \tag{9.4.10}$$

In fact, we can perform this construction for any arrow j that need not define a topology. One can show that j defines a topology iff this closure operator satisfies

$$A \subset \overline{A}^j, \ \overline{\overline{A}^j}^j = \overline{A}^j, \ \overline{A \cap B}^j = \overline{A}^j \cap \overline{B}^j \tag{9.4.11}$$

for all subobjects A. We see here a difference to the topological closure operator of Theorem 4.1.2. There, the closure commuted with unions, but here, the closure operator from a topos topology j commutes with intersections instead.

We now turn to *modal logic*. In a modal logic, ordinary logic (classical or intuitionistic) is amplified by some *modal operator*. This operator can, for instance, express knowledge (epistemic modality), belief (doxastic), obligation (deontic), or necessity and possibility (alethic). We denote such a modal operator by ∇. So, for instance, in the epistemic modality, $\nabla \alpha$ stands for "the sentence α is known". When the setting involves agents, we may also have $\nabla_i \alpha$ expressing that "agent i knows α". Here, however, we shall work with a different modal operator, and we shall read $\nabla \alpha$ as "it is locally the case that α". Here, "locally" will refer to a topology, as will be explained in a moment. The modality that we shall investigate in the sequel is also called "geometric modality".

Definition 9.4.2 A *modal operator* is an operator

$$\nabla : \Pi \to \Pi \tag{9.4.12}$$

where Π is the collection of sentences of our logic, satisfying the following conditions

$$\nabla(\alpha \supset \beta) \supset (\nabla \alpha \supset \nabla \beta) \tag{9.4.13}$$

$$\alpha \supset \nabla \alpha \tag{9.4.14}$$

$$\nabla \nabla \alpha \supset \nabla \alpha. \tag{9.4.15}$$

These axioms are different from the ones in classical modal logic for the knowledge or other operators. For instance, when $\nabla \alpha$ meant that "the agent knows α", then (9.4.13) would say that when the implication $\alpha \supset \beta$ is known, then, when the agent knows α, she also knows β. Also, by (9.4.14) she would know everything that is true, and when she knew that she knows

α, then by (9.4.15), she would also know α. These properties are not so desirable for a knowledge operator, however, and in modal logic (see e.g. [53]), rather than (9.4.14), (9.4.15), one usually requires the converse implications $\nabla\alpha \supset \alpha$ and $\nabla\alpha \supset \nabla\nabla\alpha$, that is, only true statements can be known, and when one knows something, one also knows that one knows it.

In fact, the conditions (9.4.13) and (9.4.14) can be replaced by

$$(\alpha \supset \beta) \supset (\nabla\alpha \supset \nabla\beta) \tag{9.4.16}$$

$$\nabla(\alpha \supset \alpha) . \tag{9.4.17}$$

Again, for the knowledge operator, this would mean that logical implications lead to implications between knowledges, and all tautologies would be known, resp. Again, these conditions are not desirable for a knowledge operator.

Let $\mathbf{E}$ be a topos equipped with a local operator j. If $V : \Pi_0 \to \mathrm{Hom}_{\mathbf{E}}(1, \Omega)$ is a valuation, we can extend it to Π by using the semantic rules of Definition 9.3.1 plus

$$V(\nabla\alpha) = j \circ V(\alpha) \tag{9.4.18}$$

for all sentences α.

9.5   Topologies and Sheaves

In order to understand the preceding better, we now discuss an approach to topologies that constitutes an alternative to Definition 9.4.1.

Definition 9.5.1 Let $\mathbf{C}$ be a category. A *(Grothendieck) topology* on $\mathbf{C}$ assigns to each object C a collection $J(C)$ of sieves with the following properties

(i) The total sieve $\bar{S}(C)$ of all arrows $f : D \to C$ is a member of $J(C)$, for every C.
(ii) If $S \in J(C)$ and $h : D \to C$ is an arrow, then $h^*(S) \in J(D)$.
(iii) If for some $S \in J(C)$ and some other sieve S' on C, for every arrow $h : D \to C$ in S, $h^*(S') \in J(D)$, then also $S' \in J(C)$.

When a sieve S is contained in $J(C)$ we say that S *covers* C. And we say that the sieve S on C covers the arrow $f : D \to C$ if $f^*(S) \in J(D)$. The sieve S on C is called *closed*[6] if it contains all the arrows that it covers, i.e., if $f : D \to C$ is covered by S, then $f \in S$.

A *site* is a small category equipped with a topology J.

Again, this is motivated by the example of the category $\mathcal{O}(X)$ of open subsets of a topological space X, the arrows $V \to U$ being the inclusions

---

[6]This condition is different from the closedness condition in topology and hopefully will not be confused with it.

$V \subset U$. We recall that a sieve S on U is a collection of open subsets of U such that whenever $W \subset V \in S$, then also $W \in S$. We then say that a sieve S covers U if

$$U \subset \bigcup_{V \in S} V. \qquad (9.5.1)$$

We then obtain a Grothendieck topology by letting $J(U)$ be the collection of all sieves on U that cover U. (Note that not every cover of U, that is any collection R of subsets V of U with $U \subset \bigcup_{V \in R} V$ is a sieve, but it generates the sieve of all V' with $V' \subset V$ for some $V \in R$.) Condition (i) is obvious: The collection of all open subsets of U obviously covers U. Condition (ii) means that when a sieve S covers U, then any open $U' \subset U$ is covered by the collection $V \cap U'$, $V \in S$. Finally, (iii) says that when a sieve S' covers any member V of a sieve S covering U, then S' also covers U, as $U \subset \bigcup_{V \in S} V$. Also, closedness of S then simply means that whenever $U' \subset \bigcup_{V \in S} V$, then also $U' \in S$. The latter makes the fundamental link with the sheaf concept below: Whenever something holds on all $V \in S$ in a compatible manner, that is, matches on all intersections $V_1 \cap V_2$ for $V_1, V_2 \in S$, then it also has to hold on any $U' \subset \bigcup_{V \in S} V$.

As always, there exists a trivial topology. Here, this is the topology where a sieve S covers C iff $1_C \in S$. In other words, the only sieve covering C is the total sieve $\bar{S}(C)$. Obviously, such a topology exists on every category, in particular on $\mathcal{O}(X)$ for a topological space X. In order to avoid confusion, we should point out that this is not the same as what might be called a trivial topological space in topology. For such a trivial topological space, $\mathcal{O}(X)$ would consist of X and $\emptyset$ only (this is also called the indiscrete topology on the set X, see 4.1).

In fact, the Definitions 9.4.1 and 9.5.1 are equivalent in the following situation.

Theorem 9.5.1 *Let* **C** *be a small category. Then the Grothendieck topologies J on* **C** *correspond to the Lawvere-Tierney topologies on* **Set$^{\mathbf{C^{op}}}$**.

We sketch the

Proof We have seen at the end of 9.1 that the subobject classifier for the topos **Set$^{\mathbf{C^{op}}}$** is given by

$$\Omega(C) = \{S : S \text{ a sieve on } C\}. \qquad (9.5.2)$$

Therefore, given a Grothendieck topology J on **C**, we define

$$j_C(S) := \{g : D \to C : g^*S \in J(D)\}. \qquad (9.5.3)$$

Then $j_C(S)$ is also a sieve on C, and the operator $j : \Omega \to \Omega$ satisfies the conditions of Definition 9.4.1 as is readily checked.

Conversely, a Lawvere-Tierney operator $j : \Omega \to \Omega$ classifies a subobject J of Ω, as in (9.4.4). In fact

$$S \in J(C) \text{ iff } j_C(S) = \bar{S}(C) \text{ (the total sieve on } C). \qquad (9.5.4)$$

Again, one checks that this yields a Grothendieck topology.

These two constructions are inverses of each other. In fact, when for a Grothendieck topology J, $S \in J(C)$, then by condition (ii), $j_C(S)$ in (9.5.3) is the total sieve. Conversely, if in (9.5.4), for a sieve S, $j_C(S)$ is the total sieve, then by condition (iii), S has to be in $J(C)$.

Now let J be a Grothendieck topology on $\mathbf{C}$. For an object C, consider $g_1, g_2 \in S \in J(C)$ for arrows $g_1 : D_1 \to C$, $g_2 : D_2 \to C$. We let $D_1 \times_C D_2$ be the pullback of these two arrows, i.e.,

$$
\begin{array}{ccc}
D_1 \times_C D_2 & \xrightarrow{\ g_1^2\ } & D_1 \\
\ \downarrow{\scriptstyle g_2^1} & & \ \downarrow{\scriptstyle g_1} \\
D_2 & \xrightarrow{\ g_2\ } & C
\end{array}
\qquad (9.5.5)
$$

commutes and is universal.

Definition 9.5.2 Let $F \in \mathbf{Sets}^{\mathbf{C}^{\mathrm{op}}}$ be a presheaf. F is called a *sheaf* on the site $(\mathbf{C}, J)$ if it satisfies the following condition. Whenever $S \in J(C)$, if for any collection of elements $x_i \in FD_i$ for arrows $g_i : D_i \to C$ in S, we have

$$
F g_i^j(x_i) = F g_j^i(x_j),
\qquad (9.5.6)
$$

then there exists precisely one $x \in FC$ with

$$
F g_i(x) = x_i \text{ for all } i.
\qquad (9.5.7)
$$

This x is called an *amalgamation* of the x_i, and the above family of arrows g_i and elements x_i is called compatible.

We recall the discussion of sheaves on topological spaces in Sect. 4.5. Of course, the present definition generalizes the concept introduced there. Let us consider the example of the sheaf of continuous functions on a topological space $(X, \mathcal{O}(X))$. Given $U \in \mathcal{O}(X)$ and a collection of open subsets $U_i \subset U$ and continuous functions $\phi_i : U_i \to \mathbb{R}$ satisfying

$$
\phi_i = \phi_j \text{ on } U_i \cap U_j \text{ for all } i, j
\qquad (9.5.8)
$$

then there exists a continuous function

$$
\phi : U \to \mathbb{R} \text{ with } \phi = \phi_i \text{ on each } U_i.
\qquad (9.5.9)
$$

The preceding definition abstracts this situation.

We also observe that for the trivial Grothendieck topology where the only sieve covering an object C is the total sieve $\bar{S}(C)$, every presheaf is a sheaf. This is trivially so, because the total sieve contains 1_C, and so, for this covering, we do not need to amalgamate anything, because it is already given on C itself.

Theorem 9.5.2 *The sheaves on a site $(\mathbf{C}, J)$ also form a topos, called a Grothendieck topos and denoted by* $\mathbf{Sh(C)}$ *(J, being implicitly understood,* *is omitted from the notation). The subobject classifier is similar to the classifier* Ω *for presheaves (see 9.5.2),*

$$\Omega_J(C) = \{S : S \text{ a closed sieve on } C\}, \qquad (9.5.10)$$

with $1_J(C) = 1(C)$ *again the maximal sieve on C which is obviously closed.*

Proof We shall not provide all the details, which are tedious, but not principally difficult, but only sketch the main issues. Of course, one needs to verify first of all that this Ω_J defines a presheaf. The crucial point for this is that the closedness condition is preserved under pullback, that is, whenever the sieve S on C is closed, so is then the pullback $f^*(S)$ for any arrow $f : D \to C$.

The next point to verify is that Ω_J is even a sheaf, that is, one has to verify the uniqueness and existence of amalgamations. Finally, to show that Ω_J is a subobject classifier, one checks that for a sheaf F, the classifying map $\chi_{F'} : F \to \Omega$ factors through Ω_J precisely if the subpresheaf F' of F is a sheaf itself. This is the crucial point, and so we recall the construction of the subobject classifier Ω for presheaves of (9.1.27). For $C \in \mathbf{C}$ and $x \in FC$, we put

$$S_C(x) := \{f : D \to C : f^*(x) \in F'D\}. \qquad (9.5.11)$$

Thus, $S_C(x)$ is the sieve of arrows into C that pull back the element x of the set FC to an element of the subset $F'D$ of FD. Now, let F' be a sheaf itself. We then have to verify that for every x, the sieve $S_C(x)$ is closed. The condition that F' be a sheaf means that for all sieves S covering C if $f^*(x) \in F'D$ for all arrows $f : D \to C$ in S, then also $x \in F'C$. Indeed, when we have a compatible family of arrows $g_i : D_i \to C$ and elements $x_i \in FD_i$, then, because F is a sheaf, this family possesses an amalgamation $x \in FC$. If F' now is a sheaf as well, it then follows that $x \in F'C$. This consideration shows that when $S_C(x)$ covers the identity arrow $1_C : C \to C$, then $1_C \in S_C(x)$. When we apply this argument more generally to any arrow $f : D \to C$ and to $f^*(x)$ in place of x, we see that $S_C(x)$ is indeed closed when F' is a sheaf.

Of course, for a topos, we need to verify more conditions, like the existence of power objects, but we shall not do this here, referring instead to the literature or the reader's own persistence.   □

9.6   Topoi Semantics

With the preceding concepts, we can now take a deeper look at topoi semantics. In particular, we shall be able to place the considerations of Sect. 9.3 into a more general perspective.

We begin with the requirement that, in topos logic, to each variable x, we need to assign a *type* (also called *sort*). This is an object X of the topos $\mathbf{E}$ that we want to work with, called the language. One then also says that the interpretation of the variable x of type X is the identity arrow $1_X : X \to X$.

(When the objects of **E** are sets, then one would think of the set X as the range of values for the variable x.) More generally, we also call an arrow $f : U \to X$ an interpretation of a *term* of type X. We shall not carefully distinguish between terms and their interpretations, however, and we often call an arrow $f : U \to X$ simply a term. In particular, the operations on arrows, like composition, can thus be applied to terms. In particular, for terms σ of type X and τ of type Y, we obtain a term $\langle \sigma, \tau \rangle$ of type $X \times Y$. An arrow $c : \mathbf{1} \to X$ is called a *constant* of type X. A term ϕ of type Ω (the subobject classifier of **E**) is called a *formula*. Such a formula $\phi : X \to \Omega$ then classifies a subobject $\{x : \phi(x)\}$ of X, according to the diagram

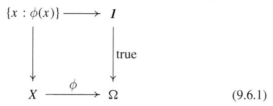

$$\tag{9.6.1}$$

(Again, if X were a set, we would think of $\{x : \phi(x)\}$ as the subset of those elements x for which $\phi(x)$ is true.)

Definition 9.6.1 For a morphism $a : U \to X$ (also called a generalized element of X) and a formula $\phi : X \to \Omega$, we say that U *forces* $\phi(a)$, in symbols

$$U \models \phi(a), \tag{9.6.2}$$

if a factors through the subobject $\{x : \phi(x)\}$ of X.

Equivalently, this can be expressed as

$$\mathrm{im}\, a \leq \{x : \phi(x)\} \tag{9.6.3}$$

in the Heyting algebra $\mathrm{Sub}(X)$ of subobjects of X, where im a is the image of a, as in Lemma 9.2.1.

In order to put the present approach into perspective, we should recall Theorem 9.3.3 of 9.3. There, we had defined Kripke validity and then identified it with validity in the sense of a topos **E** of presheafs, that is, with validity for the Heyting algebra $\mathrm{Hom}_{\mathbf{E}}(\mathbf{1}, \Omega)$. Here, now, we define validity through the forcing relation of Definition 9.6.1. We therefore now have to spell out how this validity behaves under the rules of the Heyting algebra $\mathrm{Hom}_{\mathbf{E}}(\mathbf{1}, \Omega)$. This is called Kripke-Joyal semantics (sometimes, Beth's name is also included), because Joyal had worked this out for a general topos **E**. Here, however, we shall only consider the case of a Grothendieck topos, that is, a topos $\mathbf{Sh}(\mathbf{C})$ of sheaves on a site $(\mathbf{C}, J)$, see Theorem 9.5.2. We refer to [84] for the case of a general topos, and also as a reference for our treatment here.

To set up the framework, we first consider a presheaf X on **C**. We want to define a forcing relation for an object C of **C**. We recall from the Yoneda lemma that C yields a presheaf yC, given by $\mathrm{Hom}_{\mathbf{C}}(-, C)$, and

$$\mathrm{Hom}_{\mathbf{Sets}^{\mathbf{C}^{\mathrm{op}}}}(yC, X) \cong XC, \tag{9.6.4}$$

that is, the natural transformations from $\mathrm{Hom}_{\mathbf{C}}(-, C)$ to X are in natural correspondance with the elements of the set XC. We therefore consider the elements of XC as generalized elements of X in the sense of Definition 9.6.1. Essentially, this simply means that $a \in XC$ is an element that we can insert for the variable x on the object C of the underlying category.

Henceforth, we assume that the presheaf X is a sheaf. (We shall, however, not distinguish between a sheaf and the underlying presheaf; thus, our treatment abandons some precision, hopefully contributing to an easier understanding of the key ideas.) Thus, our forcing relation for $a : yC \to X$ is

$$C \models \phi(a) \tag{9.6.5}$$

if

$$a \in \{x : \phi(x)\}(C), \tag{9.6.6}$$

that is, if

$$a \text{ factors through the subobject } \{x : \phi(x)\} \text{ of } X. \tag{9.6.7}$$

The forcing relation is then obviously monotonic in the sense that

$$\text{if } C \models \phi(a) \text{ and } f : D \to C, \text{ then } D \models \phi(af). \tag{9.6.8}$$

We now utilize the topology J and the fact that X is a sheaf. Then the forcing relation is local in the sense that

$$\text{if } C_i \models \phi(af_i) \text{ for a covering } \{f_i : C_i \to C\} \text{ w.r.t. } J, \text{ then also } C \models \phi(a). \tag{9.6.9}$$

Theorem 9.6.1 *Let* $(\mathbf{C}, J)$ *be a site and* $\mathbf{E} = \mathbf{Sh}(\mathbf{C})$ *the topos of sheaves on this site. Let* $\phi(x)$, $\psi(x)$ *be formulas in the language of* $\mathbf{E}$, *x a free variable of type* $X \in \mathbf{E}$, *$a \in XC$. Also, let* $\sigma(x, y)$ *be a formula with free variables* x, y *of types* X *and* Y, *resp. Then*

1. $C \models \phi(a) \wedge \psi(a)$ *iff* $C \models \phi(a)$ *and* $C \models \psi(a)$.
2. $C \models \phi(a) \vee \psi(a)$ *iff for some covering* $f_i : C_i \to C$, *for each index* i, $C_i \models \phi(a)$ *or* $C_i \models \psi(a)$.
3. $C \models \phi(a) \Rightarrow \psi(a)$ *iff for all* $f : D \to C$, $D \models \phi(af)$ *implies* $D \models \psi(af)$.
4. $C \models \neg\phi(a)$ *iff whenever for* $f : D \to C$, $D \models \phi(af)$, *then the empty family is a cover of* D.
5. $C \models \exists y\sigma(a, y)$ *iff there exist a covering* $f_i : C_i \to C$ *and* $b_i \in YC_i$ *such that for each index* i, $C_i \models \sigma(af_i, b_i)$.
6. $C \models \forall y\sigma(a, y)$ *iff for all* $f : D \to C$ *and* $b \in YD$, *also* $D \models \sigma(af, b)$.

Again, we do not prove this result, but rather refer to [84]. Note that, if we take the trivial Grothendieck topology on $\mathbf{C}$, then every presheaf is a sheaf, and our topos reduces to the presheaf topos $\mathbf{Sets}^{\mathbf{C}^{\mathrm{op}}}$. In this case, some of the above conditions simplify:

2' $C \models \phi(a) \vee \psi(a)$ iff $C \models \phi(a)$ or $C \models \psi(a)$.
4' $C \models \neg\phi(a)$ iff there is no $f : D \to C$ with $D \models \phi(af)$.
5' $C \models \exists y\sigma(a, y)$ iff there exists some $b \in YC$ with $C \models \sigma(a, b)$.

The reason for these simplifications is clear. For instance, to get 2', we simply observe that, for the trivial topology, we can simply take the covering $1_C : C \to C$ in 2.

In fact, Kripke's original setting was that of a site consisting of a poset **P** with the trivial topology, as described in Sect. 9.3.

A Review of Examples

10

In this chapter, we systematically describe the simplest examples of the structures that are discussed in the main text. Thus, this chapter is an anti-climax, and the preceding chapters are not a prerequisite to read it; instead you can turn to it at any point during your study of this book. Here, since the standard examples are so ubiquitous, we no longer provide margin boxes indicating their appearance.

10.1 ∅ (Nothing)

The empty set ∅ is

- a set
- not a monoid or group, because it does not possess a neutral element e
- a category with no objects or morphisms
- not the power set of any space
- a metric space without points
- a graph, a simplicial complex without vertices, a manifold without points; in particular, all Betti numbers b_p and the Euler characteristic χ vanish
- a topological space, and as such, its power set is {∅}, that is, a set with one element,

and this latter leads us on to the next example, as via the power set operation, we have created something out of nothing.

10.2 {1} (Something)

{1} is the set with a single element 1. Although it is almost completely trivial, it constitutes an example of almost all structures that we have discussed in this book. It can be equipped with the structure of

© Springer International Publishing Switzerland 2015
J. Jost, *Mathematical Concepts*, DOI 10.1007/978-3-319-20436-9_10

- a set
- a graph with a single vertex and no edges
- •
- hence also a simplicial complex with a single vertex and no other simplices; thus, $b_0 = 1$ while all other Betti numbers vanish, and $\chi=1$
- a matroid over $\emptyset$ or, more generally, over any set E, with $\{\emptyset\}$ as the only independent subset of $\emptyset$ (or E) and no vertices
- a 0-dimensional connected manifold
- an equivalence relation satisfying $F(1, 1) = 1$
- a partially ordered set (poset) and in fact even an ordered set, with $1 \leq 1$
- a lattice, with $1 \vee 1 = 1$ and $1 \wedge 1 = 1$
- a lattice with 0 and 1 and a Heyting and Boolean algebra only if we allow for $0 = 1$ (like any Boolean algebra, it is a power set, that of $\emptyset$, as noted earlier)
- a metric space, with $d(1, 1) = 0$
- a monoid and a group, with 1 being the neutral element, that is, $1 \cdot 1 = 1$ as the operation
- not a field, because for that we would also need a commutative group structure with a neutral element $0 \neq 1$
- the spectrum of any field (recall that when F is a field, its only prime ideal is (0))
- a category with a single object, the identity morphism of that object, and if we want, also other morphisms $1 \rightarrow 1$
- a topological space, and as such, its power set is $\{\emptyset, 1\}$, that is, a set with two elements, (and this is also a σ-algebra)

and this latter leads us on to the next example, as via the power set operation, we have created an alternative out of sheer existence.

10.3   {0, 1} (An Alternative)

$\{0, 1\}$ is the set with two elements 0 and 1. These elements can stand for the alternative between true (1) and false (0), between presence and absence etc. In the previous example, $\{1\}$, there was only presence, whereas absence was absent. By including the possibility of absence via the power set construction, we have created an alternative. $\{0, 1\}$ constitutes an example of essentially all the structures that we have discussed in this book. It can be equipped with the structure of

- a set
- a category with two objects, the identity morphisms of those objects, and if we want, also other morphisms, for instance $0 \rightarrow 1$
- the power set $\mathcal{P}(\{1\})$ of a space with a single element, and therefore, when we let 0 correspond to $\emptyset$ and 1 to $\{1\}$, it becomes the category with the morphisms $0 \rightarrow 0, 0 \rightarrow 1, 1 \rightarrow 1$ described in the previous item
- as the power set of a set with a single element, it also possesses a subobject classifier, $\{1\}$, and it therefore becomes a topos

- an equivalence relation satisfying $F(0, 0) = F(1, 1) = 1$ and either trivially also $F(0, 1) = F(1, 0) = 1$ or nontrivially $F(0, 1) = F(1, 0) = 0$ (in the former case, there is only a single equivalence class $\{0, 1\}$, whereas in the latter case, we have the two equivalence classes $\{0\}$ and $\{1\}$)
- a partially ordered set (poset) and in fact even an ordered set, with $0 \leq 1$
- a lattice with 0 and 1, satisfying $0 \wedge 1 = 0$ and $0 \vee 1 = 1$ as they should, and a Heyting and Boolean algebra

(like any Boolean algebra, it is a power set, that of $\{1\}$, as noted there)
- a propositional logic with alphabet $\Pi_0 = \{1, 0\}$ ("true" and "false")
- a metric space, with $d(0, 1) = d(1, 0) > 0$
- the monoid $M_2 = (\{0, 1\}, \cdot)$ with $0 \cdot 0 = 0 \cdot 1 = 1 \cdot 0 = 0$, $1 \cdot 1 = 1$ that is not a group (because 0 has no inverse)
- another monoid structure, the abelian group $\mathbb{Z}_2 = (\{0, 1\}, +)$ with $0 + 0 = 0 = 1 + 1$, $0 + 1 = 1 + 0 = 1$
- the commutative ring with 1 and the field $\mathbb{Z}_2 = (\{0, 1\}, +, \cdot)$ that combines the previous two structures (note the abuse of notation where $\mathbb{Z}_2$ first carries only a group and then also a ring structure)
- the spectrum $\mathrm{Spec}\mathbb{Z}_4$ of the ring $\mathbb{Z}_4$ because the latter has two prime ideals, the trivial one, $\{0\}$, which we naturally identify with 0, and the nontrivial $\{0, 2\}$ which we then identify with 1, with structure sheaf $\mathcal{O}\mathrm{Spec}\mathbb{Z}_4 = \mathrm{Spec}\mathbb{Z}_4$, as is the case for any ring, and equipped with this structure sheaf, it becomes an affine scheme
- a topological space, and in fact, we have several possibilities for the collection of open sets

  – the indiscrete topology with $\mathcal{O} = \{\emptyset, \{0, 1\}\}$ (that is, again with a set with two elements only, so that no further distinctions are possible)
  – the discrete topology with $\mathcal{O} = \{\emptyset, \{0\}, \{1\}, \{0, 1\}\}$, that is, equal to the power set $\mathcal{P}(\{0, 1\})$
  – or the topologies with one nontrivial open set, for instance $\mathcal{O} = \{\emptyset, \{0\}, \{0, 1\}\}$ (the latter is the topology of $\mathrm{Spec}\mathbb{Z}_4$, because the trivial ideal is not closed, but the nontrivial one is)

- a σ-algebra, either the trivial $\{\emptyset, \{0, 1\}\}$ or the full one, $\{\emptyset, \{0\}, \{1\}, \{0, 1\}\}$
- not a (0-dimensional) manifold, because it is not connected
- a matroid over $\{1\}$ with the two independent sets $\{\emptyset\}$ and $\{1\}$,

and we can use it to generate

- a graph with two vertices

0   1
•   •

that we, if we want, can connect by an edge

or by a directed edge, for instance $0 \to 1$,

- hence also a simplicial complex with two vertices that may be joined by a 1-dimensional simplex; when this is so, $\chi = 2 - 1 = 1$, and for the Betti numbers, $b_0 = 1, b_1 = 0$ and, of course, all higher Betti numbers also vanish, and thus again $\chi = 1 - 0 = 1$; when the vertices were not joined by an edge, however, we would have $b_0 = 2$ and $\chi = 2$
- as a simplicial complex with two vertices connected by an edge, it would also be a topological space, with the topology of the unit interval in $\mathbb{R}$ (for instance induced by the Euclidean metric); each vertex would then be a strong deformation retract of this space
- as such a simplicial complex consisting of two vertices connected by an edge, it could then also be equipped with the Euclidean metric on the unit interval; it would then be the convex hull of the two vertices
- as a simplicial complex with two vertices connected by an edge, then, even though connected, it would still not be a (1-dimensional) manifold, because the vertices do not possess neighborhoods that are homeomorphic to an *open* subset of $\mathbb{R}$
- finally, $\{0, 1\}$ is a set of basis vectors $e_1 = (1, 0), e_2 = (0, 1)$ of the vector space $(\mathbb{Z}_2)^2 = \mathbb{Z}_2 \times \mathbb{Z}_2$ over the field $\mathbb{Z}_2$ (of course, 0 and 1 have a different meaning now, for instance $(0, 1)$ yields the coordinates of a vector in $\mathbb{Z}_2 \times \mathbb{Z}_2$); the resulting matroid would have the independent sets $\emptyset, \{e_1\}, \{e_2\}, \{e_1, e_2\}$ and would again be the simplicial complex consisting of two vertices connected by an edge.

The bibliography lists several original references, monographs, and text-books, not all of which have been mentioned in the main text. The bibliography is definitely not exhaustive or complete, and was not intended to be so. The textbooks listed usually provide fairly accurate references, and there seems to be little point in copying existing bibliographies. All sources that I have directly consulted, however, are included in the present bibliography.

Remark In the bibliography, a superscript denotes the number of the edition; e.g., ^{2}1999 means "2nd edition, 1999".

Bibliography

1. Aleksandrov AD (1957) Über eine Verallgemeinerung der Riemannschen Geometrie. Schriften Forschungsinst Math 1:33–84
2. Aliprantis C, Border K (^{3}2006) Infinite dimensional analysis, Springer, Berlin
3. Amari S-I, Nagaoka H (2000) Methods of information geometry. Am. Math. Soc. (translated from the Japanese)
4. Artin M, Grothendieck A, Verdier J-L (1972) Théorie de topos et cohomologie étale des schémas, Seminaire Géometrie Algébraique 4, Springer LNM, Berlin, p 269, 270
5. Awodey S (2006) Category theory. Oxford University Press, Oxford
6. Ay N, Jost J, Lê HV, Schwachhöfer L Information geometry (to appear)
7. Bačák M, Hua B, Jost J, Kell M, Schikorra A A notion of nonpositive curvature for general metric spaces. Diff. Geom. Appl. (to appear)
8. Baez J, Fritz T, Leinster T (2011) A characterization of entropy in terms of information loss. Entropy 13:1945–1957
9. Bauer F, Hua BB, Jost J, Liu SP, Wang GF, The geometric meaning of curvature. Local and nonlocal aspects of Ricci curvature
10. Bell JL (2008) Toposes and local set theories. Dover, Mineola
11. Benecke A, Lesne A (2008) Feature context-dependency and complexity-reduction in probability landscapes for integrative genomics. Theor Biol Med Model 5:21
12. Berestovskij VN, Nikolaev IG (1993) Multidimensional generalized Riemannian spaces. In: Reshetnyak YuG (ed.), Geometry IV, Encyclopedia of Mathematics Sciences 70. Springer, Berlin, pp. 165–250 (translated from the Russian, original edn. VINITI, Moskva, 1989)
13. Berger M, Gostiaux B (1988) Differential geometry: manifolds, curves, and surfaces, vol 115., GTM, Springer, New York
14. Bollobás B (1998) Modern graph theory. Springer, Berlin
15. Boothby W (1975) An introduction to differentiable manifolds and Riemannian geometry. Academic Press, New York
16. Breidbach O, Jost J (2006) On the gestalt concept. Theory Biosci. 125:19–36
17. Bröcker T, tom Dieck T (1985) Representations of compact Lie groups. GTM, vol 98. Springer, New York
18. Burago D, Burago Yu, Ivanov S (2001) A course in metric geometry. American Mathematical Society, Providence
19. Busemann H (1955) The geometry of geodesics. Academic Press, New York
20. Cantor G (1932) In: Zermelo E (ed) Gesammelte Abhandlungen mathematischen und philosophischen Inhalts. Springer, Berlin (Reprint 1980)
21. Cantor M (1908) Vorlesungen über Geschichte der Mathematik, 4 vols., Teubner, Leipzig, ^{3}1907, ^{2}1900, ^{2}1901; Reprint 1965

© Springer International Publishing Switzerland 2015

J. Jost, *Mathematical Concepts*, DOI 10.1007/978-3-319-20436-9

22. Chaitin GJ (1966) On the length of programs for computing finite binary sequences. J. ACM 13(4):547–569

23. Čech E (1966) Topological spaces. Wiley, New York

24. Connes A (1995) Noncommutative geometry. Academic Press, San Diego

25. Corry L (^{2}2004) Modern algebra and the rise of mathematical structures. Birkhäuser, Basel

26. van Dalen D (^{4}2004) Logic and structure. Springer, Berlin

27. Dieudonné J (ed.) (1978) Abrégé d'histoire des mathématiques, 2 vols. Hermann, Paris, pp 1700–1900; German translation: Dieudonné J (1985) Geschichte der Mathematik, Vieweg, Braunschweig/Wiesbaden, pp 1700–1900

28. Dold A (1972) Lectures on algebraic topology. Springer, Berlin

29. Dubrovin BA, Fomenko AT, Novikov SP (1985) Modern geometry - Methods and applications. Part II: The geometry and topology of manifolds. GTM, vol 104. Springer, Berlin

30. Dubrovin BA, Fomenko AT, Novikov SP (1990) Modern geometry–methods and applications. Part II: Introduction to homology theory. GTM, vol 124. Springer, Berlin

31. Eilenberg S, Steenrod N (1952) Foundations of algebraic topology. Princeton University Press, Princeton

32. Eilenberg S, Mac Lane S (1945) General theory of natural equivalences. Trans AMS 58:231–294

33. Eisenbud D, Harris J (2000) The geometry of schemes. Springer, Berlin

34. Erdös P, Rényi A (1959) On random graphs I. Publ Math Debrecen 6:290–291

35. Eschenburg J, Jost J (^{3}2014) Differential geometrie und Minimalflächen. Springer, Berlin

36. Fedorchuk VV (1990) The fundamentals of dimension theory. In: Arkhangel'skiĭ AV, Pontryagin LS (eds.), General topology I. Encyclopaedia of Mathematical Sciences, vol 17. Springer, Berlin (translated from the Russian)

37. Ferreirós J (^{2}2007) Labyrinth of thought. A history of set theory and its role in modern mathematics. Birkhäuser, Boston

38. Forman R (1998) Morse theory for cell complexes. Adv Math 134:90–145

39. Forman R (1998) Combinatorial vector fields and dynamical systems. Math Zeit 228:629–681

40. Fulton W, Harris J (1991) Representation theory. Springer, Berlin

41. Gitchoff P, Wagner G (1996) Recombination induced hypergraphs: a new approach to mutation-recombination isomorphism. Complexity 2:37–43

42. Goldblatt R (2006) Topoi . Dover, Mineola

43. Goldblatt R (1981) Grothendieck topology as geometric modality. Zeitschr. f. Math. Logik u. Grundlagen d. Math. 27:495–529

44. Gould SJ (2002) The structure of evolutionary theory. Harvard University Press, Cambridge

45. Harris J (1992) Algebraic geometry. GTM, vol 133. Springer, Berlin

46. Hartshorne R (1977) Algebraic geometry. Springer, Berlin

47. Hatcher A (2001) Algebraic topology. Cambridge University Press, Cambridge

48. Hausdorff F (2002) Grundzüge der Mengenlehre, Von Veit, Leipzig, 1914; reprinted. In: Werke G, Bd II, Brieskorn E et al (eds) Hausdorff F. Springer, Berlin

49. Heyting A (1934) Mathematische Grundlagenforschung: Intuitionismus. Springer, Beweistheorie

50. Hilbert D (2015) Grundlagen der Geometrie, Göttingen 1899; Stuttgart ^{11}1972; the first edition has been edited with a commentary by K. Volkert, Springer, Berlin

51. Hirsch M (1976) Differential topology. Springer, Berlin

52. Horak D, Jost J (2013) Spectra of combinatorial Laplace operators on simplicial complexes. Adv. Math. 244:303–336

53. Hughes G, Cresswell M (1996) A new introduction to modal logic. Routledge, London

54. Humphreys J (1972) Introduction to Lie algebras and representation theory. Springer, Berlin

55. Johnstone P (2002) Sketches of an elephant. A topos theory compendium, 2 vols. Oxford University Press, Oxford

56. Jonsson J (2008) Simplicial complexes of graphs. LNM, vol 1928. Springer, Berlin

57. Jost J (1997) Nonpositive curvature: geometric and analytic aspects. Birkhäuser, Basel

58. Jost J (^{6}2011) Riemannian geometry and geometric analysis. Springer, Berlin

59. Jost J (^{3}2005) Postmodern analysis. Springer, Berlin

60. Jost J (2005) Dynamical systems. Springer, Berlin

61. Jost J (^{3}2006) Compact Riemann surfaces. Springer, Berlin

62. Jost J (2009) Geometry and physics. Springer, Berlin

63. Jost J (2014) Mathematical methods in biology and neurobiology. Springer, Berlin

64. Jost J, Li-Jost X (1998) Calculus of variations. Cambridge University Press, Cambridge

65. Kan D (1958) Adjoint functors. Trans AMS 87, 294–329

66. Kelley J (1955) General topology, Springer, no year (reprint of original edition, van Nostrand)

67. Klein F (1974) Vorlesungen über die Entwicklung der Mathematik im 19. Jahrhundert, Reprint in 1 vol. Springer, Berlin

68. Kline M (1972) Mathematical thought from ancient to modern times. Oxford University Press, Oxford; 3 vol. paperback edition 1990

69. Knapp A (1986) Representation theory of semisimple groups. Princeton University Press, Princeton; reprinted 2001

70. Koch H (1986) Einführung in die klassische Mathematik I. Akademieverlag and Springer, Berlin

71. Kolmogorov AN (1965) Three approaches to the quantitative definition of information. Probl Inf Trans 1(1):1–7

72. Kripke S (1962) Semantical analysis of intuitionistic logic I. In: Crossley J, Dummett M (eds) Formal systems and recursive functions. North-Holland, Amsterdam, pp 92–130

73. Kunz E (1980) Einführung in die kommutative Algebra und analytische Geometrie. Vieweg, Braiinschweig

74. Lambek J (1988) Scott PJ (1986) Introduction to higher order categorical logic. Cambridge University Press, Cambridge

75. Lang S (^{3}2002) Algebra. Springer, Berlin

76. Lawvere F (1964) An elementary theory of the category of sets. Proc Nat Acad Sci 52:1506–1511

77. Lawvere F (1971) Quantifiers as sheaves. Proccedings of the International Congress of Mathematicians 1970 Nice, vol 1, pp 329–334. Gauthiers-Villars, Paris 1971

78. Lawvere F (1975) Continuously variable sets: algebraic geometry = geometric logic. Studies in Logic and the Foundations of Mathematics, vol 80, (Proc Logic Coll Bristol, 1973), North Holland, Amsterdam, pp 135–156

79. Lawvere F, Rosebrugh R (2003) Sets for mathematics. Cambridge University Press, Cambridge

80. Leinster T (2004) Higher operads, higher categories. Cambridge University Press, Cambridge

81. Li M, Vitanyi PMB (^{2}1997) An introduction to Kolmogorov complexity and its applications. Springer, Berlin

82. MacLane S (^{2}1998) Categories for the working mathematician. Springer, Berlin

83. MacLane S (1986) Mathematics: form and function. Springer, Berlin

84. MacLane S, Moerdijk I (1992) Sheaves in geometry and logic, Springer

85. Massey W (1991) A basic course in algebraic topology. GTM, vol 127. Springer, Berlin

86. McLarty C (1995) Elementary categories, elementary toposes. Oxford University Press, Oxford

87. May JP (1999) A concise course in algebraic topology. University of Chicago Press, Chicago

88. Moise E (1977) Geometric topology in dimensions 2 and 3. Springer, Berlin

89. Mumford D (^{2}1999) The red book of varieties and schemes. LNM, vol 1358. Springer, Berlin

90. Novikov PS (1973) Grundzüge der mathematischen Logik. VEB Deutscher Verlag der Wissenschaften, Berlin (translated from the Russian, original edn. Fizmatgiz, Moskva, p 1959)

91. Oxley J (1992) Matroid theory. Oxford University Press, Oxford

92. Papadimitriou C (1994) Computational complexity. Addison-Wesley, Reading

93. Peter F, Weyl H (1927) Die Vollständigkeit der primitiven Darstellungen einer geschlossenen kontinuierlichen Gruppe. Math Ann 97:737–755

94. Pfante O, Bertschinger N, Olbrich E, Ay N, Jost J (2014) Comparison between different methods of level identification. Adv Complex Syst 17:1450007

95. Querenburg B (1973) Mengentheoretische Topologie. Springer, Berlin 1976

96. Riemann B (2013) Ueber die Hypothesen, welche der Geometrie zu Grunde liegen, Abh. Ges. Math. Kl.Gött. 13, 133–152, (1868) edited with a commentary by J. Jost, Springer

97. de Risi V (2007) Geometry and monadology: Leibniz's analysis situs and philosophy of space. Birkhäuser, Basel

98. Schubert W, Bonnekoh B, Pommer A, Philipsen L, Bockelmann R, Malykh Y, Gollnick H, Friedenberger M, Bode M, Dress A (2006) Analyzing proteome topology and function by automated multidimensional fluorescence microscopy. Nature Biotech. 24:1270–1278

99. Schur I (1905) Neue Begründung der Theorie der Gruppencharacktere. Sitzungsber. Preuss. Akad. Wiss. pp. 406–432

100. Schwarz M (1993) Morse homology. Birkhäuser, Boston

101. Serre JP (^{2}1995) Local fields. Springer, Berlin

102. Shafarevich IR (^{2}1994) Basic algebraic geometry, 2 vols. Springer, Berlin

103. Solomonoff RJ (1964) A formal theory of inductive inference: parts 1 and 2. Inf. Control 7, 1–22 and 224–254

104. Spanier E (1966) Algebraic topology. McGraw Hill, New York

105. Stadler B, Stadler P (2002) Generalized topological spaces in evolutionary theory and combinatorial chemistry. J Chem Inf Comput Sci 42:577–585

106. Stadler B, Stadler P (2003) Higher separation axioms in generalized closure spaces, Annales Societatis Mathematicae Polonae. Seria 1: Commentationes Mathematicae, pp 257–273

107. Stadler P, Stadler B (2006) Genotype-phenotype maps. Biol. Theory 1:268–279

108. Stöcker R, Zieschang H (^{2}1994) Algebraische Topologie. Teubner, Stuttgart

109. Takeuti G, Zaring W (^{2}1982) Introduction to axiomatic set theory. GTM, vol 1. Springer, Berlin

110. Tierney M (1972) Sheaf theory and the continuum hypothesis. LNM, vol 274. Springer, Berlin, pp 13–42

111. van der Waerden B (^{9}1993) Algebra I, II. Springer, Berlin

112. Wald A (1935) Begründung einer koordinatenlosen Geometrie der Flächen. Ergeb Math Koll 7:24–46

113. Welsh D (1995) Matroids: fundamental concepts. In: Graham R, Grötschel M, Lovasz L (eds) Handbook of combinatorics. Elsevier and MIT Press, Cambridge, pp 481–526

114. Weyl H (^{7}1988) In: Ehlers J (ed) Raum, Zeit, Materie. Springer, Berlin. (English translation of the 4th ed.: Space-time-matter, Dover, 1952)

115. Weyl H (^{6}1990) Philosophie der Mathematik und Naturwissenschaft, München, Oldenbourg

116. Weyl H (^{2}1946) The classical groups, their invariants and representations. Princeton University Press, Princeton

117. Whitney H (1935) On the abstract properties of linear dependence. Am. J. Math. 57:509–533

118. Wilson R (2009) The finite simple groups. Graduate Texts in Mathematics, vol 251. Springer, Berlin

119. Wußing H (2008/9) 6000 Jahre Mathematik, 2 vols., Springer, Berlin
120. Yoneda N (1954) On the homology theory of modules. J Fac Sci Tokyo Sec I 7:193–227
121. Zariski O, Samuel P (1975) Commutative algebra, 2 vols. Springer, Berlin
122. Zeidler E (2013) Springer-Handbuch der Mathematik, 4 vols., Springer, Berlin
123. Zermelo E (1908) Untersuchungen über die Grundlagen der Mengenlehre. Math Ann 65:261–281

Index

Made in the USA
Middletown, DE
28 December 2015